ALONE AT SEA

Gloucester in the Age
of the Dorymen
(1623–1939)

✕ On a calm summer's day, the trawl line has been laid on the seabed below, the trawl tubs sit empty, and two dorymen begin to haul bottom-feeding fish into the boat.

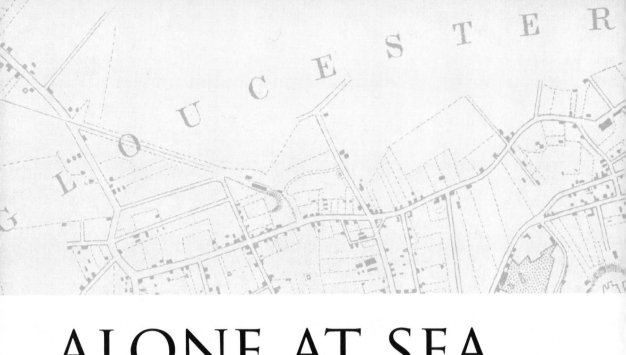

ALONE AT SEA

Gloucester in the Age
of the Dorymen
(1623–1939)

John N. Morris

*The story of how generations of brave men in small
boats, under oar and sails, built a great New England
fishing community, and in so doing were
themselves transformed into Americans.*

Commonwealth Editions

Beverly, Massachusetts

Library of Congress Cataloging-in-Publication Data

Morris, John N.
 Alone at sea : Gloucester in the age of the dorymen, 1623–1939 / John N. Morris.
    p. cm.
 Includes bibliographical references and index.
 ISBN 978-0-9819430-7-7 (alk. paper)
 1. Fisheries--Massachusetts--Gloucester--History. 2. Fishers--Massachusetts--
Gloucester--History. 3. Dories (Boats)--Massachusetts--Gloucester--History.
4. Gloucester (Mass.)--Economic conditions. 5. Gloucester (Mass.)--History. I. Title.
 SH222.M4M67 2010
 639.209744'5--dc22
                          2010000068

Book packaged by Open Book Systems, Inc. (OBS), Rockport, MA
Project management, design, and proofreading: Books By Design, Inc.
Photo editing and research: Fred Buck
Photo permissions consulting: Linda Finigan
Substantive editing: Jenny Bennett Marshall
Copyediting: Nancy Menges
Indexing: Gordon Brumm

Printed in Canada

Commonwealth Editions is the trade imprint of Memoirs Unlimited, Inc., 266 Cabot Street, Beverly, MA 01915. Visit us on the Web at www.commonwealtheditions.com.

Foreword

In the course of writing this absolutely authoritative fishing history of our Gloucester, the oldest, most dynamic and colorful such port in the Western Hemisphere, John Morris—himself a wise and salty son—has fortunately focused symbolically on our modest, flat-bottomed, slope-sided, two-man wooden dory, the inanimate hero of his fishermen heroes.

I daresay this Gloucester dory is the most versatile rowboat ever devised by man. You can row it backwards as well as forwards by pushing instead of pulling your oars, thereby actually seeing where you're going for a change. Rather tippy without a cargo, add a load of fresh-caught codfish to two dory-men and their gear, and the lower in the water almost to the "gunalls," the more stable and seaworthy. Indeed, our jack-of-all-boats can be fitted with a removable mast and sailed instead of rowed, or, in one of our old-fashioned New England winters when men were men, with a couple of temporary runners on its bottom as skates, it could be sailed back and forth across our frozen harbor fast as the wind on the ice for the pure thrill of it.

One of our fabled fishermen nicknamed Alfred "Centennial" Johnson sailed his dory, with a cabin about the size of a coffin, all alone across the wild Atlantic Ocean to England for the first time in history to mark the hundredth anniversary of Independence (and called himself a damn fool for doing it). And another, our larger-than-life man of men Howard Blackburn . . .

Well, read on, because John Morris is about to tell you all there is to be told about Gloucestermen and their wives and widows and fatherless kids, and ways of life, and of death by the thousands, of good times and of bad, in a masterpiece that's been waiting for generations to be told.

Joseph E. Garland
Gloucester, Massachusetts

Preface

On March 7, 1935, Charles Daley from St. Joseph's, Newfoundland,* and Steve Olsson from Gloucester, Massachusetts (by way of Gullholmen, Sweden), sat in a two-man, 20-foot dory trawling for halibut on St. Pierre's Bank, due south of the island of Newfoundland. Theirs was one of 12 dories that had been set on these grounds by Captain Carl Olsen of the knockabout schooner *Oretha F. Spinney*. At sea since February 16, the *Spinney* had run into unusually severe winter weather. This was one of the first times the skipper had been able to set his dories, permitting his men to drop baited trawl lines to the seabed below. But on this day, for reasons never fully explained, these two experienced fishermen disappeared from their dory—never to be seen again. Captain Olsen sailed back and forth over the grounds, but found only an overturned dory, trawl line, and Steve Olsson's glove caught on one of the fishhooks.

Steve Olsson was my grandfather. His drowning came six years before my birth, and I know of him only through the stories passed down by his children. My mother and her four siblings believed Steve to be a wise and gentle man, and they often told the story of his last trip on the *Oretha F. Spinney*. But I knew nothing of Charles Daley and had only a vague knowledge of Gloucester in the days of schooners and dorymen.

All of this changed, however, in 1999, when I scanned the website then kept by Roberta Sheedy on Gloucester fishermen lost at sea. There I found the names Steve Olsson and Charles Daley, and a brief reference to their drowning in 1935, but nothing more. There was no account of the events of the trip—nothing but their names. But, in a note on the site's chat room, the grandson of Carl Olsen, captain of the *Spinney*, offered to share information with anyone interested in the vessel. Well, I was interested, and Captain Olsen's grandson responded to my e-mail as follows: "Thank you for writing. My father, Einar, spoke very highly and fondly of your grandfather. He named me Stephen after him. I have a picture of your grandfather on board

---

*Charles's grandfather Denis was born in Dungarvan, County Waterford, Ireland. He left for Newfoundland in about 1830 from the adjacent community of East Passage, Ireland, and married Mary Flynn.

the *Spinney*, taken the summer before his drowning. The *Spinney* was sold to MGM, in 1936, for the making of the movie *Captains Courageous*. After the movie, the actor Sterling Hayden bought it. He wrote to my father that he had sold it and that it had sunk somewhere off Jamaica. My father passed on in 1985. My mother and brother live in Rockport, MA. I have an uncle in Gloucester. He has two sons."

This stunned me: The grandson of Captain Carl Olsen had been named not after his own father or grandfather, but after my grandfather. His own father, who went to sea as a teenager on the *Spinney*, "spoke very highly and fondly of" my grandfather Steve Olsson. This picture of my grandfather brought a lifetime of stories and thoughts into focus. I could see my mother telling stories of her father and the family home in East Gloucester. There was my grandfather sitting in the cellar smoking his pipe, my mother and her sisters sitting at his knee. He almost always had a small pencil in his hand, sketching images of dories, sloops, and schooners, while talking to his children of everyday life. I loved thinking of him in the coal cellar smoking his pipe. My grandmother Hilda did not want her husband to smoke in the house, and Steve loved his wife and respected her wishes. I could also see him chatting with the children from the neighborhood, handing out a penny to any child who could spin a good yarn.

The sea is where Steve made a living, but the land is where he lived his life. When he was home, the whole family celebrated. It was like Christmas when a Gloucester doryman made port. As I thought about the stories, I remembered my mother saying that "Papa" was a kind man, a gentle man, a family man. And here, for the first time, I had proof positive that someone other than his children thought highly of Steve Olsson. The picture that had been with me since childhood was not simply the revisionist memory of his children; the stories neither select nor invented—Steve Olsson was someone special, someone I wanted to know.

First, I began to try to better understand the life of this Gloucester doryman, talking with his children, reaching out to the Daley family, and gathering information on the vessels on which Olsson and Daley had sailed. A Gloucester doryman might be at sea for weeks and even months at a time, and there were thousands of dorymen sailing on hundreds of Gloucester schooners. The story went beyond these two men. It was a compelling story, and this book describes the experiences of the Gloucester dorymen in the age of sail.

I have been fortunate in finding a wide array of resources to help me in telling the story. Of printed material, the most important are the microfilm copies of Gloucester's newspapers found at the Sawyer Free Library.

Gloucester was a fishing community and her newspapers provided extensive information on the industry, information that men such as Joseph Garland and Gordon Thomas before me have used so wisely. Second in importance in telling this story has been access to the rich primary source material preserved at the library of the Cape Ann Museum (Historical Society). Then there has been my access to the children and grandchildren of the Gloucester dorymen with names such as Olson, Daley, Lee, Thomas, Arnorsson, Olsen, Lund, Files, Cluett, Rutchick, and Shields, as well as access to two of the last dorymen themselves—Brendan Daley and Billy Shields.

I also visited many of the fishing communities frequented by the Gloucestermen, getting a perspective on the places they called home. My grandfather came to Gloucester in 1893, and our maritime heritage goes back for hundreds of years in Gullholmen, a small island on the west coast of Sweden, which I was able to visit. I visited St. Joseph's, Newfoundland, the ancestral home of the Daley family, and met many of the descendants of Captain Daley. Other sites visited include the east coast fishing communities of Nova Scotia, from whence so many of the Gloucester dorymen and skippers came; the Westfjords area of Iceland, where the Gloucester halibut fishermen went in great numbers during the 1880s and 1890s; and Port Townsend, Washington, a place to which a few Gloucester skippers sailed in the 1880s and in the process began America's Northwest halibut fishery. And, of course, I grew up in Gloucester—the son of a man whose job was to cut fish and a mother whose job was to pack fish in boxes, and I have always felt close to this community.

Of all the people I met, my first contact with the kin of Charles Daley, the doryman who drowned with my grandfather, was the most unusual. I knew I had to track down the Daley family, and again the web came into play. The Canadian website "Fishing?—It Was a Way of Life" had a listing on Charles Daley posted by a Royal Canadian Mounted Policeman. He had gathered the information from Gary Daley, grandson of the Gloucester doryman. Reading this, I contacted the website and they put me in contact with the Mountie, who contacted the Daley family on his next visit to the small village of St. Joseph's. Gary was not home. He was at sea, fishing on the very banks on which our relatives had drowned. But his wife, Mercedes, was home and she radioed her husband saying I wished to speak with him. It was after nine at night, and while Gary was at sea, I was sitting in my living room. Then, out of the blue, he called me. The link had been made between the grandsons of Charles Daley and Steve Olsson—a truly magical moment.

As weeks rolled into months, and months into years, these contacts and many other wonderful sources of information revealed to me the men in the dories, on the fishing schooners, and at home in Gloucester. Two of Gloucester's last dorymen, Steve Olsson and Charles Daley, ages 63 and 53 at the time of their drowning, were calling out to me from the past, and it became obvious to me that I had to tell their story, a story that was part of my own heritage.

Acknowledgments

In writing this book I was fortunate to have access to the collections of two superb Gloucester libraries. Sawyer Free Library is the city archive, and its microfilm collection of local newspapers goes back to 1827, providing a treasure trove of information on vessels, storms, economic conditions, and the changing community. It also houses extensive collections of audio interviews from many decades earlier and historical manuscripts that speak of the people and their fishing heritage. This resource was invaluable, and I thank former director David McArdle and his staff for their support.

The second Gloucester library is maintained by the Cape Ann Museum. It sits one block from the Sawyer Free Library, but could not be more different. One enters as a member, hours are limited, but the resource is unique. The librarian during most of my period of active research was Ellen Nelson, and she became a friend and trusted informant.

Then there were the many people who spoke to me of their stories and the stories of their families. First among them are my elderly aunts, who provided rich accounts of their father, the community, and the experience of growing up as children of a doryman. Although they are now gone, I must thank Gerda Cluett, Romaine Olson, and Harriet Rutchick. Next comes Jeff Thomas, the grandson of his namesake (a great Gloucester skipper), and the son of Gordon Thomas, whose book *Fast & Able* is a Gloucester classic. Jeff had listened well to the stories of his father, and he and Gordon had systematically taped conversations with many fishermen and their families in the 1980s. Jeff shared this unique resource with me, and I cannot thank him enough for access to these accounts. Jeff became a friend, and he made himself available whenever I asked. Next are the many relatives of Charles Daley, who told me so much of Charles and the Newfoundland experience. I would especially like to thank Charles's children, Mercedes Daley Lee and Al Daley, his cousin Richard Daley, and his grandchildren Gary Daley, Jerry Lee, and Noel Daley. Finally, there is my neighbor, Lillian Lund Files, the granddaughter of a Gloucester fisherman, whose roots go back to the same Swedish island as do my relatives, and whose grandfather served with my grandfather for over 30 years.

As early drafts of this manuscript came together, I reached out to experts on the Gloucester fishing experience for comments on what I had produced and where I was going. Three people stand out, and I am greatly indebted to them for their willingness to read early copies of the text that covered the period from 1623 to 1899: They are Courtney-Ellis Peckham, who was then associated with the Essex Shipbuilding Museum; Erik Ronnberg, the premier model maker and historian; and Joseph Garland, the most prolific of local historians. All encouraged me to go on, and I thank them for their kind words. Of the three, I was most fortunate to enlist the council of Joe Garland. I can remember sitting in his Eastern Point home when we were just about to discuss my manuscript for the first time. Before Joe was willing to comment, he asked who I was and why I was writing the story. Hearing of my ties to Gloucester and the sea, and my desire to tell this story in as full and accurate a manner as possible, Joe was satisfied. In his mind, I could write the story, and from that day on he has been free with his advice and counsel.

Every book also requires those who are willing to pore over the text and provide candid responses to the author. I have been fortunate to have two such reviewers, my wife, Shirley, and Will Andrews from Idaho. Shirley encouraged me to write the full account of the dorymen: She provided textual support on an ongoing basis and challenged me to make the story flow—the same wise advice provided by Joe Garland. Will Andrews, a retired surgeon and expert model schooner maker, became the key person holding my toes to the fire. He knows as much about Gloucester and the fishing industry as any person alive. His roots go back to the soil of Gloucester, his research of the fisheries as he worked on his series of exquisite model schooners is second to none, and his ability to bring this knowledge forward is most appreciated. Will took the manuscript, chapter by chapter, draft by draft, and read and reread the text. He forwarded reviews—what worked, what needed to be cut, what had to be added, what I had missed, and how to make the story flow. His careful reviews helped shape the manuscript, and I owe him a debt of gratitude.

In producing the book, the decision was made to include a large number of photos, some seen before and others that are new to this manuscript. The photos come from a number of sources, including my own extensive collection of Gloucester images. I would like to thank the following individuals for sharing photos from their collections: Jeff Thomas, Lillian Lund Files, Johann Arnorsson, and David Cox. I thank Fred Buck from the Cape Ann Museum for his yeoman work in finding images to fill out the storyline.

Finally, I am grateful to my agent, Laura Fillmore at OBS in Rockport, for her help in steering the project over its last year, and to Webster Bull, for agreeing to bring the manuscript forward and for counseling me several years ago on the need to shorten and focus the text. Laura in turn introduced me to a peerless marine editor, Jenny Marshall, and her advice proved invaluable as the manuscript went through its final restructuring.

Work on this book took up much of my leisure time over the past ten years, and I am thankful to Shirley for putting up with my obsession; in the early years I was blessed to be able to spend time every Saturday with my three dear aunts. I am also fortunate to have worked as a research director in a most unique aging institution, a place that has deepened my commitment to digging for details and making sure that I get the story right. Over my tenure at the Institute for Aging Research at Hebrew SeniorLife, I have worked for two strong and dedicated men, Murray May and Len Fishman, and I have also been privileged to hold the Alfred A. and Gilda Slifka Chair in Social Gerontology.

Let me end by hoping that in some small way the descendants of the Gloucester dorymen get to know these men through this book. It has been my privilege to come to feel close to men such as Charles Daley, Steve Olsson, and Jeff Thomas; and to my three grandchildren—Dylan, Max, and Gaby—I hope that when the time comes for you to read this book you, too, will be proud to be descendants of a Gloucester doryman.

John N. Morris

Contents

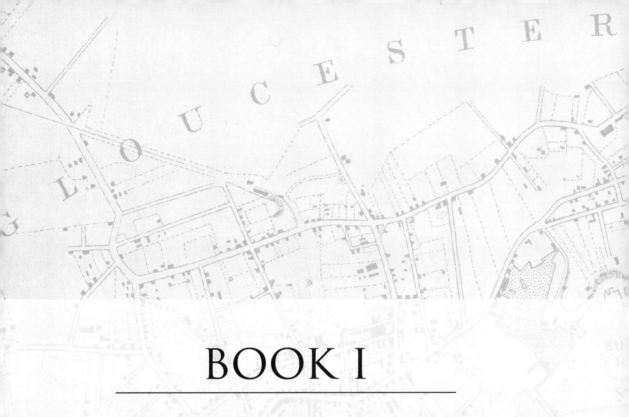

BOOK I

Climb to Greatness
(1623–1869)

*Gloucester begins as a simple fishing station,
but by the late 1860s will become the
fishing capital of North America.*

CHAPTER 1

A Time Before the Dorymen

In 1623, just three years after the Pilgrims landed at Plymouth Plantation, 14 Englishmen wintered over on Cape Ann, about 35 miles northeast of present-day Boston, to set up a fishing station on the southwest shore of the inner harbor. They chose a pristine location, protected from easterly gales by an extended peninsula, with fine beaches, fresh water, large stands of timber, and a broad expanse of flat land on which to graze their cattle, build shelters, lay out salt pans, and erect fish flakes (the raised wooden platforms on which gutted salted cod would be exposed to the sun for drying). The land was rocky, unpopulated, and ill suited to planting, but cod were plentiful, and the men eagerly anticipated the spring return of the British ships.* But the two ships came late, the catch proved small, and the price received for the fish once brought back to Europe did not cover the cost of the expedition. With a small catch and exorbitant rates for carrying the fish, no profit was made.

By 1626, the Cape Ann station had failed: Fishery proved a losing proposition. Most of the settlers, under their leader Roger Conant, who had never liked the location, picked up and moved west to Salem, driving their cattle before them. They wandered through the dense Magnolia woods, then walked along the coast of "Cape Ann side," modern-day Beverly, and finally crossed a major saltwater estuary (now called the Danvers River) as they made the trek west to start a new life.†

Cape Ann was largely deserted. A few of the original settlers remained to eke out a meager existence, and a vessel or two came up from Plymouth

---

*If the fish station had been attempted 15 to 20 years earlier, hundreds of local Native Americans (or as the Europeans would have called them, "savages") would have been found in residence. They grew grapes, peas, pumpkins, squash, and corn; their small villages stood on the shores around the harbor. In 1606, when Samuel de Champlain visited Cape Ann (Le Beau Port), he was greeted by the local chief, Quioihamenec, and a detailed account tells of the people and their reaction to the Europeans. In the intervening years, European diseases devastated Massachusetts Native Americans, killing 50 to 90 percent of the population. Different accounts describe the disease as flu, smallpox, and viral hepatitis; but whatever the name, Native Americans had no natural immunity, and tens of thousands died.

†Charles E. Mann, "The Old Planters," *Cape Ann Weekly Advertiser*, December 10, 1897. Of the 50 people on Cape Ann in 1626, more than 30 went to Salem, some of the more disgruntled men went back to England, and a few remained on Cape Ann. These few were on their own. They received no pay from the Dorchester Company that had established the fish station, nor were they supported by those who went the 25 miles west to Salem.

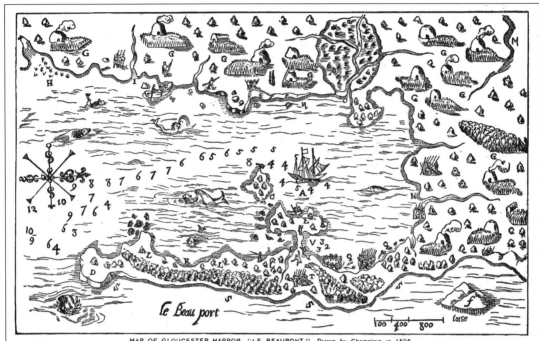

MAP OF GLOUCESTER HARBOR. "LE BEAUPORT." Drawn by Champlain in 1606
A, Place where their ship was anchored. B, Meadows   C, Little Island. (Ten Pound Island.) D, Rocky Point. (Eastern Point.) E, Rocky Neck. F, Little Rocky Island. (Salt Island.) G, Wigwams of the Savages. H, Little River and meadows. (Brook and marsh at Fresh Water Cove.) I, Brook (at Pavilion Beach.) L, Tongue of plain ground, where there are saffrons, nut-trees and vines. (On Eastern Point.) M, Where the Cape of Islands turn. (The creek at Little Good Harbor.) N, Little River. (Brook near Clay Cove.) O, Little Brook coming from meadows. P, A Brook. (At Oakes' Cove, Rocky Neck.) Q, Troop of savages coming to surprise them. (At Rocky Neck.) R, Sand Beach. (Niles' Beach.) The sea-coast. T, The Sieur de Poutrincourt in ambuscade with seven or eight arquebusiers. V, The Sieur de Champlain perceiving the savages. The figures probably denote the depth of water in metres.

)(*Champlain's 1606 map of Gloucester Harbor—the first known map of Cape Ann.*

on seasonal fishing trips. But for the next 15 years the peninsula was almost undisturbed by human endeavor. Then, in 1639, another group of settlers arrived to set up the first permanent village on Cape Ann, the First Parish, and in 1842 we find the name "Gloucester" used for the first time on a local tax record. These new immigrants were not fishermen but farmers, and for the next 60 years the people of Cape Ann turned their backs to the sea, scratching out a subsistence living from the stony soil.[1] They built no fleet, started no fishery, and as late as 1689 the seaworthy craft in the Parish consisted of only six sloops, one boat, and a shallop.

Yet, Cape Ann was on the verge of change. As the community grew from 400 in 1690 to 700 in 1700, it became more difficult to subdivide the rock-strewn land for the upcoming generation. First one, and then a trickle of families came to realize that Cape Ann's future rested with the sea. The Cape had rich stands of virgin timber with which to build vessels. The coastal inlets provided ready shelter as craft came scurrying in before an approaching storm, and the beaches and flat uplands around the harbor provided secure areas on which to cure fish and haul out boats for repair and storage. All the pieces were in place: By 1700, a number of the families on Cape Ann had made an irrevocable commitment to the sea, constructing small, two-man sloops for use in the nearby shore fisheries. Young men dropped hand-lines over the side in search of cod on their favorite spots on the shoals off Cape Ann. In 1713,* they began making larger, seven-man sloops, of 50-ton burden (with each ton equaling about 100 cubic feet of carrying space),[2] sending the craft north to the Grand Banks,† the most productive fishing grounds during the colonial era. The sloops were sturdy vessels, built to hold up under the most extreme conditions. Indeed, in the first half of the eighteenth century, the one and only report of a major loss of life occurred in 1716, when five vessels and 20 men were lost off Cape Sable Island, on the Scotian Banks, about 110 miles off the Nova Scotian coast. Recorded from this disaster is the name of the first man known to have drowned from a Gloucester vessel: a boy of 15 named Nehemiah Elwell. But the loss of the men and their craft was a unique event, leaving no lasting impression on the community.

Indeed, in less than a decade, Gloucester would send 49 large offshore craft, with 245 men as crew, to the Scotian Banks[3] off the east coast of Nova Scotia, as well as an unspecified number of vessels to the more distant Grand Banks. Fishing as a career was now common on Cape Ann.[4] At the same time, many of Cape Ann's smaller shore craft continued to fish on the nearby grounds of the Massachusetts and Ipswich bays.[5] Life on Cape Ann had changed. Experienced fishermen now moved into the community and the people of Gloucester accepted them into their midst. They almost all had English roots, as did those already in Gloucester, many moving to Cape Ann from other Massachusetts towns. Settlers carved out new homesteads on the land abutting the inner harbor. Front Street was laid out, new homes went up, rudimentary wharves were built, and flakes were set up for the curing of fish. The cornerstone of Gloucester's future was in place: Immigrant fishermen had arrived, and the community soon had a sizable fleet.

By 1740, some 70 offshore vessels were involved in the northern hand-line fishery, most making two trips per year. They left in April and remained

---

*Britain acquired the island of Newfoundland and the surrounding waters at the end of a protracted conflict with the French, known as Queen Anne's War in North America, and it was this act that opened the Grand Banks to the fishermen from Cape Ann.

†The Grand Banks are a series of interrelated plateaus off the eastern and southern coasts of Newfoundland separated from each other by deep troughs (in some places as deep as 650 feet). Gloucester captains used these troughs to help locate their position on the Banks, making soundings with a weighted line. They sailed for days and even weeks over the open ocean, guiding their vessels by a process of dead reckoning—determining their position based on an educated guess, chart readings, a compass, last known position, and the conditions of the wind, tide, and weather. Beginning with the Flemish Cap, the Grand Banks sweep from east to west to the south of Newfoundland, and include the Grand Bank, Green Bank, St. Pierre's, and Burgeo.

on the distant grounds for months on end. As they made ready to sail, wives, sweethearts, and children crowded in for a last hug, a last good-bye. As the sails filled and the boats pulled away into the harbor, the men on board craned their necks for one last glimpse of a loved one's wave.

Once at sea, the seven-man crew prepared for the long trip to the Banks, setting watches, assigning the helm, and getting down to the daily routine of a boat at sea. It took one to two weeks to reach the northern grounds (about 1,000 miles away), and the men fell into a routine of sailing the vessel, sleeping, and eating. Food was important, and they took turns cooking on a brick fireplace, using two iron pots, a tin teakettle, a cast-iron bake pan, and a spider-legged frying pan. They used the larger pot to boil meat and make fish chowders and the smaller one to make coffee. Once the meal had been cooked, they dumped it into a large tin pan from which each man took his share using a large iron spoon. The men called this "flip and dip."*

The Thursday and Sunday dinner meal consisted of boiled hearts and skirts, a salt pork dish. The Sunday main dish was boiled rice, sweetened in the big pot, with fried doughnuts for breakfast. On all days, the men had pounded leftover meat, mixed with boiled rice, sweetened with molasses, and made into pies. A tinderbox, flint, and steel kept the fire going; a matching set was kept in a safe, dry place.

Once on the Banks, weather permitting, the crew stood along the rails dropping baited handlines to the seabed below, usually two lines at a time. They waited for fish to take the bait and landed cod, one at a time—as one line came up another was dropped into the sea. Each line was a 900-foot length of strong twine, with conical lead weights to bring it to the bottom. At the end of the twine two short, lighter lines, known as gangings, held the baited hooks that enticed the bottom-feeding fish.

The skipper counted the catch that each man landed, for a man's share of the stock depended on the count. The sloops remained on the grounds until their food and bait ran out, and then sailed for home, usually with a full load of salted cod. Once the craft made the inner harbor, families helped unload the fish, and oxen teams hauled the salted cod to the grounds near the fishermen's homes.† Here the men from the sloop dried the fish hard on brush flakes§ in the fresh New England air, even as their vessel refit for a

---

*In the early days, the men sailed "on their own hook," everyone supplying his own hard crackers, coffee, sugar, tea, salt pork, and molasses. Later the men pooled their supplies, adding rye and Indian meal, beans, flour, butter, lard, and rice to their diet. Still later, the owners of the vessels supplied such necessities—although the owners' out-of-pocket costs were repaid out of the stock of the catch before any other distribution was made.

†In 1800 Gloucester had 553 dwellings.

§The salting process still required that the fish be processed and sold as soon as the vessels had returned to port. It was not until 1846 that a way was found to store salted fish in butts, so they could be held over until demand was high and market prices were at their best.

second trip to the northern grounds. Each skipper disposed of his catch with a local agent, although as time went on, a shore curer took the catch off the sloops. The curer received 1 quintal (or 112 pounds of dry, salted fish) out of every 14 he cured.

Meanwhile, cargo vessels stopped in Gloucester to pick up the processed fish for delivery to the nearby ports of Salem and Boston, as well as the distant old-world ports of Lisbon and Bilbao, and the British islands in the Caribbean. It was a profitable fishery, a family fishery, and salt cod was the key. The northern Grand Bank fleet landed about 4 million pounds per year, while the shore fleet brought in a slightly lower catch.

The fishing fleet soon had more than 100 vessels, and further growth seemed inevitable. But America's changing political fortunes over the ensuing decades would overwhelm the fisheries, and the people of Gloucester soon learned to pay close attention to the world around them. In the 1740s and 1750s, Britain waged war with France; in the 1770s Britain waged war with her American colonies; and in the early nineteenth century Britain first waged war with France and then with the United States. In each contest, the presence of the British fleet in North American waters kept the offshore fishing fleet off the grounds. Vessels rotted in the harbor and the fleet shrank, but the end of each conflict saw a new period of growth. Gloucester continually rebuilt her fleet.

The first period of decline began in 1741 and extended into the 1750s. Britain mounted expeditions against the French enclave at Louisbourg, Cape Breton Island, and the offshore fisheries on the Scotian and Grand banks were all but abandoned. It was just too dangerous to fish on these grounds, and many vessels in the Gloucester fleet wasted away on the beaches. By 1750, Gloucester had only 20 offshore craft, and the fishermen limited their efforts to the shore grounds off the Massachusetts coast. It was only when Louisbourg surrendered to the British in the late 1750s that fishermen again commissioned new offshore craft and returned to the Scotian and Grand banks.

Within 20 years, Gloucester had rebuilt its fleet to prewar levels. She had 145 vessels in fishing, a fleet second in size only to that of nearby Marblehead. Seventy were large, seven-man sloops and schooners that sailed out of the town's inner harbor. The other 75 were smaller two- and three-man shore boats that sailed largely from the Sandy Bay (modern day Rockport) area of the peninsula.

Overall, the Massachusetts cod fishery of 1775 employed 4,400 men and 565 boats. Of those, 146 sailed from Gloucester, with 900 men, and 150 from Marblehead. In that same year New England's catch came in at over

400,000 quintals of fish, most shipped out of the colonies: 178,000 went to Europe (80,000 from Marblehead, 30,000 from Gloucester) and 172,500 went to the West Indies (40,000 from Marblehead, 42,500 from Gloucester). In the winter, the fishermen of Gloucester freighted their summer catches of salted codfish to the ports of Lisbon, Bilbao, Cádiz, and the West Indies. For Gloucester, the annual salt-fish catch of the mid-1770s stood at about 8 million pounds.

But the prosperity was short-lived and came to an abrupt end when Great Britain and its American colonies went to war. Gloucester's inshore fleet was at risk every time a boat went to sea, and with the Royal Navy based in Halifax, Nova Scotia, fishing on the Scotian and Grand banks was precluded; most of Gloucester's offshore fleet wasted away on the beaches and wharves. Some, however, went out as privateers to hassle British merchant vessels. A few of the smaller boats did fish close to shore and landed their catch for home consumption. Men traded fish for salt, grain, and other staples of daily life, and occasionally sold it for the all but worthless paper money then changing hands throughout the colonies.

Along Gloucester's waterfront, wharves fell into disrepair, buildings crumbled, and the flakes and other apparatus used to cure fish disappeared. More tragic, the families of many fishermen were living in miserable huts since the men no longer had the money necessary to rent or keep up their homes. It became common to see "half starved . . . women & naked children, without wood for fuel, nor food but what they beg."[6] Children often slept in unfinished attics, with snow filtering in through the cracks between the boards, and no hope for warmth filtering up from the great room below. War was hard on the fishermen and their families.

CHAPTER 2

A New Future?

With war's end in 1783, the terms of the Treaty of Paris gave New England-ers the right once again to fish on the Scotian and Grand banks,[7] although they lost the right to do business in British markets. American trade now focused on Spain and Portugal in Europe, and the French- and Dutch-controlled islands of the Caribbean. Yet illicit trade with the British Carib-bean never stopped.

The owners of Gloucester's few remaining offshore schooners quickly fit out their craft for trips to the Banks, while a handful of individuals commis-sioned new vessels. Yet, while the northern fishing grounds had the potential to sustain a rich fishery, it failed to prosper. The seemingly endless conflicts between Great Britain and France limited access to the world markets, and the men of Gloucester ceased building new vessels for the northern offshore fishery.*

Of greater importance, however, 100 or more small shore boats remained on Cape Ann, and continued to sail to the Cashes Bank and Jeffreys fishing grounds in the Gulf of Maine. Individual skippers owned the boats, and the fleet was to carry Cape Ann's fisheries for more than a quarter century. Own-ers began to repair their long-neglected wharves, and new houses went up on the three streets near the inner-harbor section of the town.

By 1804, more than 200 boats participated in the shore fleet, 70 sailing out of Sandy Bay and others out of Lanesville and Annisquam.[8] And, as they returned from a day or two at sea, wives and children joined the skippers and crew in landing and salting the catch. Boys played about the boats and wharves, becoming familiar with the lines and gear. They spoke with their fathers and uncles about a life at sea, and boys as young as eight sailed as

---

*Gloucester was not a major offshore fishing port, and would not become one until the 1820s, with the advent of the mackerel. In the interim, Gloucester, like Salem and Boston, became a merchant maritime center. In the 1790s, square-riggers first carried mixed cargos from Cape Ann to Surinam in South America. The trade lasted for over half a century, and Gloucester benefited from her role in it. One trading firm was led by Col. William Pearce, who engaged in the Surinam trade from 1810 until 1834—others continued for several more decades. The Pearce firm imported molasses, which was converted into New England rum at a brick distillery near Center Wharf. Pearce owned 6 of Gloucester's 10 brigs in 1816 and several large schooners, and exported fish, furniture, pork, beef, hardware, and miscellaneous articles.

third hands on summer trips. But as one commented later in life, "It was not the best influence a boy could have. It was the duty of the younger to fill the pipes, mix grog and do the cooking. . . . It is scarcely to be wondered at that so many became addicted to strong drink."[9] At the same time, most people on Cape Ann led simple lives—alcohol was not a major problem. Theirs was a subsistence existence, with each family having small vegetable gardens, pigs, chickens, and fish. They ate what they produced, put up as much as possible to get through the winter, and salt fish was their only cash crop.

In those early days of the nineteenth century, as one walked about Cape Ann one would have seen the small, two-masted, gaff-rigged pinky boats of the shore fishermen. They were square-sterned or sharp-sterned boats of 10 to 30 ton burden, often less than 40 feet in length, with two or three cockpits where crew stood for fishing, a center fish hold, and a poorly lit forward cabin with two to four bunks and a brick fireplace. Pinkys were the signature craft of the day: safe, slow boats with just enough room for three-man crews to stand and fish. In April, after a winter ashore, men hauled them on the beach to caulk and "gray" their sides. Pitch tar was wiped on the bottom and burned, giving the boat a smooth, glossy surface. The men bent on sails and stowed a week's stores, including 2 quarts of molasses, 5 pounds of fat pork, 4 pounds of flour, 7 pounds of hard crackers, and a half barrel of water. They also brought on board flints to start the fire—and once at sea, the smell of brimstone was sure to make a squeamish crewman sick.

In port, Gloucester had few wharves; shore boats moored in the harbor and pinkys usually went out for a few days to a week at a time, searching for cod and haddock. On the fishing grounds, the crew stood along the waist, dropping baited handlines to the fish below. It was no different from the fishing of 100 years earlier. Spring saw the men landing fresh haddock and salted cod. The hake season began in July, and the pollock fishery ran from September to mid-November—after which the smaller craft would be hauled out, above the high-water mark, on long, greased spars; while larger craft were wintered either in Gloucester or Annisquam. Early in the year, as the season began, clams or alewives were used as bait, followed by herring once these small baitfish appeared off Cape Ann. The fishermen set herring nets at Kettle Island, Salt Island, and Stage Fort in the inner harbor, depending on the wind, and when at sea caught herring on the grounds.

As the weather improved, the pinkys went farther off the Massachusetts shore, to the eastern part of Middle Bank in Massachusetts Bay, or along the ledge extending east of Cape Ann to Tillies (18 miles out) and Jeffreys banks (42 miles out). Larger vessels went even farther, to Fippinies and Cashes banks (60 to 80 miles off the Cape Ann coast). For boats going to Middle or

Jeffreys, skippers first baited up with clams or alewives, added herring once on the grounds, and then spent up to a week in search of cod and other ground fish. The typical vessel made two trips a week during the height of the season. The three-man crew, or three men plus a boy, removed the fish entrails while at sea, but waited for their return to harbor to split and cure the fish. The salting process added a third to the weight, and once the fish had aged, it was typically brought to Boston for sale. Either the pinkys men or a local packet made the 35-mile sail.

However, by 1807 even the shore fishery was interrupted: President Thomas Jefferson imposed a two-year maritime embargo on the country,* and for 24 months only eight vessels of 30 or more tons made a fishing trip out of Gloucester. For those that did go to sea there were dangers more potentially hazardous than political interference. In 1808, a violent gale on Cashes took three small boats and 10 men. For days no one knew of their fate. Then, when news of their loss spread across the community, the grief was real and deep. Yet, as in 1716, the losses had no lasting effect on the community; the fishing continued.

When the embargo was lifted in 1809, the Cape Ann shore fisheries again met with success—fishermen sold their fish within local and regional markets, while the offshore fisheries continued to wither.[10] America's foreign trade in salted fish had all but collapsed; Newfoundland and Nova Scotia now sold the bulk of the salt cod to Spain and the islands.[11] Then, in 1818, the treaty ending the war between Great Britain and America gave U.S. skippers the legal right once again to fish on the northern banks. At last there emerged the first signs in almost a half-century of a true rebound in Gloucester's offshore fleet. In 1811, Gloucester's total fleet had numbered 208 vessels, with an average burden of 15.6 tons. Now, with new vessels coming on line, the fleet numbered 261 craft (41 of which were less than two years old), with an average vessel burden of 30.7 tons. One firm, the Gloucester Fishing Company, commissioned six large schooners and purchased one older vessel, ranging from 57 to 93 ton burden, for use in the Grand Bank salt-cod fishery.† Local men manned the vessels on their three-month northern trips,

---

*The European war between 1793 and 1832 can be traced to the execution of Louis XVI and the "reign of terror" in France. In that same year, the United States declared itself neutral regarding the war in Europe, while in 1794 the Jay Treaty with Britain tied the two countries closer together, although it did not open the British West Indies to New England salt cod. This was also a period in which the offshore craft of many countries pulled out of the Newfoundland fisheries, and the New England shore fleet continued to play a central role in the cod fisheries.

†All vessels owned by the Gloucester Fishing Company changed ownership by 1820, and the Gloucester Fishing Company ceased to exist. The schooner *Favorite* (57 tons) and *Amity* (93 tons) were sold out of town; William Babson Jr. and four partners took ownership of the schooner *Diligent* (87 tons) and *Crescent* (89 tons); Babson and one other partner took ownership of the schooner *Borneo* (88 tons); and a group of 10 took ownership of the schooner *Economy* (60 tons). The one older schooner purchased by the Gloucester Fishing Company, the schooner *Lydia*, which was 90 tons burden and 10 years old when purchased, was simply delisted by 1820.

and they returned with good fares—35,000 to 57,000 pounds of salt cod. Children of the crew gathered on the bank near the old fort as the schooners came up the harbor, calling out, "What luck? How many fish have you got?" People hoped that this once-vibrant fishery would revive, but it was not to be.[12] American fishermen had little access to foreign salt-cod markets, and the domestic markets could be served more cheaply by the shore fleet. By 1820, the experiment in the offshore fisheries was coming to an end, and in the next few years Gloucester saw springs when not a single offshore vessel sailed to the Grand Banks. Gloucester fishermen focused on the closer, highly profitable shore fishery.

CHAPTER 3

The Arrival of the Mackerel

As Gloucester entered the 1820s, there seemed little to suggest that the future would be different from the past. In addition to the small number of vessels going to the Grand Banks, in 1821 a few of the offshore craft began to fish on the more accessible Georges Bank* off the Massachusetts coast. They sailed in late spring and summer in search of cod and mackerel, and in the fall fished for pollock and herring on the more inshore grounds. Gloucester remained a small, weather-beaten village. The offshore cod fleet was modest. Few vessels were outfitted for the mackerel fishery, and even the shore fleet had begun to decline.[†] Few craft ventured beyond sight of land and earnings were meager at best. Gloucester had no large fish-processing firms, no marine railroads, no new piers, and few orders for new vessels. As one walked around the inner-harbor district the future did not seem bright.

From the Cut[§] on one end of the harbor to Rocky Neck on the other, one first saw the old ropewalk along Pavilion Beach in which rigging was made for square-riggers and fishing vessels. Also used for large political, religious, Fourth of July, and Sunday school gatherings, this building was a creaky old

---

*Georges Bank (41°N, 067°W) lies to the south and east of Cape Cod and Nantucket Shoals (with the latter south and east of Nantucket—41°N, 069°W). It is separated from Brown's Bank by the Northeast Channel. The area between the shoal area on the western side of Georges Bank and Nantucket Shoals is known as the South Channel. The greatest length north to south is 150 miles, and the greatest width is 98 miles. The depth of the water over the plateau of the bank ranges from 2 to 50 fathoms, and the bottom is chiefly sand. Its location between the Bay of Fundy and the Gulf Stream is said to cause the tides to run swifter than on other banks, moving in swirling currents, and it is the mixing of the spinning waters from the Gulf and Arctic currents that prove so productive to life. As the water moves around the bank and the sun penetrates the relatively shallow depths, huge crops of plankton are produced.

[†]Between 1818 and 1820, the total fleet of vessels of all types dropped from 304 to 249.

[§]The Canal at the Cut completes the extension of the Annisquam River (actually a tidal creek) from Ipswich Bay into Gloucester harbor, setting off most of Gloucester and all of Rockport from the mainland. The first minister, Rev. Richard Blynman, opened it as a private enterprise, the cut having a width to permit the passage of small shallops. It was kept open until 1704, when a storm filled it in with gravel. The owner cleared the sand and reopened the cut. In 1723 another high tide filled in the canal, and the Cut stayed closed for many years. In 1822 the Gloucester Canal Company created a cut that was 25 feet wide and a canal 200 yards long. They also erected a drawbridge at Western Avenue. In 1830 a stationary bridge replaced the drawbridge and stayed in place until 1840. At that time the corporation moved to fill in the canal to make a solid road, dropping stones into the opening. In 1864 the canal was cleaned and a new drawbridge was constructed. In 1907, when the wooden bridge was replaced with a steel structure, the city renamed the Cut Bridge to become the Richard Blynman bridge. The old name was felt to be uncouth.

structure that spoke of a past time. Going toward town one next met an old, eight-sided windmill, the arms of which made an "uncanny noise as they turned swiftly in the breeze." Down Beach Street, on the left, were two short cob wharves (built out of a mixture of straw, clay, and sand). On the right was a sandy beach with small fishing boats hauled out. It was here the men scrubbed, tarred, and painted their boats. Next came a few storehouses and the whitewashed ramparts of an old military fort.* Turning onto the main street, "the old Gloucester House" sat on the left, while on the right were various small buildings.† Along Front Street several short wharves extended into the harbor, one housing a distillery that took cargos of molasses to be made into New England rum. Next came Pierce's Wharf and warehouses, the center for ships engaged in foreign business. On the waterfront, Vincent Cove extended up almost to Front Street. A few yards up, at the foot of Union Hill, a wall enclosed a large field, preventing progress toward East Gloucester. From Vincent Cove to the head of the harbor there were no wharves until Rose Bank. There were almost no businesses or vessels at this end of the town; Rocky Neck itself was a sheep pasture. The inner-harbor area of Gloucester in 1820 did not present an impressive picture.

Moving to Sandy Bay (modern-day Rockport), on the exposed eastern side of Cape Ann, Dock Square was a bit more lively. Most of Cape Ann's shore fleet sailed from this harbor, and ox carts carried fish to the flake yards on the uplands, a mile from the wharves. About 15 coastal vessels called Sandy Bay home, carrying cured fish to Boston, New York, and other more distant centers. Continuing around Cape Ann, shore craft worked out of small fishing stations at Pigeon Cove, Folly Cove, Lanes Cove, and Annisquam. All sent their cured fish to the inner harbor for shipment to the commercial markets.

But with the advent of a vibrant mackerel fishery all would change: New schooners entered the offshore fleet—it was not unusual to see notices in the paper offering to build new schooners, such as that of Jacob Story of Gloucester announcing the availability for immediate sale of "a schooner (of) about 65 tons . . . to be finished by the first of April . . . for the banks and other fisheries." Wharves and support facilities were also expanded, the population grew, and craft from the outlying fishing stations moved to the inner harbor. Prior to 1819, the American mackerel fishery had languished— mackerel had been used largely as bait for cod, and sales of fresh mackerel

---

*The old fort was completed in 1743 on a hill that guarded the harbor at a cost of £527, appropriated by the General Court (Pringle, pp. 62–63). Complete with breastworks and eight 12-pound cannons, it was built to protect the harbor from the French.

†All of these had disappeared within a decade as a great fire swept through this part of the town.

were limited to local markets.* Then, in 1818, an American vessel made the first salt-mackerel trip, and everything changed: The annual U.S. mackerel catch rose from about 1,000, 200-pound barrels in 1814 to more than 100,000 barrels in 1819, and continued to rise. The country was at peace, and with a growing population and major commercial seaports up and down the coast, salt mackerel found its niche. Processed fish flowed into New York, Philadelphia, Baltimore, and New Orleans. Philadelphia accounted for one-third of the trade, and salt mackerel became a staple in the American diet.

In Gloucester, prior to 1820, the annual mackerel catch did not exceed 100,000 pounds (or 500 barrels). Small boats known as jiggers—pinkys with the pointed prow cut off and a bowsprit and jib added—fished on the nearby grounds off Cape Ann, on Cashes and Middle banks.† They stayed at sea for up to a week at a time, drifting over the grounds as the men dropped floating handlines over the sides, throwing bait on the waters to help attract the fish. The mackerel were brought up one fish at a time to be sorted by size and stored in barrels. Some were landed fresh and others came in salted. In 1820, the men moved to a more stationary fishery, anchoring over schools of mackerel and using weighted handline jigs dropped over the sides of the vessels. The whole crew now had access to the fish, and the catch exploded.

In 1822, the Gloucester fleet found large schools of mackerel on Georges Bank, the vast fishing ground east and south of Cape Cod, and with this new find the productive season now extended from late spring into early fall. An anchored fishery made all the difference—drift fishing on the shoals was dangerous, strong tides washed over the area, and vessels could be lost. But now, with a dedicated summer fishery on Georges, the catch rose: 400,000 pounds in 1821, 1.3 million pounds in 1824, and 6.8 million pounds in 1828. By the 1860s, the annual Gloucester catch came in at more than 30 million pounds, and the mackerel fishery helped carry the local economy for over half a century.

By 1827, the Gloucester fleet had grown to 238 craft. There were 5 large seagoing ships, 1 barque, 7 brigs, 46 coastal (trading) schooners, 14 ships, and 165 fishing schooners—the latter made up of pinkys and jiggers. The average burden of the fishing schooners came in at 39 tons.§ The fleet

---

*There had been a more substantial mackerel fishery in the eighteenth century, but it did not continue after the war with Britain.

†The more inshore grounds included Cashes Ledge Bank (about 80 miles due east of Cape Ann—42.8N, 069W), Jeffreys (between Gloucester and Maine—43°N, 069°W, in the Gulf of Maine), and Stellwagen (between Gloucester and Cape Cod—44°N, 060°W).

§From 1816 to 1827, while the inshore fleet dropped from more than 100 small boats to only 38, the fleet of larger ocean schooners rose from 143 to 187 vessels. The total tonnage of the fleet went from 4,347 tons to 7,256.

had at last returned to its prewar size in number and tonnage. It had taken 50 years, but Gloucester's fleet was back.

Many in the community had a stake in the fleet. Eight individuals owned three or more schooners; 15 owned two schooners; 104 owned a single vessel. It was also common for the owner of one vessel to have relatives who owned other vessels in the fleet. The fishery was a family business, and the survival of extended families depended on the success of the catch, one vessel at a time. The fishermen made $150 to $200 (about the same as an agricultural worker), and at home they had small vegetable gardens, chickens, and a pig or two to carry them through the winter.

From 1827 to 1833, the fleet increased by 72 craft, including 25 new fishing vessels.* Gloucester now sent out 190 schooners, the largest fishing fleet the town had ever seen. Gloucestermen fished on the shoal areas of Cashes, Middle, and Georges banks. They searched for mackerel and cod, and brought in larger and larger fares. Boats drifted over the shoals in the daytime and ran to the safety of deeper water at night. They returned to the grounds at daybreak, unless a head breeze sprang up. It was a safe fishery; few vessels or men were lost in these years.

On some trips, jiggers found mackerel so plentiful they filled their boats in a few days of good fishing, while on others they had a more protracted stay on the grounds. S. B. Brown described these different types of trips:

> I made one trip with Edward Wonson in August 1832. Our crew consisted of old Charlie Wonson of Rocky Neck, Reuben Rich of Cape Cod, Joseph Warren and James Green of the Point, and myself. We had bad luck, so we made but one trip. The next year I went to Gloucester, hunted up my old skipper, who was still master of the same boat, and went with him that season. I recollect well the great school of mackerel that struck Middle Bank that year. Sept. 22nd, at ten o'clock at night there were some two hundred sails at anchor, twenty-five miles southeast of Eastern Point light, in a dead calm, when our skipper sang out, "Here they are boys!" At the same moment every vessel in the fleet commenced the catch. We fished for three days and filled everything, even our boat, and struck on deck until we were knee-deep in fish. Then, a breeze springing up, we ran in and packed out two-hundred and eighty barrels, and returned to the Bank just as the wind left us.

With days such as these, the mackerel fishery of the early 1830s reached new highs—10.3 million pounds in 1830 and 14.0 million pounds in 1831.

---

*America was also changing. The opening of the Erie Canal in 1825 ushered in a new era of westward expansion and economic improvement; and the fish of Gloucester and New England followed the path inland. The best of Gloucester's salt cod, which might before have gone to Bilbao, Spain, now went to western New York, Ohio, and the Mississippi valley.

The season, too, was expanding. There were late winter and early spring trips to Georges for cod and early summer trips to the Gulf of St. Lawrence for mackerel. The Gulf grounds lay between Nova Scotia and Prince Edward Island, and extended west to the Bay of Chaleur. The fishermen called these grounds the Bay, and each year the vessels of the mackerel fleet made one or two northern Bay trips as well as their regular trips to Georges, Cashes, and Middle banks. The success of the year's effort depended on the mackerel, and the Georges cod fishery helped fill in the time. The Bay fishermen followed the fish, setting lines in deeper waters when the fish remained offshore, and coming near to shore as the fish made their annual migratory runs. Early summer fishing in the Bay could be spectacular. Hundreds of craft from Gloucester and other ports were joined by quaint old-fashioned boats of the native fishermen. The white sails of the Gloucester vessels stood out in stark contrast to the brightly colored sails of the local fishermen.

Sylvannus Smith described trips where his boat sailed into the Bay to fish on a long bar that extended off the shore of Prince Edward Island. The men were all from Cape Ann, and looked little like the images of the later offshore fishermen. They wore satinet trousers, red flannel shirts, "guernsey" frocks (oiled-wool sweaters), pea jackets, low wide shoes, and tarpaulin hats. Once on the Bay, they made handline sets from the deck, hitting many of their favorite haunts in search of the elusive mackerel. They set lines at night and fished into the morning. The summer weather in the Bay was normally fine, for it was an inshore sea with an almost total absence of fog and a wooded shore that threw forth delicious odors. Yet, the Bay could be subject to sudden and violent gales, "tho' not usually of long duration."

On a clear day, all watched for the telltale ripple mackerel made when swimming near the water's surface. Men waited with baited hooks and lines at the ready, throwing bait overboard to help concentrate the fish. Then over went the lines. Quickly they came up with mackerel attached, and a curious twist of the wrist sent the fish "slating" into a barrel. Back went the line, and so it continued as long as fish took the hook. All worked at a breakneck pace, for the school could disappear as quickly as it came, the fish plunging into the lower depths. The fishermen ended their day by dressing out and salting the catch.

Yet, handline mackerel fishing could be unpredictable—fish movement was precarious, the mackerel rising to the surface one day and unseen the next, with unpredictable seasonal shifts as the fish moved north from North Carolina to the Bay. In the second half of the 1830s, the fishery went into a brief period of decline, the catch dropping to only 3.1 million pounds in 1839. The fishermen were unprepared, not knowing why the mackerel had disappeared. Fog hid the fish, rough weather kept the fleet in port, and the

)(*In 1840 the fleet is expanding.*
A partially completed schooner
sits above the highwater mark
in Gloucester's Vincent Cove.
A horse cart delivers a load
of timber, as skilled crafts-
men work the wood, caulk the
sides, and lay down the deck.

mackerel vanished from view. Even more troubling, with the success of the previous year, few craft were ready to venture into one of the two poorer paying alternative fisheries: mid-year fishing for cod and halibut. Gloucester had become dependent on the mackerel. As vessels returned empty or with short fares,* men had no way to recover from the loss. In the best of times, many of Gloucester's mackerel fishermen spent up to five winter months ashore with no employment—only a successful mackerel season made life bearable. Now, the men looked to their future with dismay. One could either wait for the mackerel to reappear or take the drastic step of refitting for the more limited cod and halibut fisheries. Even here, there were risks—a few decades earlier the halibut market had been so unprofitable that fishermen threw their catches overboard at Eastern Point; the fish could not be sold or given away. Thus, in such troubling times owners sought alternative uses for their vessels, fishing for cod and halibut in February and March, cod and pollock (another ground fish) in October and November, and mackerel from April to September. In the 1860s, it was just such a mind-set that drove vessel owners into the dory fishery. Gloucester had the vessels, an immense fleet, and owners found ways to keep them at sea and were ever ready to consider new innovations, whether which grounds to fish, which bait to use, or how to fish.†

* The term "fare" or "hail" indicates the amount of fish landed from a trip, while the term "stock" indicates the dollars received for the fish.

†In 1821, the schooners *Three Sisters*, *Eight Brothers*, and *Two Friends* spent a few days on Georges successfully setting their handlines as they drifted over one of the shoals, afraid to anchor because of the strong current. However, it was not until the 1830s, when the mackerel fishery faltered, that larger numbers of the offshore craft conducted a productive cod fishery on Georges. The 1830s also saw the coming of a successful Georges halibut fishery. The schooner *Nautilus* under Captain John Fletcher Wonson was the first to return with a fare. In 1831 he fitted his five vessels and others followed his lead.

CHAPTER 4

Period of Expansion

In 1845 Gloucester entered a remarkable 35-year period in which the mackerel fishery was the centerpiece of life on the Banks. The catch jumped to 9.7 million pounds and held fast at that level for many years to come. Gloucester was now the center of a large offshore mackerel fishery, supplemented by smaller ground fisheries in cod, haddock, and halibut.

In 1846, the winter fleet that sailed to Georges rose to 29 vessels with an average burden of 62 tons: Georges was now Gloucester's offshore grounds, not the Grand Banks of the prewar years. The first 10 schooners left on January 5 and made an average of five trips during the season—fishing had become a year-round activity for many in the fleet. The next 11 left on February 6, while the remaining 8 started on March 2. As yet the fleet did not carry ice, and many of the boats had a compartment amidships in which to hold live halibut in salt water. Such compartments often leaked.

By season's end, Gloucester's combined fisheries had upwards of 3,000 landings. But the community was ever alert to new ideas for expansion, and two innovations were to prove important to the future growth of the fisheries: the founding of a mutual vessel insurance program[13] and the coming of the railroad.

By 1841, the vessel insurance rates of the old Marine Insurance Company (controlled by Gloucester's retired square-rigger captains, and not the fishermen) had become prohibitive, and a meeting was held in Epes Merchant's store to organize a locally based mutual insurance company. The company functioned for six years. Then, with a growing fleet, a larger, more stable concern took its place—the Gloucester Mutual Insurance Company. Owners now insured their craft under a pooled, shared-risk arrangement, and trips throughout the year became possible. The Gloucester Mutual met the needs of the fleet over the next 20 years. But when winter fishing became more common, the company was reluctant to cover the risk. Without insurance no one could go to sea, and in response business and fishing interests came together to form the Cape Ann Mutual Marine Insurance Company, a firm willing to take winter fishing risks.

With a vibrant offshore fishery and a guaranteed insurance program, the next change that helped Gloucester become a premier fishing port was the coming of the railroad.* This led to the building of fish processing plants and a diverse array of support industries. In 1845 the Massachusetts legislature approved the construction of the Gloucester branch of the railroad. By July 1847, the rail bed and six miles of the superstructure were in place. In August the first train ran from Boston to Manchester (Gloucester's nearest neighbor to the west). By September, the rail sleepers reached the new bridge over Gloucester's Little River. Finally, in January 1848, Gloucester had a railroad, and a group of citizens organized a company to buy and sell halibut. John Pew and others leased the Redding Wharf at the foot of Water Street to receive the catch, and agreed to pay skippers a uniform rate for their fish. Within a year, several other companies had been formed for the same purpose, and not a train left Gloucester without a quantity of boxed fish, up to 20 tons a day heading west to Boston. All had changed. It was a time of optimism: New processing firms had been set up, and local shipyards in both Gloucester and the neighboring towns of Essex and Rockport took orders for the next generation of vessels.

WITH NEW PROSPERITY, however, came a darker tale—newspaper headlines began to tell a recurring story of disasters at sea. The Gloucester outfitters no longer limited their craft to a simple shore fishery. Vessels could not scurry back to port at the first sign of an approaching storm. There had been losses in the past, but the community had seen them as the unusual twist of fate in a safe and secure fishery. Now there came an unfamiliar sadness.

In 1842 no fisherman had drowned; no vessel had gone to the bottom. Newspapers and private letters reveal no sense of communal concern for the safety of the fishermen. But in 1843, for the first time in recent memory, the local paper reported that the loss of a vessel had thrown a "pall of sadness" over the town. The schooner *Byron* had gone down in a late August storm. For one month the community awaited her return, hoping for the best. But on September 20, *Byron*'s owners gave her up for lost; 10 young men were gone. No one knew of their last day, of the waves that swept over the decks, of the last prayers or thoughts for loved ones left behind. Men and schooner were gone. All the men were young, some left families, and others left this world without having taken a wife or fathered a child.

---

*The first link in a North Shore railroad came in 1835, when the General Court (the state legislature) approved the leg between Boston and Salem, stopping some 20 miles from Gloucester.

The community was shocked, and the words in the local newspaper, the *Telegraph*, could not be mistaken: "Never have we known a loss that awakened a deeper interest or more sincere commiseration. It is a loss that even in the lapse of time, or amid the luxury and bustle of this changing world, will not soon be forgotten." The disaster also caused the community to focus again on the losses of individual vessels over the previous decade, severe and frequent: In 1830 the schooner *Olive* was lost with all hands. In 1837 three vessels with 26 men were lost on the return from Georges in a tremendous gale. In 1840 the *Ida* disappeared on Georges with a crew of eight. In 1841 the *Forest* went ashore on Cape Cod and all on board were lost. Something different was happening, and Gloucester's vast fleet now brought both prosperity and sadness. Bigger vessels, larger crews—7- to 10-man crews were common—and the fishing on Georges had changed the landscape.

Over the remaining years of the decade, Gloucester lost an average of four vessels and eight men a year—almost all on Georges, and often in the harsh weather of late fall or mid-winter. In 1849, 43 men drowned, 35 on four vessels that went down, and such losses became even more commonplace in the years to come. In April 1850 the *Telegraph* reported that the Georges fleet had met with severe weather two months earlier. Many of the 100 vessels, and the 800 men who served on them, had yet to return. There had been growing concern in the community, but the paper had not publicly reflected it. The practice of the day had been to avoid early speculation. Such thoughts were seen to be private, and newspapers sought not to raise concerns; they trod lightly on the possible fate of men and vessels. But now, with so many at risk, the local paper felt duty bound to tell the community of a change in policy:

> We have not thought it proper when reports have been in circulation respecting any of our vessels, to give publicity to the facts of the case until a reasonable time had elapsed, and until all hope of their safety are given up. A different course would, in many cases, create unnecessary alarm and distress in the families of those particularly interested. By many the announcement of a case of this kind in the public prints is considered as the extinguishment of all hopes, and to thus suddenly crush all hopes in a sensitive mind is a cruelty which we shall endeavor always to avoid.[14]

Henceforth, accounts of disasters would become common; people, the editor said, "had the right to know."

A report soon followed on the status of one of the missing Gloucester schooners, the *William Wallace*, Stephen D. Griffin master. The skippers of three other craft had reported seeing her in February, but there had been

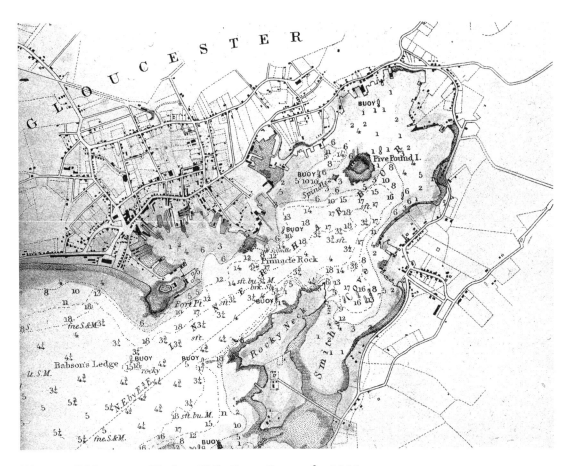

)(*Map of Gloucester Harbor, U.S. Coast Survey for 1855.*

no subsequent sightings. She had sailed with eight men, and the paper now acknowledged their loss.

Over the next two months similar stories appeared on the fate of three other vessels. The paper first reported the vessels as overdue, then published a later statement that the vessel owners had given up all hope for a safe return. By the end of 1850, Gloucester had lost six vessels and 39 men. The town had moved into an era when such losses reinforced the unique bond between community and fishermen.

CHAPTER 5

The Mackerel Prosperity of the 1850s

Through the 1850s, robust landings of mackerel and cod supported a growing fleet and town, with foreign-born residents comprising 20 percent of the population.* Schooners now went to sea with salt and ice to preserve the catch, and in the dead of winter, Newfoundland herring for use as bait.[†] Confidence was in the air. Vessel owners sold old-style pinkys to buyers from Cape Cod to Maine, even as the Essex shipyards[§] launched a superior class of replacement schooners. Called "sharpshooters" and "clippers," these schooners were designed for speed, with sleek hulls, shallow drafts, tall masts, elongated spars, and an increased sail area.[15] They sliced through the sea under an immense load of canvas, reaching the grounds in record times. There, handlines went over the side, as the skipper moved from one spot to another until the hold was filled and he again piled on sail in a race for home—top dollar went to the swift. The sharpshooter was a 65-foot, 70- to 80-ton vessel, while the faster clippers came in at about 90 feet and 100 to 110 tons. Speed, tonnage, and greater deck room made the clipper Gloucester's workhorse for years to come.

Around the Gloucester waterfront, 15 firms sprung up to process the 20 million pounds of fresh and salted fish landed each year. Vessels now went to sea with fresh ice to preserve bait and fish—a practice that had begun in the mid-1840s, but only became prevalent in the new decade, with the Georges

---

*Foreign-born made up 20 percent of the population: 435 from Ireland (a half-dozen Irish families had resided in Gloucester prior to 1840), 384 from Nova Scotia, 63 from Newfoundland, 57 from Portugal, 53 from England, 42 from Scotland, 18 from Prince Edward Island, 12 from Denmark, and 31 from other places.

[†]In the 1850s, George Blatchford, skipper of the schooner *Laurel*, demonstrated that freshwater ice could successfully preserve saltwater fish, taking the ice from the Kennebec River. The following winter, ice was housed on various wharves for the purpose of supplying the vessels. At about the same time, Captain Henry O. Smith of the schooner *Flying Cloud* returned in early January with a cargo of naturally frozen herring from Newfoundland. Herring were abundant in the bays of southern Newfoundland, and a vigorous Newfoundland herring fishery soon emerged to supply bait for the Gloucester winter fleet. Up until that time, the Georges winter fleet had relied on herries, poggies, or mackerel caught on the Georges for use as bait when going after cod and halibut.

[§]Shipbuilding in Essex began in 1668, when the Crown granted an acre of land to the inhabitants for the establishment of a yard to build vessels for the locals' use, and to employ workmen to that end. In 1852–1853, the 15 Essex shipyards built 78 of the 103 schooners added to the Gloucester fleet. An additional 16 came from yards in Gloucester.

VIEW OF GLOUCESTER, MASSACHUSETTS.

✕ *Gloucester in the 1850s: The fisheries have brought prosperity, businesses crowd in on the waterfront from the old fort (left) to Five Pound Island (right). Smoke rises from the processing plants; schooners sit at docks or head out to sea.*

fleet taking an average of 6 tons of ice per trip, while the fresh halibuters went out with an average of 16 tons of ice. Wharves were extended and firms modernized their buildings. Life revolved around the inner harbor and the streets and pathways that encircled it. The town had 23 forges, 4 bakeries, 2 small beer breweries, and 2 livery stables. A daguerreotype photographer produced 1,000 likenesses in a year, while 6 lofts made 1,270 sails. Homes on Front Street were beginning to be converted to eateries, shops, and boarding-houses; one-story buildings were common and the street itself was unpaved.

Gas and the magnetic telegraph came to the community, and the town fathers provided free gas to anyone who put up a streetlamp. They believed this helped check "the operations of evildoers," thereby increasing the safety and comfort of the people. Special note was made of the excesses of the new class of immigrant fishermen starting to come into the town, men from Portugal, Nova Scotia, and Scandinavia. Prior to the mid-1840s, the schooners were crewed by the men of Cape Ann, men of English roots who lived in the harbor area, Annisquam, Lanes Cove, and Rockport; but new men were needed, and immigrant fishermen flowed into Gloucester. The town fathers and others in the community said such newcomers drank too much, lost control when in port, and visited prostitutes. There was also a general concern with rowdyism at public concerts. One account said, "No decent person . . . (could) attend a public concert without being thoroughly disgusted with the behavior of a set of miserable fellows who ought not be suffered to be at large in any community." Another account said that rowdies followed the police when they had a prisoner under arrest. They were noisy and rude. It had to stop, and the town fathers considered appointing a vigilance committee to check the evil.

Yet there was another view of the new immigrants. While the town fathers might call them rowdies, others said they were just ordinary fishermen letting off a little steam after weeks at sea. But Gloucester was undoubtedly changing. There was a rough edge to this burgeoning community, and the more established Cape Ann citizens were having a hard time accepting the new element in their midst.

Gloucester had become a bustling, changing town, whose principal street (Front Street) was as yet unpaved. Gravel was thrown down to fill the ruts and water sprinkled to hold down the dust. Boardinghouses went up, and young, unattached fishermen walked the streets. Yet even in the 1850s, with their unparalleled prosperity and signs of social upheaval, the people of Gloucester were ever vigilant for anything that might threaten the emerging fisheries. The future rested with the fleet, and the people quickly contested any outside forces that might imperil the fishermen.

ON OCTOBER 15, 1851, almost three-quarters of Gloucester's offshore fleet, 112 vessels, sat quietly on the Bay at Prince Edward Island. Warm air flowed across the water as the mackerelers dropped their baited lines over the rails. It was still, and a thin layer of clouds slowly overspread the northern sky. Only as the sun began to slowly set did the northwest sky take on a more lurid, brassy appearance, as storm clouds moved in.

As darkness crept over the fleet, the winds rose, the seas stirred, and skippers readied their craft to ride out the storm. Torches went high in the rigging. Double watches patrolled the decks. Helms were lashed in place and skippers dropped giant banks' anchors to the seabed below. When the storm hit, all they could do was sit it out and wait.* Through the long night the winds howled and at the half-light hour of dawn still ripped through the

---

*Prince Edward Island is about 100 miles long and 30 broad, sitting 30 to 35 miles off the mainland. The northeastern coast curves inward in an arching 18- to 20-mile bend. Here, harbors such as Cardigan, Murray, Richmond, and Holland offered good anchorage from northerly gales, but in a southerly storm, such as this, these harbors were less secure, as sandbars obstructed their entrances.

rigging and masts. Many a sail had been torn apart, spars were gone, and vessels were in danger. The storm continued much of the day, and before long, anchor lines broke or men were forced to cut the cables as skippers struggled to save their craft.

Some schooners hoved over, filled with water, and disappeared below the churning waves. Others smashed together and disappeared. The dead and dying littered the waters. The howling winds and the storm-driven flood tide had carried many vessels high into the sands of Prince Edward Island's northern beaches. The land was littered with vessels, masts, barrels, clothing, and the bodies of dead fishermen. It was an appalling sight. Corpses were everywhere, mixed in with bits of clothing and the splintered wood of the schooners. In the first 24 hours, more than 50 corpses came ashore between Brackley Point and Cavandish, and more tumbled ashore on each incoming tide. One lifeless fisherman had a young boy lashed to his back; others had tied themselves in the rigging only to die as their craft smashed onto the rocks. Most of the bodies carried no identification, and newspapers carried full descriptions of the dead. Captain Tarr of Gloucester described one such man as stout, about 5 feet 10 inches tall, wearing a red flannel shirt, dungaree frock, white flannel drawers, and blue satinet pants. Captain Tarr had a lock of the dead man's long, fine brown hair, and knew in which cemetery the body could be found.

The people of Prince Edward Island buried the dead. But as each incoming tide brought new horrors, it must have seemed an impossible task. After a week's search, Captain James Wilson of Dennis found the bodies of two of his four lost sons, their clothes torn to shreds. But that was far from the end of the grief. Bodies washed ashore for weeks, and in the most deplorable of conditions. Many had no flesh on their hands, faces, and arms, and remnants of clothing were often the only way to identify the nationality of the man who had been lost. In total, about 200 men were lost, and the people soon labeled the storm as the Great Yankee Gale.

Of the 400 New England vessels in the Bay, 15 went down in the storm and many more were abandoned on the shore. Two of the lost boats came from Gloucester. The *Flirt* had a crew of 15 and the *Princeton* a crew of 10. The *Flirt* was found off Malpee—anchored, full of water, a total loss. Captain Aaron Stubbs and the crew had cut away the mainmast, which carried the foremast with it. But one could not tell if the craft had capsized or filled with water at her anchors. The cabin held 5 bodies, and the other 10 men surely rested among the many unidentified fishermen lying in Prince Edward Island graveyards. The fate of the *Princeton* probably mimicked that of the *Flirt*: She had gone to the bottom, east of Prince Edward Island. Her skipper,

Thomas Guard, 34, came from Gloucester. He left a wife. Two of the other crewmen also came from Gloucester, two came from Marblehead, three from East Boston, and two from Maine.

In all there had been 112 Gloucester vessels in the Bay: 110 made it through the storm. Of those that survived, 36 went ashore, many to be quickly sold. Among the beached craft, the *Garland* sat abandoned in 18 inches of water at high tide, her hull filled with sand. Nearby, efforts to save the *Lucy Pulcifier* began almost immediately, and they were to succeed. Further up the beach, the *Eleanor* sat high in the sand, dry and dismasted. At Tignish near the North Cape, the *Powhatten* was wrecked beyond repair, while the *Golden Rule* was high and dry but not damaged. Agents from Gloucester's mutual insurance company came north and sold off many of the stranded vessels. There was no shortage of Nova Scotian takers: Gloucester craft had superior oak timbers, not the soft pine found in Nova Scotian craft. Oak was less subject to rot, and a salvaged vessel, once repaired, could last for years.

Yet, while the devastation was stunning and the sadness real, the losses did not cause the men of Gloucester to turn from the sea; there were, after all, no real alternatives. Indeed, in 1852, 50 new craft were added to the fleet, and Gloucester now sent out 2,650 fishermen on 357 vessels.

The mutual insurance company had provided coverage for the two lost vessels and had worked to sell off many of the grounded schooners. Owners received an average of $3,000 per vessel—a fair price—and they reinvested the money in new craft. Thus did faster, more modern schooners come into the fleet.

THE CONTINUED SUCCESS of the mackerel fishery was the primary driving force behind Gloucester's investment in its fleet. Thus, it came as a shock to see the annual catch drop to only 7 million pounds from 1852 to 1854. But the decline had little to do with the presence or absence of fish on the grounds, nor was it related to the skill of the Gloucester fishermen. Rather, it came about because in 1852 the British moved to restrict American access to prime fishing grounds in the Bay of St. Lawrence. There had been sporadic interference earlier, but this was different.* British colonies sent powered vessels to patrol the Bay, while the London government sent armed Royal Navy ships.

---

*An August 1852 account in the Halifax paper contains the official record from the office of the Court of Vice Admiralty at Halifax on the number of American vessels seized since the treaty of 1818, and the numbers are not large. "The first was the *Hero*, seized June, 1838 and condemned Jan. 28, 1839. Two vessels were seized in 1838, 9 in 1839, 5 in 1840, 7 in 1841, 1 each in 1843, 48, 49, 50, and 51. Of these three were restored." The last seized from this list, the *Tiber*, taken October 29, 1851, was condemned and sold at auction, and repurchased by her American owners for $2,500.

These actions shook the Gloucester community. For most of the fleet, the year's profit depended on the fall Bay fishery. The winter and spring cod fishery brought in a good catch, but with the damage sustained in severe winter gales, such fishing brought only small gains.

The newspaper insisted the community would not stand for this new British policy. Gloucester fishermen had a right to these grounds, and it was inconceivable that the British had imposed new limits out of the blue. The people rallied behind the fleet, and they became most accomplished at reciting the American side of the story: Their rights went back to the treaty of 1783, with only slight modification by the treaty of 1818. As the mackerel fishery had expanded, so the Gloucestermen had fished throughout the Bay of St. Lawrence, following fish within sight of land and into the Bay of Chaleur. There had been instances of interference, but they had amounted to little, and Gloucester insisted on a return to the status quo. From the British and provincial viewpoint, the issue was not American access, but commercial reciprocity: permitting duty-free entry of Canadian fish and other products into the American market. If America agreed, the British government was willing "to open to American fishermen the fisheries along the coast of Nova Scotia and New Brunswick." But for now, British vessels of war had come to the Bay. The Province of Canada (Quebec) outfitted an armed vessel, giving it the authority to seize American vessels fishing within the restricted zone and imprison all on board. Nova Scotia had four armed cruisers in the Bay; New Brunswick placed a cutter in the Bay of Fundy; and, from Prince Edward Island, Her Majesty's steam frigate *Devastation* was under the instructions of the Governor of the colony.

By late July 1852, with 300 American vessels in the Bay and mackerel scarce, there had yet to be a confrontation. On arriving back in Gloucester from the Bay, Captain Whalen, of the schooner *Flying Cloud*, reported no annoyance from the British, but he had caught his fish outside the limits claimed by the British government. Soon, however, the British seized two non-Gloucester vessels, followed within days by the taking of a Gloucester schooner, the *Helen Maria*. She had sat at anchor near Sable Island, not in the Bay but on the Scotian Banks off the east coast of Nova Scotia. A British cutter had boarded her, found fresh bait, and taken her to Pubnico. The crew had not been fishing at the time and claimed they had come within the new coastal limits only to buy supplies. About a week later the British took the American schooner *Union*, and the people of Massachusetts petitioned the U.S. Congress, Secretary of State Daniel Webster, and President Franklin Pierce, asking for "prompt and efficient action" on behalf of the U.S. fishing vessels. But as this request sat in Washington, the British auctioned the seized American fishing schooner *Coral*. The American Consul made the high token bid of $155—the fishing gear bringing an extra $20. A week later, the British seized another Gloucester vessel, the *Florida*. She was one of several craft that had anchored in the Bay of Chaleur to pass the Sabbath. The boarding officer said "they had no business in that place," even though they were five miles from shore. In the officer's view, American craft could not enter Chaleur through a close headland. When the distance between opposite headlands was within a few miles (as it was in this case), the British maintained that the Bay on the other side was not open to international fishermen.

In late August, returning skippers reported they had remained at sea while in the Bay and thus experienced no trouble. Meanwhile, the Canadians released the *Helen Maria*. Returning fishermen said the officers of the cutter seemed ashamed for taking in a craft when all that was found was a bucket of fresh tinker mackerel on deck.

On September 7 the British condemned the *Florida* and sold her at an auction. With such pressure on the fleet, the mackerel catch fell. Being forced to fish farther offshore, the schooners missed the run of the fish. Losing access to the Bay of Chaleur, American vessels flocked to the waters around Prince Edward Island. They had notable late-season success, with the waters both inside and outside the three-mile limit teeming with mackerel. Then, on October 15, a severe gale went through the Bay, and 21 vessels went ashore, including 8 from Gloucester. One boy from Malpeck lost his life.

By the end of 1853 it seemed that the politicians in Washington were at last at work—a reciprocity treaty was nearing completion. But then, for reasons of both politics and timing, Congress failed to pass the bill during the 1853 session. Rules governing U.S. access to the Bay remained in force, and even though the British seized few vessels, by November Gloucester's annual mackerel catch came in at only 7.2 million pounds. Finally, however, a new treaty was agreed on in June. British cutters left the Bay of Fundy, and the molestation of American vessels ceased. Gloucestermen could once again go wherever the fish took them.

The mackerel fishery quickly rebounded, centered on the Bay. Many skippers made a few Georges trips in late winter, mackereled in the Bay over the summer and into early fall, and sometimes even made an early winter trip to Newfoundland for herring. But, more important for Gloucester, the people had shown that they could come together behind their fisheries.

CHAPTER 6

Civil War

In November 1860, the people of Gloucester, about 10,000 in total, celebrated Abraham Lincoln's election as President of their country. They illuminated their homes and joined in festive processions through the streets of the town. Lincoln had been the overwhelming choice of the people of Gloucester. Yet, by early spring, in what seemed like a mere blink of an eye, "Deep South" states had left the Union and the people of South Carolina had seized the federal garrison at Fort Sumter. The Civil War had begun.

In Gloucester the beat of the drum soon sounded as young men stepped forward to enlist. American flags fluttered high in the rigging of schooners fitting out for trips to Georges and the Bay. Patriotism was in the air, and families gathered at the train depot as young men set off for war. Meanwhile, on the waterfront, the spring fleet prepared to sail with large numbers of green hands and foreign sailors. Some even shipped fishermen from the Provinces. There was no alternative, since hundreds of fishermen had gone off to war.

Over the summer of 1861, only small shore craft remained in the harbor as the large offshore fleet made extended summer trips to the Bay. The fishermen had little access to the day-to-day accounts of the war, and on their return in October their feelings harkened back to the patriotic days of April and May. They had not been jaded, as had many, by the summer losses on the battlefield or the questions being raised regarding the leadership of the country. Hundreds more now joined up. It was a unique moment: The young fishermen of Gloucester never again showed such enthusiasm or enlisted in such numbers.

As the local fishermen entered the army and navy, Gloucester's dependence on foreign fishermen continued to grow. A key turning point had been reached.

For the rest of 1861 and into the early months of 1862, little of import happened to the fleet or on the war front. In early February, with the return of the first of the Newfoundland herring craft, the handline fleet fit out for winter trips to the Georges shoals. Men fished over the sides at slack tides, pulling up ground fish on ice-coated lines, retreating below for warmth as their mittens and nippers iced up, the schooner drifting with the tide and dropping anchor as the skipper found a place to fish. All was in readiness: Icehouses and gurry pens had been set up (the latter built on the deck near the cabin to temporarily store gear and bait after bringing it up from below, and later to hold refuse from the fish—the head, skin, viscerals, and bones— as they were landed and prepared for either icing or salting), the gear was checked, and the supplies stowed for a trip that could last for many weeks. The grub included several hundred pounds of brown sugar and butter; about 20 pounds each of tea and coffee; 10 total barrels of beef, pork, and pigs' knuckles; 1 bushel each of rice and onions; 10 bushels of potatoes; and 1 barrel each of beans, dried apples, and dried peas.

By mid-month, the first of the boats were coming back with fine fares of fresh cod and halibut. By March, more than 100 handline schooners set in close proximity on Georges' southern shoals, and good stocks seemed assured.

But then the weather turned. A nor'wester swept over both fleet and town. Chimneys came down, tin roofs lifted, blinds snapped off, and windows blew in. In the harbor, several vessels dragged ashore on Rocky Neck. On the backside of Cape Ann, where the Lanesville harbor faced the full force of the northwest gale, several boats "broke their stern fasts and swung around upon the rocks." The schooners *Helen C. Young* (35 tons) and *Zephyr* (60 tons) struck the rocks and sank, while the *Jane* (40 tons) drifted onto the ledge in front of the fish houses in the cove. The *Fountain* broke free from its stern anchor and drifted into the wharf, breaking its bowsprit. The *Martha*, too, went adrift, but the crew successfully ran her on the beach.

But this was not the worst. Out on Georges Bank the fleet was being blasted by the full fury of the storm. Without warning the clouds massed, the wind rose, and crews had no time to heave up anchors and head out to sea to escape facing the gale on the shoals. Even trimming the sails and setting double banks' anchors proved difficult.* Nearly all the schooners lost cable, anchors, boats, hatches, gurry pens, bulwarks, and sails. Icehouses and cabins filled with water, and the men looked on in amazement as the rising wind pushed the sea into towering waves.

Skippers kept a fearful eye to windward, scanning the storm-tossed waters for drifting vessels or seas that might pull their own boats down

---

*The banks' anchor was 9 feet 10 inches long and 4 feet 8 inches wide, weighed approximately 700 pounds, and was designed to hold a 110- to 150-ton banks' schooner in place. The anchor was made of iron and painted black, and at its top was a 14-foot wooden stock crosspiece. Hemp cable, some 8 to 9 inches in circumference, was used. Two coils were carried on the deck, one on each side of the forecastle companionway. Each cable was from 300 to 400 fathoms in length.

under the waves. At the windlasses, men with hatchets stood ready to cut the anchor cables, and when all hope was gone skippers gave the fateful order to cut loose the schooner and pray. The untethered boats now bore down on others still anchored; crews braced themselves, vessels collided, and men and schooners disappeared beneath the waves.

When the schooner *Borodino* went adrift, the towering waves swept clean her decks. Her stanchions broke, and the seas swept away the mainmast, booms, and sails. Only a jib set to the foremast gave the helmsman some control, as he held the schooner's head into the wind. But the ballast shifted, and the *Borodino* hove over onto her beams; the men worked frantically to trim the ballast. They pushed and pulled, knowing their lives sat in the balance, and in a matter of moments the *Borodino* rose from the sea. The men had succeeded—for now—but the turbulent seas would ultimately claim the *Borodino*. But in the last minutes her crew made the death-defying transfer to the *Peerless*. Cut adrift, the *Borodino* disappeared into the night.

By storm's end, *Borodino* was 1 of 15 Gloucester vessels lost in the storm, of which the crew of only 2 had survived. In all, 168 fishermen lost their lives. They left 70 widows and 147 children, and the community was shaken. One old-time skipper thought the lost vessels must have been deeply laden with fish: They sat low in the water, out of trim, and when forced to break adrift and drag their anchors they were apt to be knocked down and swamped by the heavy seas. How else could the town have lost so many men?

Story after story echoed across the community. Four captains, friends of the skipper, went down on the *George F. Wonson*. He had prevailed on them to make the trip, as their own schooners were yet to be prepared for the season. A young man from East Gloucester, a mere boy who had recently come of age, went down on the *May Queen*. He had gone against his parents' wishes—this was his "freedom year," he said, and he had to earn all he could to make a good year's work.

Never before had Gloucester seen such devastation, and the community rallied, forming the Fishermen's, Widows', and Orphans' Relief Committee. Money, hams, and flour poured in for the families. People recognized their civic duty to provide succor to those left behind. Gloucester established a new precedent: With great loss there would be a bold response. Town, state, and federal agencies provided no help, but no one on Cape Ann wished to see Gloucester's widows in poor farms, workhouses, or brothels.

Yet, even as the community raised funds for the widows and children of the lost fishermen, vessel outfitters fit out craft for the next trips to Georges and Western banks. One commentator in the newspaper said it all: "It was useless to repine o'er the past; it will not bring back the lives of those lost, or help those who have been left, or restore the property which has been sunk on Georges." Gloucester was a fishing community.

One month later, in April 1862, on the one-year anniversary of the beginning of the war, the people of Gloucester well understood that these losses on the fishing grounds far overshadowed the losses of its men in the war. But they also realized that the war would not be quick. Many more young men would be called before this war was over. The Battle of Bull Run, the first major conflict of the war, had not resulted in the expected Northern victory, and more than 1,000 Cape Ann men were now in the army and navy.

With the coming of spring, the Northern military strategy focused on the latest push for Richmond, the Confederate capital, but it was to fail, and it was followed by a call for 300,000 men. The people of Gloucester came together at Town Hall to meet the challenge. As there was no enthusiasm for a draft, volunteers had to be found. E. G. Friend moved a motion for the town to pay a $100 bounty to each volunteer, with the clergy and citizens in the audience raising their voices in support of the motion and the war. They poured into the streets to help recruit volunteers. A week later, recruiters reported that 39 men had enlisted. Speakers urged others to follow their example, and the throng again paraded through the streets of the town, with the Mechanics Band playing patriotic songs.

Of Gloucester's 186-man quota, 93 had soon signed up under the bounty. The town now had to wait on the return of its fishermen. The vast mackerel fleet had long since sailed to the Bay, and only with its fall return could Gloucester hope to fill the quota. In late September, following the most bloody battle of the war at Antietam, Maryland, the Lincoln administration raised the quota yet again. The Gloucester quota now stood at 251.

With many men having died on the battlefields, all waited to see what would happen when the Bay fleet returned in the fall. Would the men of the fleet help Gloucester meet its quota? Or would a draft echo through the community? The answer was quick in coming. The fishermen signed up, Gloucester easily met its quota, and as they did thoughts of war faded quickly into the background.

Talk in Gloucester now focused almost exclusively on the winter fishery. Firms had outfitted 47 craft, up from 17 in 1861, for Newfoundland herring trips. By early December these craft and two handline schooners remained at sea, and the local paper expressed the hope that the Georges fleet would not leave as early as in the past. Lives could be saved if the outfitters kept their craft in port during the worst of the winter weather. Yet they also recognized that come late January, with empty pockets, the men would feel the need to go back to sea.

In early January, Gloucester papers weighed in on the continued Northern war disasters as Union troops had pushed a campaign in Virginia, expressing disillusionment with the effort. It had been a trying year. Sad hearts grieved "over the dead who lie buried in nameless graves far away on distant battlefields." Yet, the local paper also insisted that the "question must be settled in blood." The North felt disillusioned, but the commentator claimed most in Gloucester stood ready to continue the fight. But whatever the truth of the feelings, on January 9, with the Northern army again in winter quarters, Gloucester vessel outfitters, the owners of the vessels, announced their intent to send more than 40 craft into the winter fisheries. With empty pockets, neither the outfitters nor the fishermen were dissuaded by the severe losses of last February's storm. Icehouses went up, ballast went below, and the people of Gloucester looked on as its handline fleet sailed for Georges. Only a few were put off by high insurance premiums and chose to wait until March, when the price was scheduled to go down.

Clear skies and calm weather settled over the grounds, and by April landings of 40,000 to 60,000 pounds per vessel became common.* The many outfitters who had come down on the side of their "empty pockets" won out. April also ushered in a renewal of fighting between Union and Confederate armies, but the spring offensive failed. In July, however, Northern armies at Gettysburg, Pennsylvania, and Vicksburg, Mississippi, won great victories.

This news, which awakened the hearts of the cities and towns in eastern Massachusetts and the nation, resulted in a much more muted response in Gloucester. One commentator in the local paper said Gloucester seemed "dead as far as any symptoms of rejoicing or public spirit." The people were distracted by new national conscription calls, with 285 eligible Gloucester men now subject to the draft. An extra edition of the local paper listed those who were eligible, and within two hours the 720 copies had been sold out. All levels of society were affected, although some sections of town carried more of the load. But folks made the most of it.

---

*While the storms of 1862 did not reoccur, the fleet was in danger from the Confederate raider CSS *Tacony* on the fishing grounds. In a two-week period the raider took 15 Northern vessels, including 6 Gloucester schooners. The first to be captured and burned was the merchant brig *Umpire*, followed five days later by the packet-ship *Isaac Webb* (bound to New York with 750 passengers) and the fishing schooner *Micawber*. The Confederates placed a $40,000 bond on the packet, and sent her on to New York. They burned the schooner. The *Tacony* next took the clipper ship *Byzantium*, followed one day later by the first 4 of the Gloucester fishing schooners: the *Marengo* (82 tons), the *Ripple* (64 tons), the *Elizabeth Ann* (92 tons), and the *Rufus Choate* (90 tons). They burned each craft, but did not harm the men. The last of the Gloucester fishing schooners taken and burned were the *Ada* (70 tons), captured about 40 miles northwest of Cape Cod, and the *Wanderer* (90 tons), taken northwest of Georges. The *Tacony* continued to gather in merchant vessels for several more days prior to being captured by sailors out of Portland. The New England fishing towns, concerned about further pirates, petitioned Secretary of the Navy Welles for protection. In response he sent 5 vessels to the East Coast fishing grounds, while the army announced plans to build a big fort at Gloucester's Eastern Point.

As the last of the Bay fleet fit out, fishermen found their names on the list. One captain tried to avoid the call by saying that he did not live in Gloucester. He said he spent the winter in Boston, and asked to be taken off the Gloucester draft list, but his story crumbled. He had paid taxes in Gloucester for two years, and a draft official had him arrested before he could leave for the Bay. He quickly paid the $300 fee that released him from the draft and went to sea.

By early fall, with the Northern and Southern armies stalemated in northern Virginia, the news in Gloucester focused on the fisheries. In particular, there came calls for the growing cod fleet to land and process fresh and salt cod in Gloucester, rather than in Boston. The fisheries were prospering—affected both by a growing population and the need to feed the people in a time of civil war, salt cod became a staple food commodity—and in Gloucester, jobs became the watchword. George Perkins of George Perkins & Son, a leader in this effort, said he was tired of the slipshod manner in which businesses outside Gloucester reaped the profits. Change was needed, and he started westward with 25 cases of Gloucester salted codfish comprising 11,250 pounds. He stopped at Albany, New York, and sold most of the fish to Messrs. Hong & VanHaven. When he returned, the company telegraphed for more cases of fish. On his next trip he went to Buffalo, Cleveland, Detroit, Chicago, and other cities of the Midwest. Dealers were quick to recognize that his fish were not only a good price but of superior quality: Codfish packed in clean boxes direct from the flakes in Gloucester (where the salted fish were air-dried), packed on trains at the Gloucester depot, arrived in the Midwest cities in much better condition than they had previously coming through Boston and New York. Orders began to pour in for the Gloucester product.

Such were the first tentative steps to transform Gloucester into a major independent distribution center for salt cod. In the postwar era, her dory-based cod fishery would wrestle control from the fishermen of Portland, Maine. But the first signs of change could be seen in the fall of 1863. Gloucester was beginning to grow its cod fishery while maintaining its prosperous mackerel fishery. In 1863 nearly 31 million pounds of mackerel come into the port. It was the largest catch on record, and the merchants of Front Street benefited. Returning fishermen had money in their pockets. They patronized clothing and other stores, and foreign fishermen stocked up before heading home to the Provinces. Gloucester businessmen were taking advantage of the favorable economic climate, transforming the town into a center for the salted-cod fishery.

)(*An early image of Gloucester's docks and fishing vessels, c. 1864. The Somes rigging loft is in the right foreground; Harbor Cove lies beyond.*

By the end of the fishing season in November, people noted that there had been little loss of life and vessels, and the focus of the community again turned to the draft. Of 285 men subject to the draft, all but 37 had reported to the Provost Marshal. Gloucester remained 89 men short of its quota and at risk of a draft. A committee of citizens canvassed the town, but it seemed the only real hope lay in Gloucester men currently in the service reenlisting as their terms ended. In the first local unit heard from, Company K, 12th Massachusetts, none of the Gloucester men reenlisted. However, Gloucester men in other units, like their counterparts across the North, reenlisted in large numbers. On January 8, 1864, Gloucester filled its quota.

Fourteen days later the town welcomed home its first group of returning veterans: 56 men from the 32nd and 23rd Massachusetts Regiments, 17 of whom had been fishermen. It was an Indian summer day, and by noon 5,000 people awaited the train. They stood in the streets, filled the depot, and gathered on rooftops. As the special train pulled in, big guns belched forth a salute, church bells rang out a merry chime, and the Gilmore Band struck up "Home Sweet Home." Loving friends and relatives encircled the men.

The men soon regrouped on the street, picked up their tattered flags, and marched through town. Ladies waved handkerchiefs and cheer after cheer rang out as the men progressed through the town. The first veteran fishermen had returned to Gloucester, and with a larger fleet and a steady demand for mackerel and cod, they found jobs readily available.

After the celebration, Gloucester returned to more mundane routines; fitting out vessels, unloading Newfoundland herring, and sending craft to Georges and Western banks. Then, on the morning of February 18, disaster struck. An enormous fire engulfed the downtown business district, devouring buildings only recently decked out to greet the returning veterans. On the east end of Front Street 100 buildings were destroyed. It was a bitterly cold day: The mercury stood at –6°F, strong winds blew in from the northwest, and once the fire began, it ran wild. Only when Salem's powerful steam fire engine came in on the 9 o'clock train did the firefighters get the upper hand.

The same cold front caused havoc at sea. Schooners lost cables, anchors, and boats, and most returned from Georges with only partial fares. But the owners soon replaced the lost gear, and the boats went back to sea. Again

)(*On the coldest of days, the great Gloucester fire, backed by a fierce northwest wind, swept through Front Street. Mercifully, the fleet was spared.*

they ran into a great storm, a northeaster that claimed six schooners and 54 men. But the town had little time to grieve, for the papers were again filled with news of the war.

The North had begun a coordinated two-pronged offensive in the eastern and western theaters, putting hundreds of Gloucester boys into the heat of battle. In Virginia, General Grant mounted a prolonged siege against General Lee's army in the trenches surrounding Petersburg, Virginia. In the west, General Sherman pressured the Southern army outside the strategic city of Atlanta. On both fronts, the troops became bogged down, but this changed on September 2 when General Sherman telegraphed to Lincoln, "Atlanta is ours, and fairly won." Lincoln's fall reelection was assured. In Gloucester, of 1,308 ballots cast, 1,122 went for Lincoln.

Shortly after the election, General Sherman left Atlanta in flames and gave Lincoln the city of Savannah, Georgia, as a Christmas present. Now, with the end of the war in sight, Gloucester looked to the day when her boys would come home and join in her booming fisheries. Much had changed between 1859 and 1865: The fishing fleet went from 301 vessels to 365, while the number of fishermen increased from 3,434 to 4,700, with foreign-born young men filling in for native fishermen. The annual mackerel catch went from 12 million to 30 million pounds, and the businessmen had laid the foundation for Gloucester to become the center of the all-important salt-cod fishery. Gloucester's fleet was second to none, new and expanded businesses surrounded the waterfront, and with a superb harbor on the tip of Cape Ann, Gloucester had now become America's premier fishing port.

With this prosperity, many continued to invest in the community's future. John Low had begun to build an extensive wharf complex, said to be one of the best in town, and he contracted to have a fish processing complex built on the bordering property. J. O. Procter was extending his wharf, adding 75 feet to its length, providing him with several fine deepwater berths. Charles Parkhurst began a boat-building business at the head of the harbor. On Front Street, new buildings rose from the ashes of the fire, and Leonard Walen, Joseph Rowe, and Sylvannus Smith began to fit out new craft. Success was in the air and the population exploded. For each house that came on the market, 20 to 30 ready buyers stepped forward. Even though fishermen lacked decent homes, still the fisheries expanded.

Then in January 1865, a shock wave echoed across the community. Word reached Gloucester that Congress was considering a bill to revoke the 1854 reciprocity treaty with Great Britain, and Gloucester's access to the Bay was again threatened. Four years of distinctly pro-Southern British bias had angered the Congressmen in Washington. British shipyards had built

Rebel sea raiders. Only three months earlier, in the rural town of St. Albans, Vermont, Southern soldiers crossed the Provincial border, robbed the town's three banks, and returned safely to British-controlled Canadian soil.

Now, as the war wound down, Congress prepared to act. In Gloucester the people were horrified by the prospect of returning to the chaos that had existed before the treaty. But their voices carried little weight, and in a matter of months the U.S. government announced the reciprocity treaty would end in 1866. Gloucester, it seemed, had no say in the deliberation.

Meanwhile, back at Petersburg, U.S. troops prepared for a spring campaign. March 25 saw General Lee send half of his small army to break through the Union lines. But they failed. Four days later, General Grant put the Northern troops into motion, and they did not fail. Lee evacuated Petersburg on April 2. One day later the Union army entered Richmond. Six days later Lee surrendered his army to Grant at Appomattox Court House, and although there were other skirmishes in North Carolina and Alabama, the American Civil War had effectively come to an end.*

When news of Richmond's fall reached Gloucester the people poured into the streets. Bells rang, guns fired, and flags appeared on buildings across the town. Everyone seemed jubilant and happy. On the following day, the soldiers at Eastern Point fired a 100-gun salute, and church bells tolled in celebration. Flags flew from all the vessels in port, tricolor hangings appeared in shop windows, and schoolchildren frolicked in the streets. Then came news of Lee's surrender, and again every flag in town floated in the breeze. People fired powder packets and muskets, bells rang, and a parade wound through the streets late into the evening.

Then, on April 21 the euphoria ended. President Abraham Lincoln had been assassinated at Ford Theater, and on the day of his funeral the bells in Gloucester tolled mournfully at sunrise, noon, and sunset. The fleet displayed the national colors at half-mast, and dipped flags flew across the town. Services at the Methodist church spilled onto the street as the congregation spoke words of honor and loss. Gloucester mourned the man who had led them to victory, a man whom they had only recently helped reelect as President of their country.

Yet, as in much of the North, there were those who had not voted for Lincoln. They had ridiculed his looks, his long face, his gangling body, and his presumed lack of leadership. Now, even at this dark hour, some unwisely

---

*Between 1861 and 1865, around 683,000 soldiers had lost their lives: 394,000 from the Union and 289,000 from the Confederacy. Of these, more died of disease (some 389,000) than in combat (233,000) or in prisoner of war camps (61,000).

expressed contempt for Lincoln, and the loyal men of Gloucester no longer had the good sense to ignore such fools. Offenders were "waited on" and "respectfully invited to pay due homage to the flag of their country." George Steele Jr. was the first to be "respectfully requested" to hoist a flag to half-mast on the schooner *Alhambra* as she sat at his wharf. On seeing the approaching crowd, he "expressed himself perfectly willing that it should be done," and by "special request" of the group he accompanied them to the ship. Here he personally lowered the flag to half-mast and affirmed his patriotism.

The assembled multitude next went to the Rogers Street wharf of Epes Porter and questioned his loyalty. Porter looked at the crowd and expressed his willingness to salute the flag. He covered himself with bunting from head to foot, kissed the folds of the flag, and presented a patriotic appearance. The group continued to Thomas Hall's net and twine factory on Duncan Street. No flag flew from the building, and Hall, seeing the approaching procession, hurried to the roof to hang one. The crowd next marched to Jackson Street. William T. Cooper did have a small flag, draped in mourning, flying from his shop window, but it had only seven stars, and the crowd interpreted this as a gesture of defiance, an insult to the country. Cooper got the message, and pulled down and destroyed the offending flag, kissing and saluting the more suitable flag handed to him by someone in the crowd. Robert Rowe was the next visited. On seeing the flag-waving multitude approach his shop, he cheered the Stars and Stripes, and the crowd moved on. They next stopped at the Coggeswell wheelwright shop, where they easily persuaded Coggeswell to honor the flag.

By now the crowd had become large, and dissent of any type was impossible. They arrived at the home of John Wheeler on Washington Street, who took offense at the multitude at his front door. He said he would "not be forced into obeying their request," but to no avail. Some in the crowd placed him on a rail, with the American flag wrapped around him, and carried him to the Custom House. Here, he finally complied. He saluted the flag and cheered for the Union.

With victory, Addison Gilbert, chairman of the town Selectmen, called for silence. The men had done enough for one night. He reminded them of the solemnity of the day and the tolling bells. The men heard his words and returned to their homes. By the next day, there were those who "regretted that such a spirit was manifested, especially on a day set apart for the funeral solemnities of the late President." But feelings soon returned to a more even keel, and again the town looked to its economic future.

CHAPTER 7

Dory Fishing Comes to Gloucester

With war's end, Gloucester entered an era of unprecedented prosperity, fueled by her huge fleet and fine harbor, as well as by national geographic and economic expansion. Larger and more diversified companies came to the fore. Pew, Wonson, Tarr, and Smith, among others, outfitted sizable fleets, while also owning extensive wharves and modern fish processing plants. In 1866, 40 individuals or firms owned 3 or more schooners, 17 owned 10 or more, and the total fleet stood at 478 vessels and would increase by almost 100 by the end of the decade. Fresh and salted fish were now sent to Boston, New York, and Chicago, and Gloucester soon became America's salt-cod capital. America's improved rail and waterway systems offered easy access to Gloucester's salted fish by the heartland of the country.*

The large clipper schooner had become the standard of the fleet, and to maximize one's return on these expensive vessels ways had to be found to bring in more fish. The fishing season ran year-round, from January to December, the profitable mackerel fishery going from late spring into early fall, with the ground fishery over the rest of the year. For the mackerel fishery, the fleet moved from handline to seine netting—a totally new, revolutionary approach to catching the fish. For the ground fishery, the move was first to handline fishing out of dories, which extended the ground over which the men could fish and increased the catch by 30 percent. Shore-based dory fishing had been common in Annisquam and Lanesville for most of the nineteenth century, and the addition of two-man dories to schooners became common toward the end of the 1850s. The final innovation was to move from handline to trawl fishing, and it too was a revolutionary change—laying long, baited lines across the seafloor. The process came from France, and North Atlantic fishermen had experimented with it for over a decade. By bringing together the two-man dory and the ground trawl, vessel owners were able to

---

*The fishermen of Marblehead had dominated the salt-cod fishery prior to 1845, when storms caused a significant loss of the vessels in their fleet. With her demise, the fishermen of eastern Maine stepped in for a number of years to fill the void. Wayne M. O'Leary, *Maine Sea Fisheries: The Rise and Fall of a Native Industry, 1830–1890* (Boston: Northeastern University Press, 1996).

maximize their return: Larger catches on shorter trips became the standard for the next 70 years.

As vessels now went to sea with purse-seines and dories, the catch exploded. For the mackerel fishermen, with the warming seawater of spring, purse-seine-equipped schooners sailed in search of rising fish. Men sat high in mastheads scanning the horizon for swirling mackerel, and once they were sighted, the vessel's seine-boat (a 38- to 40-foot rowed craft, whose origins go back to the whaleboat) set off in pursuit. Men bent to their oars as they closed the distance between themselves and the fish. Soon the seine-boat boss called for his men to lower the 250-fathom-long purse-seine net into the water. Corks held its top edge afloat, while weights pulled its bottom below the swirling fish.

The net now hung on one side of the fish, and the boss of the seine-boat called on the men to once again bend to the oars, drawing the net around the schooling mackerel. Just as the ends met, the men pulled on ropes to close (or purse) the bottom of the net, trapping the swirling mass of mackerel. The schooner then closed on the men, one end of the net was raised onto the deck, and the men dipped small nets into the thrashing mass of mackerel, bringing out a half barrel at a time. Soon the mackerel lay knee-deep on the deck of the schooner, and men split, salted, and stored the fish in 200-pound barrels. This was a prosperous fishery, with the eastern fleet bringing in enormous fares from grounds to the east of New England and Nova Scotia, which now proved remarkably productive in the initial spring rush and later summer season. The Bay became of secondary importance once the purse-seine came on the scene.

For Gloucester's salt cod and halibut fishermen, the year began in January, with craft fitting out for both handline fishing from the schooners and ground-line trawling from dories. They fished for halibut on Georges and the northern Scotian and Grand banks,* and for cod on Georges and Western banks. But it was on the more distant fishing grounds, Western and Grand banks, that ground-trawl dory fishing first gained its foothold, and once started, it overspread the fleet by the end of the decade.[16] Some vessels continued to engage in spring and summer handline fishing well into the next century—particularly in the treacherous shallows of Georges Bank—bringing in smaller catches from quick trips to the grounds. The fresh fish were in the best of shape and brought top dollar at the Boston market. They called

---

*In 1865, with the practical demise of the Georges halibut fishery, Gloucester skippers had moved north to the Grand Bank. In June, after a four-week trip, the schooner *Hattie M. Lyons* under Captain George W. Miner (owned by D. C. & H. Babson) returned with 75,000 pounds of halibut. This was a superb catch, and in 1866 a handful of vessels made the trip. By the end of the decade, this had become an established fishery.

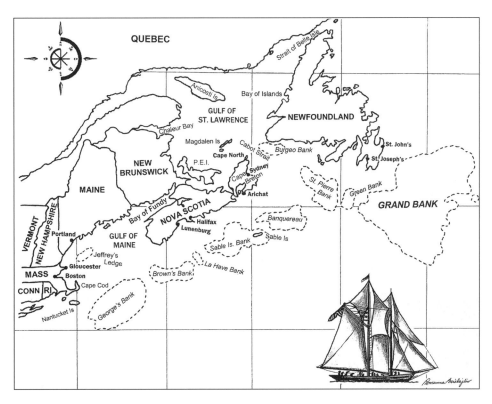

)(*The major fishing banks off the coast of New England and the Maritimes.*

it "Rips" fishing, referencing its start as an anchored fishery in the Rips area off Nantucket, followed later as a nonanchored, drifting fishery over Georges and other grounds.

But it was ground-line trawling that would revolutionize the port. In this fishery the men went out in small, 13- to 18-foot two-man dories (see the next chapter for a full description of these boats). A schooner scattered 6 to 8 dories, and later even 12 to 14, over many miles of open ocean. Each dory laid long ground lines containing hundreds of baited hooks on the seabed below (see the next chapter).* On a good day, a dory could land upwards of 1,000 to 2,000 pounds of cod, haddock, hake, or halibut.† Dory fishing became commonplace, and with the return of the soldiers and abundant capital, outfitters committed ever more schooners to the dory fishery. All the pieces were in place.

---

*While the ground trawl became widespread in the 1860s, a number of craft continued to go out in the Georges handline cod and halibut fisheries into the twentieth century.

†From 1859 to 1889, cod landings went up spectacularly: 12.8 million pounds in 1859, 28.0 million in 1869, 36.4 million in 1879, and 44.3 million in 1889. Over this same period, halibut landings also remained reasonably high: 4.5 million pounds in 1859, 9.7 million in 1875, 13.2 million in 1879, and 7.2 million in 1889.

X *The Gloucester waterfront in the mid-1860s. Sitting in the foreground is the Bennett Griffin lumberyard on Commercial Street; lying beyond are the masts of the vessels in Harbor Cove and the cupola of the Pavilion Hotel.*

In 1867 the first craft left on January 2,* while the schooner *Leonard McKenzie*, the last of Gloucester's 45-vessel Newfoundland herring fleet, returned with baitfish on February 26. The emergence of a Newfoundland herring fishery from December into March had made possible Gloucester's move into a winter salt-cod fishery; without a reliable source of bait the fleet could not have gone to sea.† By year's end, the men of the fleet looked back with favor on the fishing season that had just passed. The ground fisheries had become a year-round activity, although the summer fleet was smaller than the fall, winter, and spring fleets. In 1867, 239 vessels fished on Georges.

*The previous year, 1866, the early arrival of the Newfoundland herring fleet back in Gloucester had caught the community somewhat by surprise. Early January saw few of the ground fleet fit out, and it took weeks for the first of the dory-based fleet to sail for the grounds. But outfitters learned from this experience, and in the years to come they had their boats scraped, painted, and stocked in the days between Christmas and New Year's. They sailed to the grounds as soon as the first of the herring fleet returned to port.

†In the winter of 1856–1857, 6 vessels were engaged in the herring fishery. The next year the fleet increased to 11; in 1858 there were 13; in 1859, 16; in 1860, 19; in 1861–1862, 15; the next season, 28; in 1863–1864, 39; the next year the number was 21; in 1865–1866, 28; and now 45 (8 going to New Brunswick and 37 to Newfoundland).

One hundred and eighteen craft left early in the season and averaged trips of three weeks each. Boats entering later in the season went out on shorter trips, averaging one to two weeks on the grounds. The better weather later in the season translated into fewer days sitting idle on the grounds, waiting for an opportunity to set trawl or drop handlines, and thus less time to fill the hold. The 85 vessels that went out all year averaged six to eight trips to the grounds.* In addition to fishing on Georges, 28 craft made one or two dory trips to Western Bank. Five vessels also sailed to the Grand Bank in the spring and brought back good fares. Finally, many vessels went out for mackerel during the height of that season.

Over the 1867 season, 871 ground fishing trips came in from Georges, while another 46 came in from the northern banks. The ground fishermen had become a powerful force, dory fishing was sweeping through the fleet, and, by 1869, the fleet made more than 100 trips to the Grand and Scotian banks, while the Georges dory fleet continued to expand. With fewer and fewer vessels tied up over the winter months, the combined salt cod and halibut catch rose to more than 30 million pounds per year. Gloucester had become the country's premier fishing port, with a diverse fishery and rising profits.

New vessels poured into the fleet: 46 in 1867, 47 in 1868, and 39 in 1869. The Essex yards rapidly turned out new craft, with easy terms offered to skippers—one-quarter down and the rest in three annual payments. Schooners went out of the yards at $65 a ton.

Change was everywhere. New stores, banks, and insurance offices were opened on Front Street. The waterfront saw new wharves, storehouses, marine railways, and mechanic shops. New firms opened to outfit vessels for the fleet. As schooners returned to port and the men crowded onto Front Street, the police closed liquor shops in the evening to prevent "suppressed spirits" from getting out of hand. Excess alcohol consumption was common: The police processed 370 liquor-related warrants in 1868 alone, a 50 percent increase over the previous year. Streetwalkers were said to frequent Middle Street, and the churchgoing community came together to support its temperance league. Yet, the overwhelming spirit of the day was pride in what Gloucester had become, not concern with these side issues. The number of outfitting firms sending out three or more vessels increased from 39 in 1866 to 47 in 1869. The fleet grew from 365 schooners in 1865 to 506 in 1870; the number of fishermen rose from 4,700 to 6,084. Gloucester had reached its peak, and would stay on top for many years to come.

*One vessel made 8 trips to Georges, 21 made 7, 27 made 6, 42 made 5, 30 made 4, 35 made 3, 36 made 2, and 47 made but 1 trip each.

CHAPTER 8

To Be a Dory Fisherman

By the late 1860s, with more than 450 vessels in the Gloucester fleet, there was always a berth for a vigorous young man. Many of the returning Civil War veterans went back to sea, as did the immigrants who poured in from Canada, Scandinavia, and Portugal. More than 6,100 men served as fishermen on Gloucester vessels.

Once the crew had signed on for the season, and the skipper had filled out the customs declaration for alien men in his crew, the dorymen set about fitting out the vessel for sea. They checked the condition of the schooner and dories, helped the cook store supplies, and waited for the last few men to come aboard before preparing to sail. Those already on board could be found talking among themselves, stowing luggage, or sitting about before the order came to unfurl the sails. Among the last to board would often be a man who was badly off from drink—a good sailor but one who could not hold his liquor. He may have been drinking from the moment he reached port, spending all of his money on alcohol. The skipper, or one of the men, might even have dragged the man from his favorite barroom. Soon it was time to go. In a surviving account of one such trip to the grounds, a novice fisherman tells of a mid-July voyage on the schooner *Oliver Eldridge*.[17] He went out as a doryman, and on a clear summer's day Captain John Scott had the men unfurl the sails, heave the anchor, and then begin to drift away from the dock. Soon the schooner was sailing past Ten Pound Island, then rounding Eastern Point Light, and finally beating out for her long trip to the Banks.

The men congregated on deck to catch a last glimpse of Cape Ann, even as the veterans began to take the measure of the new man, and the new man looked to the veterans for advice and guidance. Among the crew was a boy of 14, the son of one of the dorymen. He, too, would pull his weight, helping in preparing bait, sending off dories, landing fish, and doing whatever the skipper or cook might assign him to do. But as a lad on his first trip he did not go out in the dories, nor did he expect to receive a full share for his work—a quarter share would be his take. Such a boy was called "a catchee," named after his catching the painter of the dories as the men returned to the schooner.

Once under way, the skipper set the compass heading for the watch: With a fair wind, they sailed north by northeast, at 8 knots on this first night out. He called all hands aft to "thumb the hat," by which means he set the order of the watch. One man pulled off his hat, and each of the crew placed a thumb and finger on the rim. The captain started at a random spot and counted to a fixed number, placing his hand on a finger. The first man at whom he stopped had the first watch, and immediately took himself to the wheel to stand his turn at the helm. Meanwhile, the counting went on, men dropping out as the skipper assigned their watch. Some turned in and slept, others smoked or played cribbage, still others read.

The skipper also told the men when they were to start overhauling the trawls, and at the appointed hour on the designated day the men brought the gear on deck. They straightened and sharpened hooks, and attached the gangings to the ground line to hold the hooks. When necessary, they replaced hooks and worn ground lines.

Often, and especially in summer, the sail to the grounds was slow and easy. Sails flapped lazily to and fro as the craft rose and fell on gentle swells. Men looked to the west for more wind, hoping a stiff breeze would spring up and push them to the grounds. But dorymen quickly learned that the schooner sailed at her own pace. As each watch came on deck, the men exchanged the orders left by the skipper, and the helmsman went below to turn in. As his head hit the pillow, he hoped the first sound to disturb his rest would be the welcome cry of "fish, ho."

But not all trips to the grounds were as peaceful. Passage to Georges or the northern banks could be troubling, even in summer. As clouds rose on the western horizon, and the weather took on a menacing appearance, the skipper's voice would ring out, calling the men to lower and reef the main and foresails, and set double watches. In extreme gales, he lashed the wheel and made soundings to ensure his craft did not run afoul of shoals and reefs. Shallow sailing often spelled doom to the Gloucestermen. Water swept over the deck, and as her head dove and rose, she shook herself free—like a dog coming out of the water—and surged forward for another dip. The men bunking on the leeward side of the vessel spent hours listening to the sound of rushing water overhead. Time and again the heavy crash of the waves hit the vessel broadside.

When the onrushing schooner approached shallow water, the skipper called all men on deck to tack the boat: Here was real danger, especially if there was a thick fog over the water. He unlashed the wheel to bring the ship about, calling on the helmsman to set a new heading toward the safety of the open ocean. But tremendous seas could hold the schooner on a dangerous course, and skipper and men worked ceaselessly to trim the sails. With no drop in speed, the helmsman could not turn the rudder. Then the moment arrived. The pace had finally come down, and hard over went the wheel. The rudder began to turn, the schooner paused, and then one heard the much-anticipated shout, "She's over!" Perhaps the reefed foresail had caught a breeze; the vessel was now headed to the safety of deeper water. The skipper again lashed the wheel and the men could at last go below.

Eventually the weather let up, and as the vessel approached its destination, the skipper called on the men to prepare to set two, three, or four tubs of baited trawl. At last on the fishing grounds, the skipper made soundings to check on the depth of the water and the presence of a sand-and-pebble bottom—the ideal habitat for bottom-feeding fish. He looked to the weather and the tides, with an eye to setting his dories on a rising tide with clear skies and mild winds. Soon the skipper shouted, "Two men to a dory! Ready! Throw your anchor!" The gripes (heavy ropes) that fastened the two stacks of dories to the deck were removed. From the rigging, hooks came down to grab the dories and lift them off the cradles—over the rail went the first dory, splashing into the water. . . . But first let us look at life on board as the schooner made for the fishing grounds.

At sea, life on board fell into a comfortable routine. In the forecastle, the cook kept the coal ablaze in the stove, providing all the food a man could want. By the 1860s, each schooner had its own cook: The men no longer took turns preparing the food and keeping the fire ablaze. A good cook was almost as important as a good skipper in attracting the best dorymen to a boat. As the schooner had prepared for sea, the cook made sure the larder was filled with all of the stores required for a multiple-week trip—flour, tea, sugar, molasses, pork, lard, cheese, eggs, salt, and so much more. He personally stored the supplies safely below, and checked on the condition of the galley stove, pans, knives, forks, spoons, dishes, crockery, wood, coal, and kerosene. Once the stove was lit, he made sure it never went out until the schooner was back in Gloucester, where he would clean it up as he left the forecastle in shape for the next trip. But now it was time to go to bed, and in a few hours at the first hint of a lightening sky, he would rise, prepare breakfast, and call out the men at about 3:30 to 4:00 A.M. with a blast of a whistle. The cook was a key man on a schooner, second only to the skipper.

Most of the men were sleeping on bunks that encircled the forecastle, stacked two high, and when they were not on watch this was their home. The rows of bunks became narrower toward the bow and a bed curtain shut off the sleeping man from his mates, who would be sitting about—eating a meal, grabbing a mug-up, or playing a game of cards. The men ate, read, talked, and played cards around the permanently installed table in the middle of the forecastle, a table with hinged leaves that added a couple of feet to its width. The mast passed through the forward end of the table, and there was little room to move about; thus the leaves of the table were usually raised only at mealtime. A kerosene lamp swung over the table, another was fastened to the pawl-post, and they would be lit until lights out was called at 9 P.M.

The 3:30 A.M. breakfast consisted of oatmeal, eggs, bacon, bread, and doughnuts. At midmorning, a meal of cooked meat was put on the table, followed by a big midday dinner at about 3:30 P.M. and a hot supper at day's end. Food and the company of the other men were all-important on board a schooner. While the men were baiting up their trawls, the cook could be heard chopping up the ingredients for the later meals. He decided what the men ate, and pots of coffee and tea where always available for the returning fishermen. The forecastle was a warm place for the men to come to dry out. As soon as they came back from manning their dories, men went below for a mug of coffee and a bite to eat (bread, cake, pie, or a piece of meat). They entered the forecastle without changing, with wet clothes, oilskins on, and gurry on their hands. They washed off the gurry and sat down to enjoy the food, rest, and talk with those around the table.

While the cook used a hand pump to draw drinking water from a tank beneath the floor, and the tank was filled when the vessels made port in the Maritimes, there was no toilet. An oak draw bucket served the purpose; bodily excretions going into the sea over the leeward rail. For a new man, it could take time to get used to the raw nature of a life at sea. There was also no tub, and the men rarely bathed—their showers came as they worked the fish from their dories.

There were also bunks in the captain's cabin. Here the skipper and a few of the men slept. It was a better place in a storm, but most dorymen preferred to be up front in the forecastle, where they were free to criticize the skipper outside of his earshot. Although the skipper would have worked his way up from a doryman, once he was in charge it was his word that governed the day-to-day activities of the boat: He decided where to fish, when to set, and when to go home. He was in control, and on a good boat no one dared question his authority.

Each bunk had a shelf to store gloves, nippers (a circular gripping device that went around the hand when fishing), books, cards, tobacco, and the other odds and ends of everyday life. Bedding consisted of a straw mattress, wool blankets, and a pillow. Each man had a small locker with a hook on which to hang his oilskins, boots, and street (or off-boat) clothes. Men came on board in their street clothes and quickly switched to their work clothes. They dressed much the same in winter and summer: long underwear, wool pants, heavy flannel shirts underneath, and a wool shirt on top—with an extra flannel shirt, mittens, and wool scarf in the winter. Over the inner clothes the men wore oiled "skins," lined suits treated with varnish and sometimes fish oil. Headgear was either a cotton or woolen cap, topped by a sou'wester. Boots, called red jacks, came halfway up the leg, with a little flap across the knee. They were usually worn large, so the men could kick them off if they went overboard.

Skippers often left port with a first baiting of frozen herring, cut-up mackerel, or even clams, but they could also leave with no bait at all, in which case they stopped at Cape Cod, Maine, or Nova Scotia to buy it. Second and third baitings were often caught on the grounds, the men jigging to bring in a catch of fresh squid or herring. Alternatively, they might again buy bait in a port in Maine, Nova Scotia, or Newfoundland.

In his personal account, Lon Littleton reports coming on a Georges man laying at anchor on the Maine coast; when asked, the skipper of the anchored vessel said he had bought bait from small boats coming out of Bath. The bait consisted mostly of small mackerel, averaging 9 inches long. This statement considered favorable, Littleton's skipper anchored, furled the sails, and in a few minutes let the crew engage in their respective amusements. Next morning, as the bait ship pulled alongside the Littleton vessel, the skipper called out, "All hands on deck! Get ready to take bait!" The men put on their oil-clothes, opened the main hatch, and began to hoist ice from below. One man in the hold hooked it on, another detached it from the tongs as it came on deck, and the rest of the crew picked it into small pieces, using pitchforks for the purpose. They kept it in hogshead tubs until needed, ready to preserve the bait. In no time at all, the 40 bushels of bait that had been purchased from the Maine bait ship had been iced and stored below—3 to 4 inches of bait covered by a layer of ice, and then more bait and more ice. The bait pen now full, they were ready to fish once they reached the grounds.

The schooner made for the fishing grounds, and once there, with the weather favorable for a day of fishing, the skipper bellowed out the call to "bait up." Soon half barrels of coiled cotton line encircled the stern deck of the vessel; their numbers depended on the number of "skates" the skipper wanted each dory to set. A skate equaled 300 feet of tarred cotton trawl line. A single skate set might be fished off an 18- to 36-inch square piece of canvas with two 9- to 10-foot ropes coming up from the corners and crossing to tie down the line. The men took to calling the piece of canvas a skate. In the early days, all were placed in half barrels, cut down from their previous use as flour containers. As time passed, this practice continued in the cod and haddock fishery, but in the halibut fishery, where the fish were larger, the ground line was left on the skate; but no matter, they still called it a "tub of trawl." Six 300-foot sections of ground line were tied together to make up a standard 1,800-foot barrel of trawl, the skipper typically calling for two to four tubs of trawl per dory.

The men had previously tarred the 3/16-inch-diameter trawl line to protect against the rigors of the sea. They had also attached 50 or more lighter 30-inch to 5-foot gangings to the line, connected at 6- to 12-foot intervals—the longer gangings and spacing was for halibut, the shorter gangings and spacing for haddock and cod. The gangings were either fastened to short 1-foot pieces of line attached to the ground line, or a small spike was used to push the ganging through the trawl line, and then it was made fast with a knot. Each such ganging held its own hook.

Using iced herring, tinker mackerel, fresh squid, alewives, menhaden, or clams, an experienced doryman baited eight to nine hooks a minute. Most of the men worked on either side of the schooner's cabin trunk. A few stood beside the gurry pen, in front of the cabin.* No matter where they worked, they all used a butcher knife to chop a typical 10-inch herring into four pieces of bait, throwing away the head and tail.

Although most men were adept in baiting the trawl, on many trips skippers characterized three or four of the men as "fussy old women," because they were "chock full of whims." Some fishermen had to have things exactly right in order to bring out their best efforts—the tub at exactly the right height from the deck, the bait to one side and within inches of the man's reach.

To bait 500 hooks (one tub—a half barrel) took about 30 minutes, or about two hours for a four-tub set. But baiting a trawl line involved more than simply sticking pieces of herring on hooks. When the men last recovered the trawl line they would have coiled it in the tub or on the skate; some hooks were now caught on either the tub or the line. It all had to be untangled, ensuring that the hooks, ground line, and gangings would later run free.

In preparing the trawl line, the doryman made three or four small loops in one hand as he ran his free hand down the ganging to the hook. He passed the hook through both skin sides of the bait and placed the hook over the loops in his other hand. The man then dropped the loops and hooks into the

---

*Men might also bait up in the fish hold. Candles lit up their work and fit on holders that sat on the edges of the trawl tubs.

trawl tub. Men quickly became adept at baiting up, and skippers had no use for men whose trawls tangled and refused to play out. They called such bunglers "mollycods," "tommycods," or "nincompoops," according to the degree of fussing.

With his jib to windward, and a rising tide over the fishing ground to ease the work of the men, the skipper would at last call for the men to make their dories ready. The dory was a simple boat: Orange-buff in color, it measured 14 to 16 feet at the waterline and 18 to 21 feet at the gunwale. The larger boat was used in the halibut fishery. It was just 21 inches deep, with a flat bottom and rounded sides, and a raked stem that rose 6 inches higher than the midsection of the boat.* When ready to launch, the two men of the top dory in the stack of six on either side of the deck arranged their gear, including buoys, buoy lines, trawls, trawl-line anchors, and the like. They placed a birch "dory plug" in the drain hole in the floor of the dory. A manila line passed through a hole in the middle of the plug. An eye loop of 1 to 2 feet in length was fixed at the end of the line, and was fastened onto the dory riser on which the thwarts sat. If the dory flipped over into a "turtle" position, the doryman did his best to grab this rope and pull himself onto the overturned boat.

The cook helped the dorymen lift the dories over the rail, one at a time. There the individual dory hung until the captain gave the word to lower. The stern of the dory went in first, hitting the water with a bang and sending up a splash of water. One of the two dorymen jumped into the dory and unhooked the stern line from the hook and used an oar to push off from the schooner, while the men on deck unhooked the bowline and the second man scrambled into the dory. The dory remained attached to the schooner by either the dory's 18-foot bow painter or a tail rope dragging out behind the schooner, as one after another of the other dories joined them behind the boat. When the skipper thought the time best—usually when he had the schooner moving across the fishing ground—he ordered the painters cast off at fixed intervals, one after another. This was called a "flying set."

In a "standing set," the small boats were lowered one at a time from a stationary schooner, and on the skipper's order went out in a star or other fixed pattern—the schooner holding position. Here, the mainsail came down, the forward sails were sheeted in, the boom was lashed amidships, and the skipper set a small triangular riding sail set to hold position. He turned the schooner into the wind. Dories went out on each side of the vessel, one over

---

*Bent-over brads held the closely shaved side planking in place, the pointed shaft of the brads sunk into the wood of the plank. Only the dory floor needed caulking. Dories had about a two-year useful life, having to withstand the constant pounding of the sea and dashing into the sides of the schooner.

✕ *Dories streaming astern of a schooner prior to being released to make a flying set in the early 1920s.*

the bow, another two points off the bow, another amidships, a fourth off the quarter, and a fifth off the stern. Depending on the wind, the dorymen either rowed or sailed away from the schooner, a few miles at most, to set their trawl lines.

Under the flying set, the skipper released his dories one after another, up to one-half mile apart, as the schooner sailed slowly across the grounds. The skipper set a straight course. The dories were lowered on the lee side of the schooner, each one going out a mile or two from the imaginary setline. As the dory left the vessel, the skipper called to the men, "Heave out your buoy!" As the buoy line ran out and the men cast off the painter, the skipper told them in which direction to set their trawl—generally to leeward, at right angles to the course of the vessel. Dorymen counted the rowing strokes as they went out, and noted the count at the point where they set the outside buoy, the buoys keeping track of where the trawl line had been set and the heading to return to the schooner. They also counted the rowing strokes when they returned. With these precautions, on a foggy day they knew how far to go to return to the beginning of the trawl line.

No matter the set, the captain used his compass to mark the position of each dory. One never knew when fog would roll in and obscure the dories from view. In fine weather the men spaced the dories far apart; in bad weather they kept them close together. By keeping the dories together in rising seas

Rowing away from a schooner to make a set. The men's full trawl tubs are in the stern of the dory and the boat's freeboard is still generous—by the time the dory returned to the schooner it might have only a plank or two above water.

and overspreading fog, the skipper could more easily find the men, and the men the schooner. The last thing a skipper wanted was for the men to be scattered all over the grounds in foul weather, but in fine weather a skipper scattered his dories, opening them up in the hopes of making a catch.

The ground line of each trawl had to lie on the bottom of the sea (halibut, cod, cusk, hake, and haddock were all ground-feeding fish). Attached to the end of the ground line was a small 10-pound anchor that took the line to the bottom. A buoy line (or cable) went from the anchor to the surface. The buoy—really a watertight barrel bearing a unique identification mark, often with a black ball on top—indicated where the trawl line began. The manner of setting the trawl varied by fishery. In the cod and haddock fishery, with both men standing, the bowman played out the trawl line with the rising current, while the second man stood in the stern, controlling the dory. The end of the trawl line was connected to the buoy line, and the bowman fed out the baited trawl line. To ensure the hooks were clear, the bowman sometimes used a short, two-pronged stick to throw the line over the gunwale. By lifting the line straight up before casting it out, the bowman made sure there was rarely a snarl. The bowman also paced the throw to ensure the line sank without getting tangled. Once in a while a hook caught in the tub, and the bowman had to pause to release its grip. On a halibut trip, the bowman rowed, while the more experienced of the two stood in the back throwing out the line. He lifted the skate of trawl onto the thwart in front of him, made the end fast to the anchor, and played out the line over the side.

When the line in the first tub neared its end, the doryman bent the line and tied it to the one in the next tub of trawl, and so on, until he reached the last skate the skipper had said to set. He then attached a second anchor to the far end of the trawl line, as well as a second buoy line and buoy—the outer buoy. The ground line of the trawl now rested on the bottom with its baited hooks. Assuming the dory set four barrels with 500 hooks for each ground line, the trawl line extended over a mile with 2,000 hooks. Depending on the weather, it took up to two hours to make this set.

The men might now return to the schooner for a meal, awaiting the skipper's command to retrieve the trawl. They might also wait at the outer buoy, listening for the skipper to blow a horn calling for them to pull the trawl—a 30- to 60-minute wait. If they had come back to the schooner, the wait time could be as little as two hours, but it was often longer. But no matter, the cook had prepared coffee, tea, and a bit of food for their well-earned mug-up. The skipper waited for the tide to turn before retrieving the trawl line, neutralizing the force of the tide, keeping the men from having to pull too hard. In night sets, the call to pull the trawls might not come until the next afternoon.

Bringing in the trawl line could take from a few to as many as eight hours, with multiple sets possible each day. One man stood in the bow and grabbed the buoy to haul up the end of the trawl line. He wore "nippers," which resembled wristbands, around the hollows of both hands to help grip the line. Every time he gripped the line the nipper closed and held on. The bowman hauled up the ground line, with its fish, if indeed fish had taken the bait. In the cod and halibut fishery, the line was often pulled directly over the side; in the halibut fishery a "hurdy-gurdy" hand winch was attached into the gunwale 3 to 4 feet from the stem, and the line was more easily pulled over a roller from the seabed below. In the stern, the standing mate coiled the ground line, ganging lines, and hooks into the barrel, while at the same time removing the fish and the bait from the hooks.

The bowman used the swell or chop to help him. When the dory rose he held the ground line fast. When the dory fell between the swells, he hauled the ground line as fast as possible: To make a mistake in when he held and released the line could swamp the boat, but these were experienced men and this seldom happened. The man in the stern developed a rhythm for taking off the fish. For the smaller cod and haddock, he snapped or jerked the fish off the ganging and heaved them into the open space between the two men. For fish that did not come off, with hooks set deep in the mouth, he used a "gob-stick" to reach down and release the hook. Old bait on the hooks always came off with a simple snap of the wrist and hit against the gunwale. For larger fish such as halibut, the man in the stern used a gaff to lift the catch into the boat. He used the barbless hand tool to first hook the fish in the eye or someplace in the head, then he stunned or killed the larger halibut by clubbing it with a 2½-foot club before hauling it into the boat. It was a tiring job, muscles became strained and sore, and the men switched positions every tub or so.

In bad weather, the men recovered trawl in winds that rose to 15 to 20 knots and seas of up to 8 or 9 feet, the two men constantly switching positions. On a heavy fishing day, they might change places every 5 or 6 lines, 260 to 320 hooks.

The stern man had one other responsibility: to watch for rogue waves or other changes in the condition of the sea. He kept the dory headed into the wind, for if a heavy sea hit the dory on its side the boat might overturn. Thus, the man in the stern kept an oar hanging in the becket, ready to spin the dory into the wind at a moment's notice. In good weather, the dory could safely hold up to 2,000 pounds of fish, but in poor weather or heavy seas, it carried only half as much. The men might even throw fish overboard to ensure their safety.*

To keep the men and the fish in some semblance of order, dories had two sets of removable boards, the thwarts and kid or kick boards, which fit under the second and stern thwarts, creating an area for the storage of the fish as the men pulled them from the bottom. By keeping the fish in one area, dorymen had enough foot room to pull in gear and remove and store the fish brought up on the trawl line. When the storage area was full, the men headed back to the schooner. With a fair summer breeze, they might use the cotton-cloth spritsail to run home to the schooner. It was triangular in shape and was held into the wind by a light portable mast. In light winds, on

---

*Goode and Collins, in a U.S. government fish bulletin from the era, describe another process, used sometimes in very heavy weather : "When fishing on George's Bank, the Gloucester haddock vessels are obliged by the force of the tide to resort to another method of setting, which is called 'double-banking the trawl.' The tide is so strong that the trawls cannot be set in the ordinary way, for the buoys would be carried beneath the surface. Two dories are therefore lowered at once, and jointly perform the act of setting; only two tubs are set by each pair of dories. The set is made in the following manner: The men in one of the dories hold fast to the weather-buoy while the men in the other dory set the trawl. After the trawl is out, the dory which sets it holds fast to the lee buoy until by some signal, such as lowering the jib, the skipper of the schooner gives the order to haul. The trawls are left on the bottom 15 or 20 minutes before they are hauled. The men in the two dories begin to haul simultaneously; the anchors are thus first raised from the bottom and presently the bight of the trawl and the two boats drift along with the tide, the distance between them gradually narrowing as they haul." G. Browne Goode and Captain J. W. Collins, *Bulletin of the U.S. Fish Commission*, 1881.

)(*The set is complete. Cod sit in the middle of the dory, ready to be off-loaded onto the schooner deck, after which they will be sorted, gutted, and preserved in salt.*

a foggy day, and from fall to spring, they usually rowed back. On a clear day, the schooner often moved back and forth over the grounds to reach the full dories. The men held up an oar at the stern as a signal that they were ready, and the captain sailed to them to unload the fish. As the dory reached the schooner, the dorymen threw the bow and stern painters onto the deck. The cook, catchee, and captain brought the dory close—the cook often remaining at the wheel to hold the schooner in position—working hard to stop it from smashing into the side of the vessel. The dorymen then moved the fish from the dory to the schooner. One man jumped onto the deck of the schooner while the other remained in the dory. Often they used pitchforks and gaff hooks to move the fish, forking through the head to maintain the value of the fish. For large fish, such as halibut, that could weigh up to 700 pounds, the man in the dory grabbed the fish with the gaff hook and handed it up to his mate on deck to haul onto the schooner. Smaller fish were tossed up overhead. Pulling a giant halibut from the dory over the schooner's rail was timed to take advantage of the rolling of the vessel: The man who hauled in the fish watched the motion, and by a sudden exertion of strength at the right moment was able to take a fish on deck.

After unloading the fish, assuming they did not have to go back out to finish hauling their trawl, the dorymen removed the equipment in the dory—the small anchors, bait bucket, trawl buckets, and water jug. They knocked down the thwarts and dividing kick boards, storing them on the floor of the dory. Next they pulled the block and tackle line down from the fore- and aft-rigging to raise the dory onto the schooner. The boat was guided over the rail and onto the other dories stacked beside the main hatch, where it waited for the skipper's call to again set the dories.

Now, with fish piled high on the deck, the men spent hours sorting, cutting, gutting, and salting (or icing) the fish, storing the processed fish in the hold and throwing the gurry over the side. The hatches in mid-deck opened wide to accept the catch. The work seemed never ending, and when the skipper succeeded in getting the schooner on the fish, the men rarely slept. Some were able to go without sleep for a day or two, some slept standing up, and some rubbed wet tobacco in their eyes in hopes the pain would keep them awake. It was a labor-intensive, exhausting way of life, but it would be at the heart of Gloucester's prosperity for almost 70 years. The technology changed little over this period. Dorymen became masters in handling the new gear and in maneuvering their small boats, their strength and endurance were tested on every set, they fished in all conditions and in all seasons, they could face the risks of a life at sea, and for days on end they truly sat alone at sea.

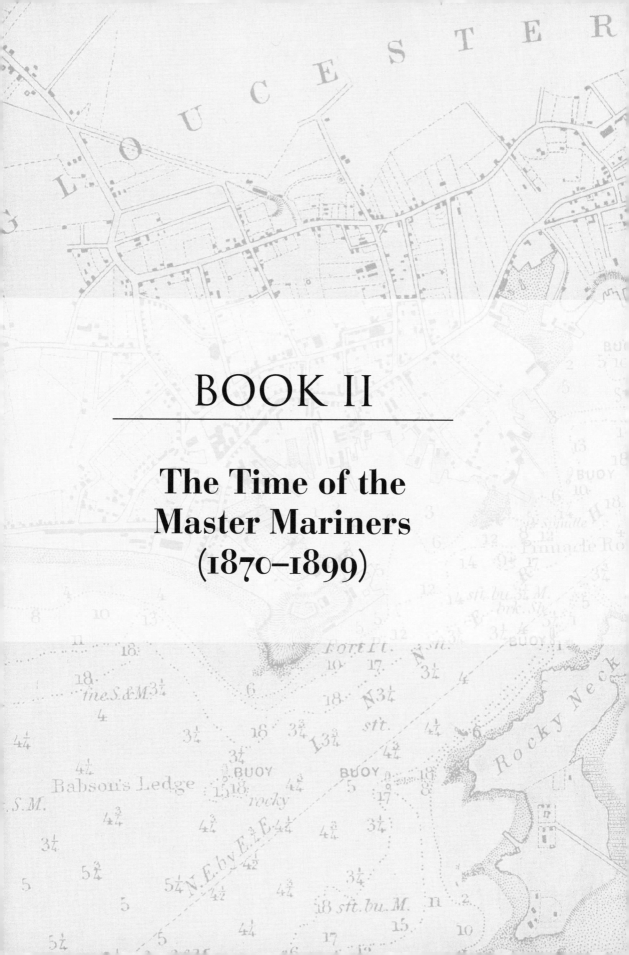

BOOK II

The Time of the Master Mariners (1870–1899)

Over the next 20 years, little changed in Gloucester. Schooners went to sea, dorymen set trawl lines, and mackerelers dropped seine nets. Gloucester remained the fishing capital of North America, attracting young immigrants from Maine, Canada, Newfoundland, Scandinavia, and Portugal's Azores Islands. A veritable babel of tongues floated over the waterfront. Even so, having risen to their lofty position, Gloucester people fretted about the future. Fishermen felt threatened by the entry of cheap, duty-free Canadian fish into the American marketplace, while the loss of large numbers of fine clipper schooners on the North Atlantic fishing grounds left lives in danger and families without hope. Yet, despite the fretting of the people—and there were bumps in the road that had to be addressed—it was a time of unbounded optimism and great prosperity.

CHAPTER 9

Reciprocity and the Bay

In 1854, Gloucester's handline mackerel fishermen had championed the reciprocity treaty with Great Britain; they supported entry of duty-free Canadian fish in exchange for access to the inshore waters of the Bay. Even when the treaty ended in 1866, they continued to support the status quo—access to the Bay still took precedence. For its part, Canada initially allowed the men of Gloucester to fish the inshore waters of the Bay for a nominal fee of 50 cents per vessel ton—about $35 a year for the typical mackerel handliner. Every schooner in the Gloucester mackerel fleet paid the fee, and little changed in the annual cycle of life on the Bay.

For the Canadian fishermen, however, the end of reciprocity had a devastating effect: They could no longer send duty-free fish south into the United States. So Canada put pressure on the American mackerel fishermen in the hopes of bringing Washington to the bargaining table. Their tactic was simple: They raised the tonnage fee, first to $1 per ton, then to $2. It would surely be only a matter of time before the mackerelers refused to pay, and when that happened, there would finally be a need for the two countries to come together. America would have to sit at the bargaining table. And, indeed, at $2 per vessel ton, the Dominion got its wish—Gloucester's mackerelers refused to pay, but they still headed north to the Bay, even without the all-important Canadian license.

The Canadians feigned outrage and the game was set in motion. The Canadians looked to British cruisers to seize all American craft that dared fish without a license, but the British sat on the sidelines. London was unwilling to enter a contest with the Americans, and no Gloucester vessel was seized. But for the Canadians there was a principle at stake: Gloucestermen had shown up in large numbers, they had defied Canadian law, and newspapers below and above the border wrote articles describing the emerging struggle. Most important of all, the issue now sat front and center in both London and Washington.

In 1870, Canada raised the ante, and its own patrol craft were sent into the Bay. They were not prepared to sit on the sidelines as the British had the previous year, and the *Halifax Chronicle* wrote, "Just think of it, seven well-appointed schooners and one excellent steamship! Hear it and tremble ye Gloucester fishermen. When British gunboats only cruised on the fishing grounds you treated Dominion laws with contempt and fished without license. . . . [But this] will stop your little game."

Thus had the Dominion acted, and confrontation was in the air. Then something strange happened, something that was to alter the history of the Gloucester fisheries. For the first time in recent memory, large numbers of Gloucester outfitters abandoned the Bay. For Gloucestermen, the issue of trade reciprocity and access to the Bay was stood on its head. Rather than confront the Canadians, many firms fit out their craft with purse-seines and fished in international waters off the New England coast—on Georges Bank and the South Channel. They called the Canadians' bluff, and in so doing found that they did not need the Bay. Who needed the Bay when seining worked so well off the New England coast? Meanwhile, other mackerel outfitters fit out craft with dories. They moved them into the shore and Georges ground fisheries, and they too found a prosperous alternative fishery: The salt fishery had emerged as a major player in the Gloucester economy, and dory-based trawling had contributed to an altered landscape.

Even so, a few mackerelers did sail to the Bay, and they called on Washington to send American patrol boats. But President Grant sent none; instead, he brought trade reciprocity back to the diplomatic table.

On the inshore waters of the Bay, little of any consequence happened during the 1870 season. Canadian patrol craft seized only three American vessels, although they stopped and searched many others. But the intrusions had their intended effect: Gloucester's small fleet of mackerelers in the Bay remained on edge. Newspaper articles described the situation, and of greatest import, the federal government was back in the game.

As 1871 dawned, American, British, and Canadian commissioners looked for ways to address the issues. But even as discussions were under way, the delegates had no idea that Gloucester had lost its fascination with the Bay. If the price of future access to the Bay was the entry of duty-free Canadian fish, the people of Gloucester said to hell with the Bay. But Gloucester had no access to the commissioners, and, indeed, some commissioners even said they wished to hear no more "fish stories."

It came as no real surprise, therefore, when later in the year the Washington Treaty both opened the Bay to American fishermen and allowed duty-free Canadian fish into the United States. Many in Gloucester felt betrayed. Everyone now knew that they had to be better prepared when the treaty came up for renewal. But that would not be for a decade, and for now the Gloucester fishermen resolved to look to their future, to continue to thrive.

With 435 vessels in its fleet, Gloucester would more than hold its own, although as the 1873 season opened, the town's mutual insurance company refused to cover the fleet in the difficult January period, and thus early season returns were limited to 16 trips of cod and halibut. In February, the pace picked up—25 trips from the Banks and 32 from Georges. By March the Georges and Banks fleets were in full operation—the waterfront was humming.

By late April, the first 20 schooners of the southern mackerel fleet had sailed, and they soon found the run of the mackerel, opening a lively southern season. By July Fourth, the southern fleet had largely returned, and with gear checked and supplies stowed, 126 handliners and 12 seiners were soon away for a summer on the Bay. With the treaty in place, there was no reason not to sail to the Bay. Most first fished in Chaleur, where large mackerel were in abundance—craft putting up 40 to 100 barrels. Next they sailed east to the bend of Prince Edward Island, and here too had great success.

Toward the end of July, a severe thunderstorm broke on the fishing grounds, which cost the men a day of fishing. It was followed by a three-day northerly in which many craft lost cables and anchors, while others had split foresails and jibs. Nevertheless, with immense schools of mackerel, the men were off to a fast start. But on August 24, as the fleet sat at the northern bend of Prince Edward Island, the winds intensified and the seas rose. A blow was coming in from the northeast, but it was no mere northeaster. It was the western flank of a full-fledged Atlantic hurricane: Vessels and lives now hung in the balance. The storm surge ranged from 5 to 8 feet above normal. Many double-anchored schooners dragged their anchors and found themselves beached in the sand above the normal high-water mark. Most of Gloucester's fleet had been on the north shore of Prince Edward Island when the storm struck, and judging the danger of staying in so exposed a position, many skippers sailed north 50 miles for Amherst, the southernmost of the Magdalen Islands. Captain Ingersoll of the schooner *Carleton* counted more than 40 double-anchored schooners, many from Gloucester, and by storm's end, 18 Gloucester vessels had come ashore in the Pleasant Bay area. A few were a total loss, but most could be relaunched and saved. Damages were slight, although getting the craft out of the sand took time: Ways had to

)(*The vast mackerel fleet of the 1860s at anchor inside Eastern Point on the pancake ground—the town must have been busy with idle men spending their hard-earned money.*

be dug in the sand and steamers brought on the scene to drag the vessels through the sand trenches into the sea.

Meanwhile, in Gloucester, families began to come together in the reading room of the Mutual Fishing Insurance Company to see the latest postings on vessels that had been lost and those that had weathered the storm. It was said that even as families waited, little children on Cape Ann asked, "Why does not father come home?"

Word quickly followed on the many other Gloucester vessels that were ashore, including those at the Magdalens, and the insurance company sent representatives to aid the fishermen. In a final accounting, eight Gloucester vessels had been lost with all of their crew, and six men had been washed overboard and drowned.* Yet, despite the losses, Gloucester's mackerel fleet had largely weathered the hurricane, and by season's end, several of the Bay fleet returned with more than 400 barrels of mackerel, while the average was

*The list of lost craft included the *Charles C. Dame*, wrecked off Cape North, 18 men lost; the *Angie S. Friend* foundered at anchor off Port Hood, 14 men were lost (although 2 of the crew had left the vessel in the Bay, being replaced by men from the Provinces); the *James G. Tarr* was lost at North Cape with 18 men; the *Royal Arch* was lost while homeward bound at White Head, N.S., with a crew of 14; the *Samuel Crowell*, Captain Hamilton, left Canso on the day before the gale, setting a course for Gloucester, but foundered at sea, 15 men lost; the *El Dorado*, on a codfishing trip to the Provincial shores, drifted out of Canso in the gale and was lost on White Head with a crew of 7. As the storm first worked up the coast, 2 craft were lost on Georges: the *Center Point* with a crew of 11, and the *A. H. Wonson* with a crew of 10. Three days before the gale, the *Henry Clay* left the Grand Banks with a crew of 10 and a full load of salted fish, but was lost on the passage home.

about 250 barrels per craft. Gloucester had returned to the Bay. The schooner *Col Cook*, under Captain George Bearse, was highline of the Bay: He used all of his salt and barrels, and returned with more than 700 barrels of mackerel. October's Bay receipts were the largest in years, and for the year, Gloucester's total catch of all fish species came in at 81.6 million pounds, up from 72.7 million pounds in the previous year. Of this, with a renewed Bay fishery, the mackerel catch rose from 14.2 to 17.3 million pounds, while the cod catch rose from 43 to 51.5 million pounds.

Three years later, the total catch passed 100 million pounds for the first time, and the fishermen received good prices for their catch. Even so, the people of Gloucester continued to fret. The ever-present specter of duty-free Canadian ground fish seemed a threat to the burgeoning dory-based economy. Fishermen saw more Nova Scotian craft on the Scotian grounds, and the economic balance seemed weighted in favor of the Canadians. The Nova Scotians were closer to the grounds, they benefited from lower construction and operating costs, and the people of Gloucester could only wonder what would happen in the 1880s and beyond as Canada grew its fleet. Could Gloucester maintain its fleet, or would the fishing capital of North America

In a view from the 1870s, mackerel barrels are lined up on wharf after wharf.

shift to the north? If Canadian fish were cheaper than American, who could blame the processing firms for going with their pocketbooks?

Such a prospect was troubling, and the threat seemed real indeed. The concern with trade reciprocity never waned. In preparing for the political battle to come in the early 1880s, Gloucestermen looked for instances where the Canadian and British North American Provinces failed to live up to the terms of the 1871 treaty. They would use examples of boorish Canadian behavior in their campaign to revoke the fishing clauses of the treaty. In 1877 Gloucester found just what she was looking for.

THE WASHINGTON TREATY gave Gloucestermen the right to fish in Canadian and Provincial inshore waters. They could use whatever method they chose, and whatever combination of American and Provincial crews they saw fit. Of these northern inshore waters, Gloucester most prized the herring bays of Nova Scotia and Newfoundland—the source of baitfish for Gloucester's winter cod and haddock fleet. For years, outfitters had sent vessels north in November and December to engage in a cooperative fishery with the local fishermen of Newfoundland. They sailed north with skeleton crews, and on arriving in the Provincial bays, the Gloucester skippers hired local fishermen to round out their crews and bring in the catch. Then, in 1877, in an intentionally provocative act, Gloucester's herring fleet sailed north with purse-seines and full crews. As they arrived at Fortune Bay, the local fishermen looked on in amazement. Their livelihood depended on the annual herring fishery, and as the local paper stated, "starvation under the new state of things stared the natives in the face."

The Newfoundland fishermen acted: The future of their families depended on their response, and they came together to contest the actions of the Americans. Men congregated on the shore and went out in small boats, threatening the Gloucestermen as they encircled the herring with seine nets. The incensed locals rushed forward, tripping the seines and releasing the fish. They looked the Americans in the eye and threatened even worse should the Gloucestermen continue to fish in this manner. Only the schooner *Moses Adams*, under Captain Solomon Jacobs, put up a stiff resistance. Jacobs armed his men with revolvers and threatened to shoot the first Newfoundland fisherman who dared to interfere. The locals kept their distance, and Jacobs brought in a catch.

He was the exception, as the other Gloucester craft had been intimidated and sailed home in ballast. They had no herring, and in the years that followed they returned to the old ways. It was more important to supply bait to Gloucester's winter ground fleet than to fight with Newfoundland natives over issues of principle. Yet, Gloucester had made its point. American fishermen could not trust the Provincials to live up to the treaty.

It would not be until 1881 that Gloucester could act politically, and then on October 14 the leaders of Gloucester's fishing community came together to meet with their congressional representative. They peppered him with the "Gloucester story," including a rehash of the Fortune Bay incident—or riot, as the Gloucestermen liked to call it. They stressed the new realities of Gloucester's dory-based and purse-seine economy: The fleet pursued fish on the ocean banks, in international waters. Of all the mackerel landed in the first nine months of 1881, only 43 barrels had come from the Bay. The 1873 success in the Bay had proved to be short-lived. The handline mackerel fishery had died a natural death. Inshore fishing on the Bay had no relevance, and the fishermen of Gloucester called for an end to the fishery clauses of the 1871 Washington Treaty.

In January 1882, their efforts quickly bore fruit. Congress gave notice of the country's intent to terminate the fishing clauses of the treaty, and with the schedule specified in the treaty, in mid-1885 Nova Scotia lost its right to send duty-free fish into the United States. The government and its fishermen had stood together; the future seemed bright. The next two years, 1882 and 1883, saw vessel outfitters add 91 new craft to the fleet—more than twice the number of the previous four years combined. For the first time in Gloucester's history, the fleet exceeded 500 vessels, and it was still growing.

Around the waterfront, companies began to expand, with new cod processing facilities put up by John Pew & Son,* and mackerel handling facilities added by James G. Tarr & Brothers. John Pew & Son was fast becoming the most important firm in town. It owned one of the largest fleets, sending 20 vessels to the Banks. It also produced nearly as much boned fish as the whole Boston trade and was a major processor of packaged mackerel. Pew & Son's wharves had a 650-foot frontage on the inner harbor. Two of the wharves handled the firm's vast salt-importation business (with 25 or more barques and schooners landing salt brought in largely from the salt pans of Trapani, Sicily), while the other two were for the fishing business. A brick building at the head of one of the wharves served as the corporate headquarters, while also including rooms for reconciling accounts (with a brick vault under the counting room floor) and fitting out the vessels. Pew even had a press to print the labels they affixed to the ends of the boxed fish, either their own or a customer's brand name, and sent the fish to customers all over the country—with most of the customers having their own house brand.

---

*Captain John Pew, who started the firm in 1849, died at 83 on March 7, 1890.

 *Schooners prepare to sail in the 1870s. On the right, a new wharf is under construction and fish flakes lie covered in salted cod. In the immediate foreground lies David Allen Lumber Company; on the left the George P. Trigg Fisheries can be seen on Collins Wharf.*

With such impressive new plants and an expanding fleet, Gloucester continued to attract a diverse group of skippers and dorymen. The extant notes of Frank and Fred Keene of Bremen, Maine, to Tarr & Brothers of Rocky Neck show the spirit that was then alive among the fishermen.[18] In the note from Frank Keene, he asks Tarr & Brothers: "Have you a vessel that you will send pollocking on the Fifih [sic] Bay? I have most of my crew shifted here. It is hard to get men to go on the halves, my good men will. [We will] leave as soon as they think there is good fishing if you have one to send. I would like to have her. Would like to start in ten days to get market fishing. Please answer as soon as you get this and oblige." His brother, Fred W. Keene, wrote: "I will drop you a line in regard to [mackerel] seining. If you have a vessel you are going to send to the Cape Shore I should like to have her. If you have not and should know of anyone that has, will you kindly inform them that I am looking for a vessel."

Such notes reflect the community's overwhelming sense of optimism. But then, with a blink of an eye, the future once more seemed in jeopardy. Word reached Gloucester that the incoming President, Grover Cleveland, saw little value in building layers of protection around American industries and American workers. Cleveland was a Democrat,* and in his view the American consumer benefited from the duty-free importation of Canadian fish. It shook many in Gloucester to the core—a national tariff was seen as central to Gloucester's well-being. For such reasons the town would always be Republican: The Republicans were for tariffs; the Democrats were not.

As one of his first acts as President, Cleveland recommended the appointment of a commission to consider the terms of a treaty to settle the fishery rights question between the United States and Great Britain. The commission would have the fullest latitude in its deliberations. But, in the view of the fishermen, there was nothing to settle; the Gloucester fishermen had won their case. The town had friends, and the Republican Party still held a slim majority in the Senate (39 to 37). It was to the Senate that the fishermen turned for help. The community swung into action, and by the end of December it seemed likely that Congress would listen to the voices of the New England fishermen. The fishing firms pitched in to create a new organization, the National Fisheries Association (NFA), appointing the ex-collector of the port of Gloucester, Captain Fitz James Babson, as its president. The message was simple: The "erroneous and absurd claims" of those who would sacrifice American fisheries had to be exposed. As Babson "spun" the story, the issue of concern was Canada's new and expanding fleet. America needed fish, and there was no question that it had to come to port in American vessels, crewed by American fishermen.

The message hit a nerve, and in a February 5, 1886, telegram back to Gloucester, Captain Babson said that the Senate would not back Mr. West, the British Minister. Two months later he was proved right; the Senate voted 35 to 10 against setting up an international commission. The issue of trade reciprocity was dead, and Captain Andrew Leighton, speaking for many in Gloucester, took a punch at some in Washington, while praising others. Leighton said he "wouldn't give a barrel of fish gurry for most of them fellers down there in Washington. Taint a question of politics. Mr. Bayard [the Secretary of State] probably thought that the Eastern fishermen were like those who fish off the wharf down there in Delaware. Senator Frye [of Maine] seems to have a strong grip. Senator Morgan [a Democrat from Alabama]

*For a quarter of a century, America had been ruled by a series of Republican Presidents beginning with Abraham Lincoln, but following the election of 1884, the Presidency and the House of Representatives passed into the hands of the Democrats.

may be a durned rebel, but he's got big brains, and he seems on the right side of the matter. Mr. Dingley, of Maine, has taken a hand in the business, and that's all that I think of. How about our representative from the Cape Ann district? He might as well be a graven image so far as doing anything goes. His name is Stone, by the way."

The euphoria did not last long. First, Canada retaliated, seizing a Gloucester vessel that had entered Canadian waters to buy bait, but it had little effect on the Gloucester fishermen. They had little interest in the inshore waters of the Bay, and they raised no call for Washington to intervene.

Then President Cleveland acted. If the Senate would not establish the commission, then he, as President, invoked his right to enter negotiations, convening the closed meeting Fisheries Conference of 1887. It was to be the President against the fishermen all over again. But, one week later, the *New York Daily Tribune*, drawing on the words of Captain Babson, rebuked the secret commission. Other papers questioned the President's failure to stand with the fishermen of New England; the message was soon spreading across the land.

The President's apologists, including the editorial writers at the *Boston Herald*, did not help his cause in their clumsy attacks on the Gloucestermen. The paper put forward "back of the envelope calculations" to show that it was "probable that the difference in the result obtained to American fishermen, with or without Canadian competition in our markets, even under the most adverse conditions, would not amount to $500,000 a year." Then, having created a mythical figure, they asked rhetorically whether America was prepared to let so little a sum stand in the way of an extension of trade that would amount to hundreds of millions of dollars. They suggested it would be cheaper to buy all the fishing vessels of Gloucester and let them rot at the wharves, pensioning every Gloucester fisherman for the rest of his life. But Gloucestermen had no wish to be pensioned off. Nor did they or most of their countrymen want Gloucester vessels to rot at the wharves. The threat seemed all too real, and on February 8, 1887, Captain Joseph Smith brought the skippers together to form an association that would "result in good." He noted that almost every other body of men had their association, a place where they could meet socially and talk over their business affairs. In his words: "Why not the masters of the Gloucester fishing fleet?"

While the skippers billed the meeting as a fraternal meeting of vessel masters, one needs only to note the name of the honoree to recognize the political focus of this new effort. It was none other than the hero of the Fortune Bay "outrage," Captain Solomon Jacobs, a man soon to embark on an experimental halibut fishery trip to America's Northwest Coast. All present—

some 40 skippers, 15 vessel outfitters, and several fish processors—praised Captain Jacobs. Captain Joseph Smith, "the founder of the feast," said he was glad to attend such a representative gathering of the skippers, and called for more events of the type. He added, "As a class of men we have made a mistake in the years which have passed, in not having an organization . . . and if we were organized it would make us feel that we were skippers, and prove conducive to our welfare." Captain James L. Anderson then alluded to the fishing skippers as a class of men to be found nowhere outside of Gloucester: "[You are] men who risk . . . [your] lives on the ocean and follow the most arduous pursuit known among men; and to add to this to have to put up with the annoyance of being chased by British gunboats, the same as Captain Jacobs and I have; but they haven't caught us yet."

So was born the Gloucester Master Mariners' Association, and its members quickly condemned the proposed fishing treaty. As they saw it, "the whole aim of Canada is to get our market free. This is the object sought. Free-fish into the United States without rendering any equivalent. The three-mile limit clause amounts to nothing. It's an old chestnut and concerns the mackerel fishery only. We would like the privilege of purchasing bait and paying for it (in a Canadian port), but if the price is to be a free market, it would be far better for us to find some other way of obtaining this article here on our own shores, where the traps would furnish a good supply, as also would the clam flats, thus aiding these industries and reaping the profits."

With the first meeting behind them, the Association and the citizens of Gloucester appealed once more to the Senate. America's press joined the fight. The *Norwich Bulletin* believed that "the interests of our New England fishermen [are being] sacrificed to the free trade principles of a Democratic Administration." In Worcester, the *Spy* was even more direct: "There need never be any difficulty in making a treaty if you are willing to abandon all your own claims and allow your adversary all he asks." The *Worcester Telegram* called the treaty "a great diplomatic victory for the Canadians." In Ottawa, the *Journal* agreed, saying, "The new treaty will be an eye-opener to the New England Fish Pirates, and it is ten to one there will be a howl in Gloucester against its ratification."

Hearing the chatter, the Master Mariners' Association stepped forward to speak in a clear, unified voice on the treaty. If it was a howl the Canadians expected, it was a howl they got. The Association had grown to more than 130 skippers, and on April 7, 1888, at its then annual meeting, Fitz J. Babson delivered a speech on the fishery situation in Washington. He stated, "The provisions, as far as they can be understood, are a vindication of the barbarism of Canadian local laws, in restricting our treaty rights, and a complete bar to any claim for indemnity for [past] wrongs committed."

The Association sent a petition to the Senate, signed by 146 skippers, the words written by Babson. Babson soon telegraphed the following message to his Gloucester colleague George Steele: "Every point is covered. The Senate understands the matter fully, have no fears we are all right and can manage it. Tell the people."[19] One month later, on August 21, 1888, by a vote of 30 to 27, the Senate declined to approve the treaty. They defeated it along strict party lines, and the fall election proved a disaster for the Democrats: Cleveland lost office and the Republicans swept both houses of Congress. Duty-free fish no longer threatened the Gloucester fisheries, and it would be 20 years before the issue again raised its contentious head.

CHAPTER 10

Disaster on the Banks

Having helped defeat the proponents of duty-free fish, the people of Gloucester shifted their attention back to the fleet and the men on the grounds. With more than 530 vessels and 5,300 fishermen, landings remained high and vessel owners prospered. But with big rewards came big risks. The winter gales that moved up the coast ripped through the fleet, resulting in an average annual loss in Gloucester of 18 clipper schooners and 122 men. While some vessels might pull out for a winter's rest, vast numbers of schooners were at sea through the most dangerous winter months, and the vessels were getting larger, with more dories and more men. The devastation echoed across the community, and people looked for answers even as families grieved for their lost loved ones and relatives sought to comfort those left behind.

A series of great storms in the 1870s and 1880s revealed the dangers. The first came up the coast in December 1876, catching more than 100 schooners of the winter fleet in mid-ocean on the Scotian and Grand banks. Survivors called it a "perfect hurricane," with roaring waves and howling winds. As it bore down on the fleet, men rushed to windlasses to let out cable even as huge waves broke over the decks. On some schooners, giant "comers" pulled the bows under, and they went down without warning, never to rise again. On others, the giant waves washed over, ripping away hatches, ice-houses, and everything movable on the deck. Masts and rigging came down, and the force of the sea forced schooners onto their beam ends, into the turbulent sea. Men could only hope their boats righted themselves.

On the schooner *Augusta H. Johnson*, when the storm first struck, Captain George A. Johnson and his men set anchor and prepared to ride out the blow. The cable soon parted and the men set their second, and last, anchor. The schooner bid up and for hours stayed in place. Then the second cable parted, leaving the schooner at the mercy of the wind. A tremendous sea broke over the boat, stove in dories, broke the fore-boom and fore-gaff, and carried 300 fathoms of cable into the sea. Yet, Captain Johnson retained control of the helm, maneuvering before the storm until it subsided late in the

)(*The 60-ton schooner* Howard Steele *is encased in ice in the middle of the inner harbor, c. 1880.*

afternoon. Johnson and his crew made it through, but others did not. In all, ten craft and 98 men disappeared that day.

In Gloucester, the Fishermen's, Widows', and Orphans' Relief Committee sprang into action. Calls for support poured in, and an urgent appeal for help went out. By year's end the committee had provided $5,275 in aid to 136 families—this at a time when the average fisherman made about $400 a year, and a top highliner made $1,100.

The next great storm struck in mid-February 1879, taking 15 schooners and 162 men on the Georges shoals. As the storm approached, vessels were anchored close in to the shoals, and men were handlining to bring in the best fish for the Boston market. The fish were plentiful, and as soon as a schooner met its quota, based on the amount of bait and ice on board, the skipper hauled anchor and headed for home. On this, as on any other day, many craft

)(*Gloucester winters were typically hard, and it was not uncommon for the Georges fleet to be frozen in, as seen here in 1875.*

sat on the shoals. The wind came up quickly and the sea became rough. As the snow-filled winds built into a full-fledged hurricane, skippers set double watches, and for the next 50 hours schooners were buffeted and tossed in the shallow waters. The wind roared through the rigging, and for nine hours blinding snow hid the focsles from the view of the helmsmen.

In the early part of the long ordeal the fleet remained safely at anchor, skippers taking bearings on the nearby schooners even as they disappeared into the gloom of the night. Helmsmen were told to hold their position into the wind, to watch the cable, to look out for wayward vessels, and to call out the crew at the first sign of trouble. Those below sat fitfully in their bunks, hurled about by the severe lurching of the vessels. There would be little sleep that first night as the men waited for their turn to go above, huddling near the cookstove, sipping hot coffee. Above, the masts, spars, and rigging groaned and creaked. Heavy seas pushed the vessel ever leeward, the helmsmen fighting to keep the bow into the wind. With every lurch the cables were strained, and after many trying hours began to break, sending vessels wandering among the fleet. There was no hope of regaining control; sails had been furled, the second anchor could not be set, and a sense of despair set in as the men contemplated meeting their maker.

Lookouts on other vessels called out the warning: A schooner was loose. The response was electric. Men poured out of the companionways, and as they reached deck the noise of the wind was deafening. On came the wayward vessel through the gloom of the storm. Whether you survived or went down was now in the hands of God and nature, and as the danger drew closer skippers stood ready to cut their own cables. The drifting vessel ran defenselessly before the wind, rising to the top of one giant wave, only to dip unseen into the trough of the next. On and on she came, and for many a doryman it seemed their fate was all but sealed. There was no time to think; only a slight shift in the wind could prevent a head-on crash and force the drifting schooner to the side of the anchored vessel. Every eye was strained as life itself hung in the balance. Vessels adrift in these conditions were totally at the mercy of wind and wave; if not sunk in a collision with another vessel, they would either be driven onto a lee shoal or would broach, turning broadside to the sea, and either capsize or sink. As a skipper watched a wayward vessel bear down on him, he often had no option but to cut his own cable, doing all in his power to escape the oncoming schooner. Certain death was averted, but for how long?

Other skippers waited, axes in hand, and the men would let out an unimaginable sigh of relief as a puff of wind pushed the wrecker to their lee. They were safe, but even as they rejoiced and the storm-tossed schooner disappeared into the gloom, they knew that others were now in danger. Death was inevitable.

By the time the storm had passed, 15 schooners had been lost, and almost all the surviving vessels were heavily damaged. Giant seas had washed over decks, seams had been opened, and rudders had been torn from their fastenings. Men now worked unceasingly at the pumps, keeping their battered craft afloat.

As the first of the fleet limped into Gloucester, families strained to learn the names of the survivors. Everyone knew that many would not return, and hands were clasped in prayer for the safety of fathers, brothers, and sons. Among the first to return was the schooner *Hyperion*. She had lost her anchor and 250 fathoms of cable; the *George G. Hovey* had lost a string of cable, two anchors, and her gurry-kids; the *Midnight* had stoved bulwarks, lost dories, and a split foresail; the *Tidal Wave* had a broken windlass. But the crews were safe, and eager families embraced their loved ones as they stepped onto the docks.

One returning skipper said the gale was the "heaviest he had ever known except once when he was caught in a hurricane." Another said that on the day before the storm, while fishing on South Shoals, he saw 30 craft working within plain sight, and he had enjoyed their companionship. As the gale increased, thick falling snow hid the craft from sight, and when the weather cleared, only one remained; all the others had cut cable and the skipper knew not their fate.

By March 14, the extent of the disaster had become clear. Six weeks had passed, and the community no longer had grounds for hope—the food on the vessels would have run out after five weeks, at the most. Yet, some clung to the belief that their loved ones were still alive on disabled schooners blown far out to sea. They held out hope that even then they were working their way in slowly. But as days turned into weeks, all had to face the stark reality that men and craft would not return.

In all, Gloucester lost 15 schooners and 14 crews, some 162 men. It was a staggering total. Many sobering stories appeared in the press. On the *George Loring*, the skipper, George W. Lane, had gone to sea with his brother Joseph M. Lane. Both were married and between them they left 10 children. The *Maud and Effie* lost 14 men, 8 of whom left wives and children; 6 were brothers-in-law. On the *Otis D. Dana*, John Garvey was on his first trip to sea. He left a wife and 7 children. The *Sea Queen* went down with 10 young, unmarried men from Norway and Denmark; it would be weeks before their families knew of their fate. The *Jennie Linwood* had been seen just before the gale broke: She had a full load of fish and was preparing to return to Glouces-ter. After an absence of six weeks, her owner declared her lost.

In all, the lost fishermen left 57 wives and 150 children, and as in 1876, the voluntary relief committees of Cape Ann mobilized to visit the families. One volunteer recounted:

> We had our list of the destitute families with instructions to apply immedi-ate relief, and to do everything within our power to bring comfort into the homes made desolate. . . . Said one of them with the tears streaming down her cheeks, "Since I have given up my husband and have made up my mind that I shall never see him on earth again, it has troubled me much as to how I should maintain myself and children. I have been very careful of the little money he left by me (when he went out on his last trip); but it has gradually melted away and now is nearly all gone. My landlord is a kind, good-hearted man, and has reduced my rent, and a little assistance

just now will be of great benefit. God bless the kind hearts who think of us stricken ones by this visit." Said another, "I have never known what it was to want; mine has been a life of constant labor, but husband and I pulled together and we led a happy life. Now he is gone and I will do my best to maintain the children, as I know this will be pleasing to him; but I do miss my good man—it cannot be possible that I will never hear his footsteps on the walk and welcome him home from his trip."

Families were, indeed, destitute and their plight touched the people not only of Cape Ann, but also of New England and many other communities across the country. Food, clothing, and money poured into Gloucester. The total amount received by the various societies came to $26,500, the equivalent annual wages of 65 dorymen.

For the next few years the fleet was to make it through the terrible winter season with no great gale or loss of schooners. But in 1883 came the "twin gales" of August and November—Gloucester lost 12 vessels and 149 men, and the combined tragedy provided an impetus for change: Private questioning became public, and vessel owners at last acted. In late August, as the first gale swept over Georges and the Banks, 3 vessels and 34 men went to the bottom. Returning skippers told chilling tales of a sudden storm and desperate survival at sea. Captain Frederick Hillier of the schooner *Wachusett* remembered the sea becoming heavy and the sky threatening, with dense masses of dark clouds to windward. All the vessels had dories out, skippers having paid little attention to the threatening weather. Then, in what seemed like a moment, a white bank moved in from the north and a squall struck the fleet. It was a fight between life and death as dorymen bent to their oars to reach their vessels. Some made it back. Others were overrun by the squall, and desperate men disappeared into the turbulent seas.

For five hours the cabin of the schooner *Somes* lay in the water, her bulwarks stoved in. Captain McDonald of the *W. E. McDonald* described it as the severest storm he had experienced in 17 years at sea. Its suddenness had heightened the impact; no one had time to prepare. All vessels lost cable, fish, boats, and trawl buckets. Schooners pitched, bow under, and many had to cut their anchor lines and run before the gale. As the *Wachusett* scudded along before the hurricane, she rushed past many dories whose occupants had "given up to despair." Next came the dories that were bottom up and abandoned; they seemed as "plentiful as seagulls," with spars, oars, water casks, and deck gear floating about in profusion. The wreckage lay across 30 miles.

The second gale hit on November 12 and 13, with schooners disappearing as the storm passed over Georges before heading north to the Maritimes. Nine vessels and 115 men went down. But as quickly as they disappeared,

unseen on the Banks, so the news of their loss traveled slowly to Gloucester. Families waited for schooners that never returned. A few days before
Christmas, the owners of the last missing vessel declared her lost. The papers
made their final tally: 115 men had been drowned; they left 34 wives and 62
children. For the year, Gloucester had lost 17 vessels and 209 men,* and the
men left 40 wives and 68 children.

With these new losses a clamor of public voices began to speak out.
What needed to change? The schooners? The way men fished? When the
fleet went to sea? As early as April 1879, an article in the *Salem Register*
suggested that the design of the modern clipper schooner was central to the
terrible losses suffered on the fishing grounds. Losses had become common
only after the current generation of larger schooners entered the fleet in the
1860s. While these schooners had beautiful lines for sailing, they lacked the
"qualities for riding at anchor in heavy, rough Georges seas." It was time to
rethink vessel design, but the article lacked sophistication in the technical
aspects of design, and the account carried no real authority.

But others were not so sure. To many the clipper schooner was a "staunch
and true" vessel, and it was only in the most powerful of gales that lives and
vessels were put in jeopardy. Might not the fatal mistake be the custom of
riding out the storm on the shoals instead of pulling anchor and heading
for the relative safety of the open sea? The accounts from the 1870s told of
broken anchors and chaffed cables, not of poorly designed hulls or badly set
rigging. As for the owners, they remained silent: The mutual insurance company covered their losses, replacement vessels were readily available, and
they saw no reason to advocate change.

Yet, as stories of losses continued to fill the local papers, the fishermen
were on edge, and in March 1883, for the very first time, their private concerns became public when the men refused to go to sea in the face of an
anticipated Atlantic gale. A widely circulated almanac said there would be an
immense mid-March gale off the New England coast, and as the forecasted
day approached the greater portion of the fleet settled into port. The fishermen would not go out—the storm and sturdiness of the clippers were the
all-absorbing subjects of conversation across the waterfront. One man spoke
for many when he said that all the money in Gloucester would not tempt him

---

*This was also the year in which a Gloucester fisherman was to establish a record of human endurance as he
became separated in his dory on Newfoundland's Burgeo Bank. The man was Howard Blackburn, and his schooner was the *Grace L. Fears*. He and his dorymate, Thomas Welch, went astray while tending their trawls. Welch
died from exposure, Blackburn froze feet and hands, but after the most trying of circumstances came ashore in
Newfoundland, where he was cared for by Frank Lishman and his family. The full account of his saga is told by
Joseph E. Garland in his book *Lone Voyager*.

to go out before the appointed time of the storm: "Give the fish a rest for one week any way, and stay at home; it won't harm anybody." Another said, "Life is as sweet to me as to anybody, and I promised my wife and children to be in good season, and not to go out, and I'll stay until this thing happens or the time for which it is predicted passes over."

The men were jittery; if the owners and outfitters would not act, then it fell to the men—but in March 1883 there was no great storm. A small seasonal blow crossed over Cape Ann, followed by several days of clear skies and brisk breezes. Soon the men were refitting for late winter fishing on Georges.

But they had spoken, and within the year two individuals with both knowledge and credibility took up the challenge. In a series of letters to the *Cape Ann Advertiser*, Captain Joseph Collins, a respected former Gloucester skipper and now a member of the U.S. Fish Commission, and Dennison Lawlor, a successful and respected vessel designer, put the matter of blame to rest by pointing the finger squarely at the flaws in the clipper design. The vessels were basically top-heavy; rigging that had been piled on to increase speed and the excessive spread of canvas were often a schooner's undoing. When the clippers broached they were prone to capsize, and when they did they seldom righted themselves and all were lost. Lawlor and Collins sketched out a practical plan of correction: At a very basic level, the changes were not difficult to grasp. The heavily canvassed clipper schooner had to have a deeper draft to effectively lower the vessel's center of gravity, and the hull had to be recontoured to give the schooner "a better hold on the water and superior, safer riding and handling characteristics in a seaway where she needed it."

The first vessels to move in this direction, all designed by Lawlor and built in 1885 and 1886, were the fishing research schooner *Grampus* and three fishermen, one of which was the Gloucester schooner *Arthur D. Story*. Others followed. In the most simple of terms, these new schooners had about a two-foot-deeper draft, and were thus less likely to broach when a sudden gale caught them broadside. In addition, their sail area was reduced by lowering the height of the foremast (which had previously been the same height as the mainmast) and permitting the use of topmasts in summer only. The new hull design was said to give the schooner a larger lifting capacity when full of fish, and the deeper keel improved responsiveness and maneuverability. At the same time, the new schooners preserved the virtues of the clippers, save the shallow harbor capability when in Gloucester, while adding virtues of their own: seaworthiness, handling and maneuverability, windward capability, and even greater speed.

In the short run, with so large a fleet of existing vessels, it would take years for there to be a big drop in the count of lost vessels and men. Yet, the problems had been identified, and the new schooners coming into the fleet began to incorporate the improved design.*

The succeeding generations of "Gloucestermen," as they became known, were acknowledged to be the finest working sailing vessels ever built, before or since. But even with the best of vessels, losses continued—Gloucestermen were out in all weather, navigation was a challenge, collisions occurred, dories went missing, men were lost, and fishing remained the most hazardous occupation in the world.

---

*[Personal communication from Will Andrews.] The "revolution" that took place from 1884–1885 to 1890 ultimately involved far more than the change in hull form alone. It witnessed the replacement of the bowsprit-and-jib boom with the pole bowsprit; the introduction of iron ballast, iron diaphragm pumps, and iron anchor davits; and substantial alterations in rig. Most importantly, it marked a "changing of the guard" regarding fishing vessel design. Formerly, the clipper models were essentially cookie-cutter vessels, requiring only a minimum of expertise to lay out— the "designer" was most often a shipyard owner or foreman with no special training, who simply had a knack for this sort of thing. Henceforth, schooner design would be the province of real designers, some of them internationally known such as Edward Burgess and B. B. Crowninshield, others primarily designers of fishermen such as George Melvin McClain and Thomas McManus (actually, the "revolution" would not be complete and totally accepted until Edward Burgess—then riding a crest of popularity for his three successful America's Cup defenders of the mid-1880s—came out with his *Fredonia* models). These men, along with others, put fishermen design on a scientific footing, and would continually refine and improve the genre for the next 40 years—quite a change from the previous 40, in which essentially nothing at all was done. This revolution ushered in "the golden age of the Gloucestermen," resulting in evolution through several generations of schooners that were unsurpassed anywhere for speed, seaworthiness, safety, easy handling, utility, and beauty.

CHAPTER 11

Immigrant Fishermen Come to Gloucester

In early February each year, workers at the marine railways busily scraped hulls, caulked seams, replaced worn rigging, and applied fresh paint. A great forest of masts encircled the harbor, schooners sitting two to three deep at docks from Harbor Cove to Rocky Neck. Everywhere one looked, fishermen stowed supplies for the upcoming trip, bringing on board food, salt, ice, bait, and personal belongings. Newly bent sails flapped gently in the late winter breeze as skippers slowly inched their craft out into the inner harbor, calling on the sloop *Wanderer* to come alongside to fill the vessel's giant freshwater tanks. Soon schooners tacked back and forth off Pavilion Beach in search of a favorable breeze, even as the last tardy fishermen rowed out to join their mates. Then, as the wind at last filled the giant, snow-white sails, the helmsmen turned their craft toward Eastern Point and the long trip to Georges and the Scotian Banks.

There were now more than 400 schooners in the fleet, and Gloucester opened its arms to the immigrants who arrived daily on the Boston trains. They came from the northern Provinces of Nova Scotia and Newfoundland, as well as the more distant old-world countries of Norway, Sweden, and Portugal—and 50 years later, the last group of immigrants would arrive from Sicily. The men were young and eager to work, and after dropping off their meager belongings at one of the town's small ethnic boardinghouses, they walked to the nearby wharves in search of a berth. For some, it was the beginning of a new way of life—a new home, new friends, new opportunities. For others, it was but a seasonal occupation, the men returning later in the year to their northern homes in Nova Scotia and Newfoundland. They had been coming since the mid-1800s. Amongst the earliest arrivals were young men and women from the small coastal villages of Portugal's Azores Islands.* For them, America promised a better life and better opportunities. One such immigrant, a young man named Francis Bernard, had arrived on Cape Ann in 1845. He came by whaler to New Bedford, followed by a train ride north

---

*The Azores lie in the mid-Atlantic, 970 miles from Portugal and 2,400 miles from Cape Ann.

to Boston and then east on the Cape Ann branch of the railroad.* Bernard secured a berth on one of the local schooners, and within two years he married an immigrant from Nova Scotia. He settled into a life as a fisherman and sail maker.

By 1850, a dozen other young Azorean men had joined Bernard on Gloucester vessels, and their number grew tenfold over the following decade. The men worked hard, saved what they could, and even sent money home to pay for the passage of loved ones to Gloucester. Over time, they bought small house lots on a rough, unsettled hill overlooking Gloucester's inner harbor known locally as Lookout Hill. Everyone was to soon call it "Portagee Hill"— the home of Gloucester's Azorean immigrants. In preparing their lots on the Hill, the men used simple tools to split rocks, prepare the ground, and frame their homes. They planted trees and vines—pears, apples, and grapes. They laid out small vegetable and flower beds: No home was complete without its morning glories, lilacs, roses, and elderberries. Chickens walked about the grounds, their coops adjoining the backyard pens for the many pigs kept on the Hill. In the fall, families slaughtered the pigs, putting up salt pork, lard, and Portuguese sausages for the long New England winter. Later in the year, as the snow lay on the ground, the people of the Hill settled down to meals of fried eggs and sizzling sausage.†

The Azorean community prospered, and in the spring of 1879 they welcomed a young immigrant from Pico into their midst—19-year-old Joseph Mesquita, destined to become Gloucester's greatest Portuguese master. Mesquita arrived with only 50 cents in his pocket. He boarded with Captain Joe Silva on Taylor Street and served under Silva as a doryman on the *George A. Upton*.§ He lay and hauled trawls from a single-man dory, helping Captain Silva bring his fresh-fish catch into Boston and Gloucester.[20] When Captain Silva moved to John Pew & Son Company, Joe Mesquita moved with him, and he learned his trade while serving as a doryman on the *Champion*, *Ann D*, *Eastern Queen*, and *Chocorua*. All were fresh fishermen, averaging about 72 feet in length with a burden of about 60 tons.

---

*According to Frank L. Cox (mimeo, 1938), Bernard was from St. Michael, Azores. In 1880, at age 56, he worked as a sail maker; he and his wife had two children. My own relatives, on my father's side, left the Azores about the same time. They too came to New Bedford. They changed the name Silva to Morris. My father came to Gloucester in the 1930s. He had a seventh-grade education and spent his life as a fish cutter.

†Either linguica, a mild pork sausage flavored with garlic, paprika, pepper, and onions, or morcella, a blood sausage made with pork fat, onions, and spices. (*Gloucester Daily Times*, Nov. 12, 1990, Portagee Hill.)

§Old Gloucester families had owned most of the vessels on which these immigrant fishermen sailed, but by the 1880s, Portuguese masters commanded many: Manual Silva Jr. in the schooners *Lizzie* and *J. W. Collins* (for the Daniel Allen & Son Company), Tom Lewis in the schooner *New England* (for the John Pew & Son Company), and Jesse Sears in the schooner *Alice* (also a Pew vessel).

X *A view across the harbor toward Portagee Hill, c. 1876. Sayward Brothers wharf is in the foreground, Smokey Point and the Shute & Merchant Co. in the middle.*

Mesquita quickly learned English and in 1887 became a U.S. citizen.* He married Captain Silva's daughter, Mary, nine years his junior. Finally, in 1891, with a firm grounding as a doryman and a superior knowledge of the fishing grounds, Joe Mesquita assumed his first command: the 57-foot 7-inch, 34-ton schooner *Abby A. Snow*, 1 of 11 vessels outfitted by the George Steele Company. In 1892, Mesquita moved to a second Steele vessel, the newly launched *Almeida*, a 65-ton, 86-foot schooner. He proved an excellent fisherman. Seven years later, partnering with the D. B. Smith Company, Captain Mesquita commissioned his first vessel, the 65-ton, 75-foot schooner *Mary P. Mesquita*, named after his beloved wife.

Captain Mesquita was now a pillar of the Portuguese colony, and as his career blossomed so too did his sense of community and commitment to his Catholic faith. In 1893, at the opening of Our Lady of Good Voyage Church, he and Mary stood with more than 200 other Hill families to watch their

*At City Hall the clerk said that his Azorean last name, Pereira, was too complicated for the American tongue, and suggested that he take the equivalent Anglo name in its place, Perry. But Joe had other ideas. He took Perry as his middle name, but for his last name he chose to use a version of his Azorean nickname. Joe had always loved roses, and in Pico there was a variety called Misquite; it was from this association that Joe acquired his nickname. (Cecile Pimental, *The Mary P. Mesquita: Rundown at Sea*, 1998; and *Mary and Francis Mesquita*, the Jeff and Gordon Thomas Oral History Collection.)

dream become a reality—Gloucester had a Portuguese church. For Joe and Mary, who lived across from the church, prayer was a central part of their lives: A Mesquita schooner went to sea with an altar in the cabin. There were statues of the patron saint of fishermen and Our Lady of Good Voyage,[21] and if Joe heard a man swear, he was quick to get rid of him. In his words, "If you are going to swear here, pack up and get off this boat. There will be no swearing on my boat."

Men such as Mesquita lived in a world governed both by the Catholic Church and by traditional Azorean superstitions.* Their faith was reflected in the building of the church, while their superstitions could be seen in many aspects of daily life.[22] People attributed bad luck at sea to a bewildering assortment of superstitions. Bad luck followed if a woman set foot on a fishing vessel, or if a fisherman met a woman on the way to the ship, or even if the ship left on a Wednesday or Friday. Bad luck also befell a vessel if one of the crew mentioned the word "priest" or "pig" while fishing on the Banks. On the other hand, men believed that the throwing of rosary beads into rough waters could calm the raging sea. The saying on the wall of his beloved church speaks to the worldview of the Azorean fishermen. It reads: "There is no prayer like the sea, its grandeur lifts you up to the creator, its perils send you on your knees to the savior." Call it faith, call it superstition, but in the end it helped the men have a clear sense of themselves and their place in the wider community.

AS JOE MESQUITA made his way from the Azores, young men in the west coast fishing villages of Sweden and Norway also felt compelled to leave their ancestral homes. Some moved to the large cities of their homelands—Oslo, Göteborg, and Stockholm—while others boarded giant steamers for the trip across the Atlantic. For those who took up residence in Gloucester, Boston, and Portland, the move was almost always permanent. They left everything in the hopes of finding a better life. One such group came from the small west coast fishing village of Gullholmen, Sweden. Within a generation,

---

*Old-world customs had great meaning, including the Portuguese "Crowning" ceremony, with its introduction to Gloucester being largely due to the efforts of Captain Mesquita. On a particularly trying voyage, Captain Mesquita told God that if he and his men survived the trip, Joe would introduce this traditional Portuguese ceremony to Gloucester. Joe had faith in God, and when the men survived, Joe kept his pledge. He made an annual public expression of faith during the feast of Pentecost, introducing the Azorean Holy Ghost Festival (*Festa do Espirito Santo*) to Gloucester. In this festival the priest placed an ornate crown on the head of each of the crewmen as a sign of their devotion to the Holy Spirit (the third "person" of the Christian Trinity). The great silver crown (*Grande Coroacao*) came from Portugal. Pope Leo XIII blessed the crown before it made the trip across the Atlantic. Captain Mesquita sponsored the religious celebration at the church and at a banquet at his Prospect Street home. He gave *rusquilbas* (loaves of sweet bread in the form of an enormous doughnut) to every man, woman, and child at the ceremony, as well as to any of the "deserving poor" who came to his home.

5 percent of Gullholmen's population had come to Gloucester—some to stay, others to return.

Gullholmen was one of the oldest fishing villages on Sweden's west coast, its roots going back to the twelfth century, but it was little more than a hamlet. By 1870 the village had only 170 families and 650 people, and with its rocky environment, the island was ill suited to farming and animal husbandry. Only fishing was possible, and the community had eight deep-sea fishing vessels. Men crewed the vessels with no distinction as to family or age, and when they returned from the grounds, extended families worked together to prepare and sell their share of the catch. One vessel might come in with a poor catch, while another arrived "deep in the water" with a bountiful harvest. By pooling the fish and sharing the tasks, each family was able to survive.

Before the mid-1840s, Gullholmen had had a vibrant herring fishery. But when the herring catch dwindled, the village entered a prolonged period of economic decline. The men continued to go to sea, but the catch was sparse, and not until the 1880s did the herring return. So poor were local catches that in 1876 seven of Gloucester's best schooners brought cargos of herring to Sweden's large west coast port city of Göteborg. Craft cleared Gloucester Harbor in April, secured herring at the Magdalen Islands, and sailed across the North Atlantic. Once in Sweden, the schooners made a fine impression on the young men of Gullholmen, and fishing out of Gloucester then became an option as they considered their futures.

In 1880, Arvid Olsson became the first to leave for America. He went to San Francisco, but after a brief stay returned to Gullholmen. Isack Larsson left for New England. He took up residence in Boston in 1880 and went to sea on a Boston schooner. He left his mother, wife, and young child behind in Gullholmen, and wrote letters home describing life on the other side of the Atlantic. The barrier of the North Atlantic had been broken, and many young men (and later, young women) followed Larsson's example. The 1880s saw 18 young people leave for America: 10 for Gloucester, 3 for Boston, and 5 for other American cities.

In 1882, John Lund, 22, became one of the first to reach Gloucester. He had traveled to Göteborg, and after a two-week wait took a steamer to Hull, England. From there, he traveled by train to Liverpool, where he boarded a steamer to North America. One year later his younger brother Albin (b. 1863) left for Gloucester, followed three years later by the youngest of the three Lund brothers, Nils Albert (b. 1868). During this same period, two other young Gullholmen men also made the trip, the Lunds' first cousin, Herman Olsson (b. 1863), and Lars (Oscar) Olsson (b. 1863).

Once in Gloucester, the Lunds took rooms on Short Street, just off Gloucester's main thoroughfare, and began their long service as Gloucester dorymen. We know nothing more about these men in this early period: not the vessels on which they served, nor the nature of their daily lives, nor their fears and expectations. What we do know is that they spent most of their time at sea: on watch on deck, resting in the forecastle, and fishing from dories.

No letters back to Sweden survive from this period. But write home they did, for two of the Lund brothers later returned to Sweden to marry and bring their brides back to Gloucester. The brides were the Svensson sisters, Cecilia (b. 1868, wife of Nils) and Josefina (b. 1867, wife of Albin). Cape Ann was to their liking, and they suggested that others might want to join them in this land of opportunity.

It was in this atmosphere in the early 1890s that six other young men from Gullholmen considered their futures. One was Olof Johan Christiansson, a cousin of Lars Olsson, who by then had been in Gloucester for seven years, and five others, ages 21 to 24, who left in the early spring of 1893 in the largest single migration in the history of the village. They began the five-week passage from Gullholmen to Gloucester. There was nothing to hold them back. Jobs were scarce, they had already fulfilled their duty as Swedish citizens to serve in the navy, and Gloucester seemed just the place to go.* Packing their bags and bidding adieu to mothers, fathers, siblings, and sweethearts, they left for Göteborg. Here, emigration agents helped prepare the young men for the journey. They first boarded a British ship, the Wilson Line's SS *Romeo*, for the trip to Helsingborg in southwestern Sweden. From Helsingborg they took the *Bohemia* to Ellis Island, New York, arriving on April 13, 1893, and there their names were recorded for posterity.

From New York, they took a train to Boston and then another to Gloucester. Here they sought out the Lunds—women of the Lund family now managed a boardinghouse for Swedish fishermen,† a house said to have the

---

*The naval record of Stephen Olsson indicated he was inducted as a seaman on January 14, 1892, and spent one year as a military sailor. Steve belonged to the 5th Seaman company and performed his weapon exercises at the station of the navy at Koskona, and was judged fit to be a common duty person. (Translated from the Swedish by Gunnar Ljunngren.)

†Gordon Thomas recounts the history of these boardinghouses (Jeff and Gordon Thomas Oral History Collection): "Chris told me about the Scandinavian boardinghouses in Gloucester. The first in town was the Magnussons on the corner of Short and Middle St. Also located on Short Street was Strawberry Johnsson. It was $5 a week, room and board. They served the best of food. Ham and eggs for breakfast and so forth. They also sold beer and whiskey on the QT. Davis Bottling Works used to deliver the beer to the boardinghouses. Red Olson boardinghouse on Vincent Street. Another Swedish boardinghouse, Mrs. Lund up on the top of Union Hill. She had a boardinghouse. They were as clean as a whistle. If you lived there and you got up in the morning to go to your vessel you would leave the door to your room open. No one would steal into your room. Nothing was ever taken. They would come up there change the sheets, straighten the room, there was no problem. We had a lot of Scandinavian fishermen. Among the skippers were: Carl Olsen, Iva Carlson, and Eric Carlson, they were all halibuters."

The earliest known photograph of Steve Olsson, immigrant fisherman; taken in the 1890s. Steve sometimes used the Swedish spelling of his last name—Olsson—and sometimes the American spelling, Olson. All of his children used the name Olson.

best smorgasbords in town. The Lunds welcomed their countrymen with open arms, and so began a long-term relationship among the Gullholmen immigrants of the 1880s and 1890s. With the 6 newcomers, 15 Gullholmen seamen now fished out of Gloucester.

Of these men, only Steve Olsson remained a Gloucester doryman for his entire life, and the documented information on his experiences provides a unique glimpse into the lives of these men.* Steve "Stephanus" was born in 1871, the son of Olof Josef Andersson (1842–1919), a fisherman, and

*Others who came in 1893 included the brothers Oskar Johansson (Oscar Johnson, b. 1871) and Johan Johansson (John Johnson, b. 1869), sons of John Johansson (1839–1902), a fisherman; their mother was Mathilda Olsdotter (1840–1914). They were the oldest of five children, and none of the others joined them in America. Both would have successful Gloucester careers, eventually skippering local vessels. Olof Alfred Christiansson (b. 1870) was the son of Christian Olsson (1836–1877) and Sofia Svensdotter (b. 1837). Olof had four living siblings in 1893; none ever came to America. Samuel Andersson (b. 1869) was the son of Anders Samuelsson (b. 1841) and Johanna Andreasdotter (1842–1916). Samuel had seven siblings, none of whom came to America. Unlike the other five 1893 immigrants to Gloucester, Samuel would return to Gullholmen in October 1900. He talked of his Gloucester experiences for the rest of his life. Olof Johan (Christiansson) Jacobsson (b. 1873) was the son of Jacob Christiansson (1843–1911) and Olivia Larsdotter (b. 1843), cousin of Lars Olsson and a more distant cousin of the Lunds. Olof had seven siblings, of whom three traveled to Gloucester. Of those who remained behind, Anna married Samuel Andersson one month after his return in 1900. Olof's sisters Beata (Bertha, b. 1868) and Hilda (b. 1878) and his brother John Jakobsson (b. 1885) came to Gloucester 8 to 10 years later. The two young women stayed in Gloucester for the rest of their lives, Hilda eventually marrying Steve Olsson. John returned to Gullholmen after a three-year stay in Gloucester, and within the decade drowned while skippering a vessel on the North Sea.

✕ Photographs taken inside Gloucester boardinghouses were rare, and few survive. This image, taken before 1930, is of a meal at Josefina Lund's dining table in her boardinghouse on Main Street. In the center of the picture is Albin Lund, who came to Gloucester in 1883. The Lunds did much to welcome and support immigrant fishermen from Gullholmen, Sweden.

Elisabeth Carlsdotter (1847–1935). When he emigrated he left four siblings behind, two other siblings having died in childhood. He never returned to Sweden, nor did his siblings ever join him in Gloucester. For 42 years he served as a Gloucester doryman.

The Lunds' Swedish boardinghouse stood on Main Street. It was on the edge of the Portagee Hill section of the city, within two blocks of the waterfront. Men lived one or two to a room, but it mattered little, as their stay in port was brief. They worked at their trade and learned a new language. And, although men such as Stephen Olsson eventually became fluent in English, they loved to speak in their native tongue. Many years later, Steve's daughter Gerda remembered her parents having "their heads together, speaking in Swedish, particularly when there was something the children were not supposed to know." In fact, fishermen could be quite selective in what they shared with their families. They had a unique lifestyle, and it has been suggested that their ability to forget past dangers helped save them from much wear and tear as they lived their daily lives.[23]

There was also a certain cliquishness in being a Gloucester fisherman. From the beginning, their lives revolved around fishing and other men of the sea. They spent years on vessels and in boardinghouses with their mates. For the Gullholmen immigrants, having a connection to the Lunds provided a ready introduction to the captains of the Gloucester fleet. The Lunds spoke for the integrity of the new men and their skill as deep-sea dorymen.* For captains the message was clear—these men worked hard, they knew the sea, and one would be pleased to have them as dory mates. They would not be distressed by the prospect of rowing a dory, setting and pulling a trawl, or standing watch as the wind whipped and the seas rolled. They wanted to go to sea and with the Lunds' help quickly found jobs as dorymen, proving their worth on long northern voyages. Thus, from their arrival, the 1893 immigrants from Gullholmen joined the Lunds as dorymen in the small fleet of halibuters spending the summer fishing off the Icelandic coast. Later, they served under many of Gloucester's greatest master mariners, including Lemuel Spinney, John Stream, Carl Olsen, and Charles Colson.

In the same era that the Lunds left Gullholmen and Joe Mesquita left the Azores, young men from Nova Scotia and Newfoundland boarded steamers for the trip to Gloucester, some to serve as dorymen, others as skippers. Many had made the trip before, and once in Gloucester, they pitched in to help fit out the craft on which they would serve, stowing gear and renewing friendships. Young men who had never before come south first secured a room in one of the boardinghouses and then sought out the skippers of the fleet, asking, "Are you a man short? Is a new crew being formed? Is there a place for an experienced doryman?" Before long, the new immigrants found berths and their season began.

For some it was a seasonal migration. They sailed to the Banks in the dead of winter, saved their wages, and returned home to the Provinces in time for the spring planting. Others came to Gloucester in the spring, served on a mackerel seiner or fresh haddocker until fall, and returned home for the winter. In either case, most men did their best to send a few dollars home during the course of the season. Some even delivered the money in person when their schooners pulled into a northern bay for shelter, repair, or supplies. Two such northern bays were St. Mary's Bay in Newfoundland and Arichat Harbor in Nova Scotia.

---

*For these men, it was the Lunds' decision to open a new boardinghouse on Vincent Street that first triggered a shift in residence. By 1898–1899, most of the Lunds and several (but not all) of the 1893 immigrants had moved to 5 Vincent. At the same time, the roster of those remaining at the old Lund residence at 345 Main Street in 1900 included many with Swedish roots. There were Steve Olsson, Ida and John Lund, one Johnson (Alex E.), one Reinston (Gustave), and three Anderssons (Olof, Gustave, and Samuel).

The people of St. Mary's Bay faced the harshest of winters. Coastal inlets iced over for months on end, and the local fishermen hauled out in early fall. Before long the first of the winter storms would sweep in and cover their modest sheds, docks, and vessels with a mantle of snow. Harsh winters and the absence of other job opportunities prompted many men to make the trek south. St. Joseph's was an isolated Irish-Catholic village of 50 to 60 sturdy homes, with its own small dock and fish flakes for drying and salting cod.

A single dirt road ran through the village, heading north toward the Provincial capital of St. John's, 60 miles away. But few of the 400 villagers had horse-drawn carriages, preferring to make the trip to St. John's by boat.* Before hauling out for the winter, they made a last trip for supplies. In the spring, when the ice left the Salmonier River, at whose mouth St. Joseph's lies, the men again sailed to St. John's. They restocked basics such as flour and tea, and took on replacement lines, nets, hooks, and rigging for the upcoming fishing season.

To supplement the "store-bought" goods, people kept sheep, cows, pigs, and a few horses, and during the summer, all families planted vegetables. They used the horses to haul wood in the winter, lay manure in the spring, and plow the fields in the summer.

Among the family names of St. Joseph's one finds a rich Irish heritage, and, like the men from the Azores, the Irish from St. Joseph's had a deep sense of faith. Much of village life centered on the Catholic Church and the local priest. The church ran the school, although the government paid the lay teachers, and the day began at eight and ended at four. When the men went south to Gloucester they could read and write.

In the summer, boys and men fished, while women tended the gardens and shepherded the animals and their families. In winter, when large numbers had gone south fishing (or were working in the Sydney, Nova Scotia, mines), the wives ran the families. Brendan Daley, the last remaining Gloucester doryman from St. Joseph's, remembered those days as a time when the "wives were responsible for the whole thing. Mothers mostly ruled the roost even when the fathers came home." For how else could one raise a family and tend a home? Men might be gone for weeks and even months at a time.†

---

*Even by the early twentieth century automobiles were a rarity there. One member of the community remembers "that only the merchants had cars. Just maybe a couple of cars in the settlement, so we walked to church every morning (rain or shine) and also to school—it was a half a mile to school and a mile and a quarter to church." (Mercedes Daley Lee.)

†These personal remembrances by the people of St. Joseph's were gathered in a series of interviews with the author.

For those left behind, there was a personal connectedness—one person to the other, one family to the other. Mercedes Daley Lee, daughter of Captain Charles Daley, said that in the silence of winter it was religion that gave her family strength. "I remember we knelt at evening time and prayed the rosary with Mom—always for a safe return of my father, Charles, who was away in Gloucester."

Thus, among the families left behind, there developed a unique sense of open mutual affection and support, as well as reverence for their men who now served on Gloucester schooners. Mercedes went on to say, "It was more like a big family, everyone knew the other persons' business. At the end of the summer, we harvested the crops, the staples for the winter then buried in big cellars. In this way, the summer crop of vegetables was available to the family during the short days of winter. We also reared pigs and cows to have for meat in the fall. However, with no modern conveniences, when one killed a pig or a cow it was usually shared with a neighbor. Or, if it happened you were 'short of anything' you would borrow from the person next door. I hear so much now of 'Christian unity,' I think perhaps we practiced it without realizing it. It was just what helped us survive the very difficult times."

When school was out, the boys of St. Joseph's learned their trade, and Brendan Daley described how they learned of the sea and prepared for a life as a Gloucester doryman:

> Like other lads in St. Joseph's I went out with my father and another man in a small boat. It was off Cape St. Mary's (at the headlands of the Bay), and it was pretty rough. I was young and I could take just about anything. My father was well seasoned, he had dory fished out of Boston, and it was the father who would teach the son, and it was the father who would tell stories about the opportunities in Boston and Gloucester. The men would get together to talk, and we boys would listen. The sea was at our doorstep and to the sea we turned. You would go out there in your dory and anchor up a certain distance from the schooner. On the fishing grounds off the Bay we had two dories. You went out in different directions from the boat. One man would stay on the skiff from which the two dories were set. The skiff was quite small, no more than 45 to 50 feet long. Fog was a constant problem, and the man staying on the boat would keep blowing the horn. He would also hit the big brass winch with a ball hammer. Meanwhile, you would be working in the dory, listening for the blowing horn and the clanging brass—for as the fog set in, you would lose your bearings. You wouldn't go too far when you began to set your trawl, but in the fog you had the sense of being isolated, and I suppose there was a real danger. You would always listen, and when your dory was full, you would follow the sound back to the skiff.

My father made sure we knew what it was like to serve on the big vessels on the banks—to be out in the fog, to be separated from the schooner.
We were told the big vessels of Boston and Gloucester would send out
their dories in different directions, and each dory would anchor its trawl.
Just like our experience, the fog would come up without warning, and the
skipper of the schooner "was not your father," and many a man was left on
the banks by an unthinking skipper. Thus, if you were on the wrong side
of the vessel you might never hear the horn blow or the bell ring. The men
in such a dory would not know when the signal was sounded to return to
the schooner and that is where many a skipper would leave a dory. You
would be truly alone. If you didn't hear that bell or the horn you just had
the compass to go by when it was time to head back—after you had pulled
your trawls. You would have taken a bearing when it was clear, and all
might have been OK if the vessel did not move. But that would not have
been the case, and my father made it crystal clear that the skipper would
often pull up the anchor and jog around to pick up the dories as the fog set
in. So if you were on the wrong side of the schooner they would sail away,
and you could be easily missed and left to fend for yourself on the banks.

Thus, the young men of St. Joseph's learned to survive in the most severe
of weather, until life as a North American doryman became second nature.
Then, and only then, were they prepared to take the next step of manning a
dory on a Gloucester schooner. Their experiences in the Bay had prepared
them to find the best grounds on which to fish, to set a sail, and, for some
later, to skipper a Gloucester vessel.

Patrick Kelley, from Newfoundland's southern coast, left an account of
how he first went to Gloucester at 20 years old.[24] He felt that his childhood
ended when he went to sea at 16, and by the time he was 20 he had the
urge to fish out of Gloucester. Two of his uncles had previously fished out
of Gloucester, although one had drowned while pulling trawl. The surviving uncle and a cousin came home in the winter and talked up fishing out
of Gloucester. As Kelley recalled: "You would start fishing in Newfoundland
in April and you would fish all summer. But you had to wait until the fish
had been dried, cured as they called it, and taken to Europe, to Portugal or
to wherever, to find out how much money you would make. And it would
be Christmas before you would see the dollar." But this was not the case in
Gloucester; you were paid after each trip.

So, Kelley started thinking about going to Gloucester. "You'd have been
home, local for a month or a couple of months in the winter and you'd begin
to wonder about what you were going to do . . . There was never any thought
of making a fortune. For you see people were coming and going all the time.
Men all around us were going to Gloucester . . . and it was just a question

of going . . . It only cost $12 to go from St. John's to Boston. My parents didn't discourage me. There were quite a few friends of the family up here in Gloucester. Oh though to be honest, I never bothered much with them."

The migration to Gloucester became second nature to the men from southern Newfoundland, and in St. Joseph's most of the men left just before the Bay froze. At least one member of each family went south, disappearing for months on end. There was little communication, perhaps a letter now and then, with a portion of a man's wages for his mother or wife.

One of the St. Mary's men, Captain Joseph (Joe) Bonia, became one of the great skippers of his generation. He came from North Harbour, a village of no more than 50 houses, an easy hour's row from St. Joseph's. It was a poor community, with a strong farming and fishing tradition. Joe was born in 1867, and like other boys in his village, learned handline fishing for cod and gill net fishing for herring. He learned his lessons well and by the final decade of the nineteenth century had become an acclaimed Gloucester skipper. He was said to be a kind, bright man, "a pretty knowledgeable fella." He had many comical expressions and was renowned as having "the quick sayings of the Bonia's." He also took his Catholic faith seriously, although he could see the whimsy that overshadowed a life on the Banks. He had no fear of being at sea for months at a time, or fishing on the edge of the far northern ice pack, or challenging the elements off the coast of Iceland.

In time, Captain Bonia's presence in Gloucester proved helpful to others from the Bay as they made their way south. Among them were his two brothers, his nephews Charles and Joseph Daley (the sons of his sister Mary), Charles's best friend William Cole from Penny Harbor, and three Fagan brothers (John, Dennis, and Andrew).

Charles was born in St. Joseph's in 1882 and first served on a Gloucester halibut schooner under his uncle Joe Bonia in 1902. He learned to fish under his father, John Daley, who spent his life as a fisherman in the Bay and on the Banks.* Charles later skippered at least three Gloucester craft, one of which he lost off the west coast of Newfoundland. Still later he returned to Newfoundland to skipper a number of local craft, and at the end of his career

---

*Charles's mother, Mary (b. 1850), Joe Bonia's sister, lived to 93, and Charles's daughter Mercedes tells us that in her later years, "Grandmother always wore a long black dress and a white apron, she always looked old, and for the last months of her life, Grandmother came to live with us. She had lost her memory. She was senile, and the family was afraid to leave her alone. So Mom took her over to live with us, and we children had to watch Grandmother Daley. She would walk to church, and she would walk and walk and walk. But, she wouldn't let us walk with her. So we had to go behind her, to make sure that nothing would happen to her, and when she went to church, we would have to go as well, whether we liked it or not. We would sit there and say our prayers, and in those few months we said a lot of prayers." As a child she had been educated by the Sisters of Mercy, and her last words were a call for Sister Mary Xavier Gray.

he came back to Gloucester, serving with Steve Olsson from Gullholmen on the schooner *Oretha F. Spinney*.

ANOTHER NORTHERN BAY community that sent men to Gloucester was Arichat, on the southern side of Madame Island.[25] Arichat Harbor provided a fine deepwater anchorage, extending 5 kilometers inland from its headlands. It was a French community, tracing its origins back to 1713, and prior to Canadian Confederation in the 1860s, it had been a thriving port. But the economy declined, and by the 1880s, 259 people from this area had moved south to Gloucester,[26] and by 1903, 180 sons of Arichat had drowned while serving on a Gloucester vessel.

Arichat's immigration experience was unique in that many families came south together. Men moved with wives and children and were soon followed by cousins and siblings. On reaching Gloucester they moved into boarding-houses run by French-speaking Nova Scotians, including houses managed by William Boucher on Hancock Street, Edouard Girroir on Duncan Street, and Julia LeBlanc (Boudrot) on Main Street. One of these Arichat families was to produce two of Gloucester's greatest master mariners, the Thomas brothers, Jeff and Billy, with Billy serving a stint as president of the Master Mariners' Association.* Both men became consistent highliners and daring skippers, and the son of one (Gordon Thomas) became the premier author to document the Gloucester fleet and skippers.

As Gloucester mariners, the Thomas brothers fished on Georges for haddock in winter. In late spring they moved into the southern mackerel or northern shacking fisheries. By fall they had returned to the haddock grounds, ending the season in late November or early December, always trying to be home for Christmas with their families.

Jeff was born in 1873, the second youngest of the 11 children born to Thomas Thomas and Martha (Marthe) LeBlanc. He came to Gloucester with his parents at the age of 3. Billy, 16 years Jeff's senior, spent his formative years in Arichat, but by 1880 he was living in a Gloucester boardinghouse run by Julia Boudrot. He soon married one of Julia's nieces, Marie-Anne LeBlanc, and by the time of his marriage he had become fluent in English. He was to spend the next 3 years as a Gloucester doryman. In 1883, the

---

*The Master Mariners' Association was now a positive force that took public positions on a wide variety of matters that might impact the interests of the fishing industry. The Association worked with Representative Taft to introduce a resolution into the legislature to compel ocean steamships to cease crossing the Georges fishing banks; and when the government began to construct a new lighthouse at Eastern Point, they circulated a petition asking that the light be changed from fourth- to third-class and that a fog whistle be established. In the years ahead, as the Association grew to more than 200 regular and 65 associate members, it became a central power in all matters that pertained to the welfare of the fishing industry. Nearly every question of moment was submitted for its review, and its approval or disapproval carried great weight.

George Steele Company gave Billy his first command: the *Dido*—a 79-foot, 81-ton, two-masted schooner, 1 of 11 vessels owned by the company. The firm had also given Captain Joe Mesquita his first command in 1891.

Captain Billy next skippered the *Mystic*, a 79-foot, 83-ton vessel, for the same firm, after which he moved to command vessels for the John Chisholm Company. There he skippered the newest Chisholm schooner, the *Horace B. Parker*, a vessel that he jointly owned with Captain John Chisholm. Billy served as her skipper for 14 years, bringing in record catches, and it was on her that he established his reputation for pluck and daring when he rescued 18 men from a storm-tossed, waterlogged vessel on Georges. It was in December 1890, and as a storm raged across the bank, Billy came across the *Eurydices*, a British vessel, laying on her beam ends. The crew clung to whatever they could, and Captain Billy saw the danger and responded. He fought through the storm to reach the men of the *Eurydices*, more than once thinking his own craft might overturn in the high seas. At last, just four hours before the British vessel went down, Billy and his crew rescued the 18 men on the *Eurydices*. In recognition of this deed of unusual fearlessness, the Canadian government presented him with a gold watch and gave $10 to each of his 19-man crew.

In 1905 Captain Billy took on a new command, the *Thomas S. Gorton*. The schooner measured 106.6 feet on the waterline and displaced 140.2 gross tons. She was an enormous vessel, and Captain Billy set record after record as her master. He continued as a skipper until 1916, when he finally came ashore to manage the Gloucester Cold Storage and Warehouse Company plant.

Unlike many great Gloucester skippers from this era, Jeff Thomas grew up on Cape Ann. In 1893 he married Lulu Norwood (whose Gloucester roots went back for generations), becoming a deliveryman for the Smith Bakery, on Portagee Hill. He followed this up with a job as a fish skinner for the Reed & Gamage Company in East Gloucester.

But the sea was, eventually, to beckon Jeff. He first went out not with his brother, but with his father-in-law, Clifford Norwood, with whom he served as a doryman on Captain Syd McCloud's little schooner *Northern Eagle*. It proved a good arrangement for all concerned. The *Northern Eagle* was a 61-foot shore vessel that went to sea for only a few days at a time, landing fresh haddock and other ground fish. Jeff learned quickly, and he soon graduated to the offshore fleet under Captain Marty Welch, one of Gloucester's greatest skippers. He served on several Welch vessels, including the clipper-bow schooner *Navahoe*, and he was on her when she went down on Pancake Island off Beaver Harbor, Nova Scotia. In a strong breeze, as Captain

Welch tried to make harbor at 4 o'clock in the morning, but before he found the light on Beaver Harbor head, the vessel struck, turned on her side, and caught fire. The men scurried to the dories and had a hard pull through three miles of pounding surf: It was a close call, but all made it.

Jeff continued to serve under Captain Welch, and his son Gordon Thomas says his father's relationship with Captain Welch was central to Jeff's later success. "Captain Welch took a liking to Jeff and when Captain Welch stayed ashore due to sickness or for a planned vacation . . . he let my father take out the vessel. Jeff was then known as a transient skipper, and he produced for the Sylvannus Smith Company, bringing in good sized trips."[27] In recognition of his performance, the Smith Company gave Jeff command of the *Arcadia*, and in 1905 he was highliner of the winter haddocking fleet. In March 1906, the Smith Company transferred Jeff to its new schooner, the *Cynthia*, and he took her south for mackerel. He was now a Smith Company skipper.

Jeff's advancement to command followed an old Gloucester maxim that success makes success. "As soon as a new skipper makes a successful voyage, as a general rule, he will find many vessel owners ready and anxious to place him in command of the best vessels of the port. There are those skippers, who, from their good judgment and splendid abilities, can be always depended on to finish their season's work in the front rank of their respective branches of the fisheries."[28] Jeff was such a skipper, and he soon fixed his reputation as one of Gloucester's great "sail carriers"—after days of good fishing his crew knew when Jeff was going to make a fast passage. They would have just about caught the limit of their fish, when, with the wind picking up, Jeff would climb the stairs from his cabin all dressed in foul weather gear. He would look to the men on watch and say something like, "Breezing up men. Breezing up now," and all knew that he was going to make a fast passage; he was going to put on all sails and head for home as fast possible. It was all about making money, getting a good share.[29] As his son Gordon recalled, he would now stay "on deck watching everything, taking everything in. He never went down (below)."[30] On one such trip:

> It was breezing like the devil (off Cape Ann) and Billy Thomas didn't want to bring his vessel into port. He had his schooner hoved to off Eastern Point (with her bow facing into the gale), for any right thinking skipper knew that it was best to stay offshore in a heavy gale of wind. You don't like to get in on the shore too much under those conditions. Then one of the crew called to Billy and said, "Look there is a vessel coming up on us. Look at the sails she is carrying." Billy sized up the approaching craft and turned to one of his crew with a broad grin and said, "There is only one damn fool

that would carry all that sail in these conditions, and it's Jeff." And sure enough, Jeff sped by like a greyhound. Billy was a sail carrier, but under these extreme conditions he was hoved to. Billy was more cautious. Jeff was going to Boston with a full load of fish, and he was bound to make the market. And, as Billy said, only one damn fool was carrying that amount of sail under those conditions, and his name was Jeff Thomas. A little wind couldn't stop Captain Jeff.

CHAPTER 12

A Scarcity of Fish

In the last quarter of the nineteenth century, landings of mackerel and flitched (salted) halibut were subject to wide swings—while the fish might be readily available one year, they were rare the next. Beginning in the late 1870s, mackerel landings dropped by almost half, from 19 million pounds in 1870 to 11 million pounds per year in 1875. The catch briefly rebounded in the early 1880s, to an astounding 31 million pounds per year, only to plummet to an anemic 6 million pounds per year in 1891 and 1892. Halibut landings also followed an uncertain course. Stocks had long been low on Georges, and by the 1880s the halibut catch on the Scotian Banks had also become less certain. Skippers now had to go farther and farther afield for these valuable fish.

When the mackerel catch crashed in the 1870s, its effect rippled through the community. From 1870 to 1872, the fleet lost 81 schooners, 600 fewer fishermen had jobs, and outfitters all but stopped ordering new vessels. Fishermen were especially hard hit. They had little or no savings. Their livelihood depended on what they received for the current season's work: When the fish disappeared, what else was there? So they did what they had always done, and even though the stocks fell, the mackerelers continued to fish. They made long trips to grounds as far apart as Virginia and Nova Scotia. Lookouts sat for weeks on end searching for nonexistent fish, and when the season came to a close, they and their mates had little to show for the effort.

Captain Phillip Alfred Merchant made two such trips on the schooner *Sarah C. Wharff*.[31] On the first trip he left in August on a fine, clear morning. Men rigged out the seine-boat and did the "other little knick knacks required for the business." On the next day, as the sun rose, Captain Merchant sent a man to the masthead to look for schooling mackerel. But the man spied no fish, neither on this day nor on the next. Of 18 nearby seiners, not a one had any fish. Captain Merchant wrote that he had received no reports of mackerel from the Cape Shore to the southern New England coast. Within a matter of days, he counted 89 nearby schooners, all sailing round and round with no sight of mackerel.

Then the fog closed in, and his vessel joined 37 others in a safe anchorage at Southwest Harbor, Mount Desert Island, Maine. Two days later, the fog still holding, the crew of most of the fleet went ashore to pick berries. They were gone all day and were in a foul mood when they returned. As they came back, vulgar songs sprang from their lips.

The next day the fog lifted, and Merchant put his men in the seine-boat to tow the *Wharff* out of the harbor. But fishing did not improve. After 18 uneventful days, his seine-boat finally went out after the first of several small schools, but the first school dove under the seine and the men landed only a single barrel of mackerel. After several more days, in which the men made marginally successful sets, with his supplies running low, Captain Merchant sailed home for Gloucester. There he unloaded his 120-barrel catch—less than half of what he had hoped for, but much better than the schooner *Hattie Winship* of the same firm, which came in with only 20 barrels. The skipper felt a little better, restocked, and sailed on his second summer seining trip. On reaching the fishing grounds, he spoke with Captain Solomon Jacobs, leader of Gloucester's mackerel fleet, who said he had made 12 sets on one day and came up dry each time. Again the men found that they had time on their hands, and poor weather drove them and other seiners into a Maine port. The skipper wrote that "half the crew, me including, went ashore in the seine-boat and came (back) on board at 9 P.M. Some of the boys felt happy and strong by the effect of the strong drink that they got on shore. But, anyhow we got on board alright but if I had not been in the boat, I don't know if they had got along so well. After we got on board we all turn in and everything went alright." In any case, the seining trip proved a failure, the conditions never improved, and when the season ended on October 5, Captain Merchant, like many other mackerel skippers in the late 1870s, left the fishery.

Stunned by such a rapid decline, many in Gloucester blamed overfishing. One sage suggested, "If all the vessels engaged in seining mackerel could be induced to abandon that method of fishing, we might hope for the restoration of this industry to its old time place." It did not happen, but for whatever reason, the catch levels rose in 1880. For some it was too late, and many of the families suffered. They found themselves hard-pressed to avoid "the bitterness of cold and hunger." Worthy families needed money, food, and clothing of all kinds, and the late 1870s saw hundreds lined up for handouts of soup and bread at community aid stations.

For the city as a whole, however, the diversity of the local fisheries kept vessels at sea and helped cushion the community from the worst of the mackerel decline. Outfitters sent out large fleets in search of cod, haddock, and herring. Flexibility proved key to Gloucester's continued success, and with the passage of time even the mackerel fishermen found a better day.

FROM 1886 TO 1889, Gloucester's total catch of all types of fish dropped from 123 million pounds to 75.8 million pounds. Mackerel came in at an average of 2.3 million pounds per year, while the halibut catch dropped by 5 million pounds per year. For the halibuters to thrive, skippers had to find alternative fishing grounds. While Georges, Banquereau, and the Grand banks provided superb stocks of cod and haddock, halibut were scarce. The 50-vessel halibut fleet brought in small fares, and many took a dark view of the future.

Outfitters considered two unique alternative fisheries, one in the Pacific Northwest, the other on the west coast of Iceland.

At the 1887 organizational meeting of the Master Mariners' Association, Captain Sol Jacobs stood before his fellow skippers receiving their praise and best wishes for the success of his contemplated flitched halibut* experiment in the Pacific Northwest. He had prepared two craft for the long voyage, and if he succeeded, other Gloucester skippers would follow, and Gloucester's history would change forever. In May 1886, the U.S. House of Representatives voted that for a period of five years no mackerel could be caught and landed between March 1 and June 1. This convinced Jacobs of the need to find an alternative fishery.†

Captain Jacobs owned two first-class vessels—the *Edward E. Webster* (83.5 feet and 93.9 tons) and the *Mollie Adams* (117.3 tons). On October 27, 1887, both set sail for Washington Territory carrying 23 young, experienced Gloucester mariners. Captain Jacobs made the cross-country trip by rail with an additional group of Newfoundlanders and New Englanders to round out the crews: fishermen and seal hunters.

On the trip around the Horn, Captain Charles J. Johnson commanded the *Mollie Adams* and Captain John L. Harris the *Edward E. Webster*. The schooners carried seventeen 18-foot-long boats built by Higgins & Gifford from East Gloucester for use in the sealing business, mackerel seines, lines and hooks, and other fishing gear. Each had three suits of sails, spare gaffs and booms, rigging of all kinds, paint, and all the supplies needed for a year. Finally, they carried articles for trading: dry goods, boots, shoes, rubber over-

---

*The appropriate term for this type of filleted, salted halibut is "flitched," but it was also known as fletched. Both terms are used in the text.

†No mackerel caught between March 1 and June 1 "could be imported into the United States or landed upon its shores from any portion of the world except upon production of satisfactory proof, to show that the mackerel were caught during the period mentioned with hook and line from boats, or in traps and weirs connected with the shore."

shoes, hardware, and the like. It cost Sol $13,000 to outfit the schooners, a vast amount for the time.

The previous summer, Jacobs had traveled to Washington Territory, scouting out the fishery at Seattle and other ports. Business leaders from the Northwest encouraged his venture. They questioned only the adequateness of the railroad for shipping fish east. One of the Pacific coast fishing firms, however, cautioned that the cod fisheries and seal hunting had been running at a loss in recent years—to them, Jacobs's calculations did not make sense. Captain Jacobs, on the other hand, believed the future would take care of itself.

The schooner *Mollie Adams* arrived at Port Townsend (Neah Bay) on March 3, after a passage of 123 days and 10,000 miles. As they entered Puget Sound, the crew unfurled a banner from the foremast displaying the name of the craft, and from the mainmast they flew the Stars and Stripes: This was an American vessel. Once ashore, the 12-man crew—joined by Captain Jacobs and 30 of the New England and Newfoundland fishermen who had completed the overland passage—prepared the *Adams* for her first trip in the waters of the Pacific Northwest. On March 9 the schooner left on a sealing trip, intending to go as far north as the Aleutian Islands. The Seattle newspaper, the *Daily Post-Intelligencer*, said that "under a favorable wind the *Mollie Adams*, the pioneer schooner of the Gloucester fishing fleet, plowed . . . through the white caps bound for the . . . sealing expedition." The plan called for a two- to three-month trip.

Meanwhile, the *Webster* had yet to arrive. As she had made the trip south, all but two of her crew had come down with swollen legs, probably scurvy. She lost her foremast and had to put into Montevideo for repairs. When she got back to sea, rough weather caused her to take six weeks to go around the Horn. Her meat spoiled and the crew dumped it overboard; her flour became musty. On entering San Francisco Harbor she struck on the rocks off Meiggs Wharf and went into dry dock for repairs. When she finally made Port Townsend she looked as if she had been lying on the beach for months on end.

By the end of June the *Adams*, after an unsuccessful try at sealing, went out on two fresh-halibut trips to Cape Flattery Grounds. She brought in fine catches. Captain Jacobs sent the fish east by rail and had the *Adams* refitted for a third trip. The halibut banks were near the shore and the men found abundant fish weighing 50 to 150 pounds each. Wherever the men made soundings, from the Strait of Juan de Fuca to the Aleutian Islands, they found halibut. Next came a successful flitching trip to the Two Peaks fishing grounds off northern British Columbia.

Yet, the venture failed, killed by high freight prices. The railroad charged up to 14 cents per pound to carry fish from the Northwest east to Chicago and New York—several times the cost of landing the same fish in Gloucester.

In October the *Post-Intelligencer* misleadingly reported that Captain Solomon Jacobs had sent east halibut that yielded a gross return of $33,000. What the report failed to mention was that whatever the price received for the fish, one first had to net out the cost of the two-boat venture. Shipping the fish east cost $1,000 per carload for freight and ice, $11,000 in all. This left a balance of only $22,000 to meet the expenses of the crew, the supplies, and the wear and tear on the vessels (which costs were not itemized). To anyone who understood the business, it was clear that Captain Jacobs had received no dividend for his two-boat effort.

In fact, with the high shipping costs, the skipper would have felt lucky had he just broken even. The newspaper's estimate of the stock proved to be grossly inflated. George Richards, a Boston fish dealer who visited with Jacobs on the West Coast, reported that Sol told him the *Adams* had stocked $18,000 and the *Webster* had not done as well, stocking only about $3,000.

Jacobs's luck did not improve in 1889. On January 17 the *Webster* came into Coal Bay, Unga Island, Alaska, under full sail, flying before a stiff breeze, and in heavy seas. Her skipper was unfamiliar with the grounds and had little real control of his vessel. The *Webster* smashed into a sunken reef and went to the bottom, a total loss. All of the men were saved.

Four months later Captain Jacobs sold the *Mollie Adams*, ending his West Coast experiment. He had hoped that a Northwest halibut fishery would transform Port Townsend into "the Gloucester of the Pacific." In his view of the future, large numbers of Gloucester craft would have made the trip west. It did not happen, at least not on his watch. Gloucester outfitters never did send craft to the Northwest, and fishermen and their families did not migrate to the other coast, at least not in this era. But a successful new halibut fishery was discovered at this time—not to the west, but to the east, off the coast of Iceland.

CHAPTER 13

Fishing for Halibut Off Iceland

In the spring of 1884, the schooners *Alice M. Williams*, *David A. Story*, and *Concord* set sail for Iceland's western fjords in search of halibut.* They sailed north across the Scotian Banks, turned east, passed along the southern expanse of the Grand Banks, and finally headed northeast for Iceland's Westfjords. The trip took three weeks, and once they arrived their success or failure depended almost entirely on the conditions at sea. With clear skies and a fair wind, the typical 18-man crew could land 60 to 100 fine halibut a day—weighing up to 300 pounds a fish. But the weather could change without notice, and for weeks on end dorymen might set no trawl and land no fish. Severe gales lashed the coast in April and May, and the northwest winds brought ice floes to the grounds. Even in July, whirling snowstorms could arise without warning, closing down the fisheries for days on end. Vessels seemed to be ever scampering to the safety of the nearby fjords, where steep volcanic cliffs provided some measure of protection against the raging storms.

Yet, by the end of that first summer, all three Gloucester vessels had many fine days on the grounds, and upon their return in mid-September they landed excellent fares of fletched halibut. The schooners *Williams* and *Story* each brought back 155,000 pounds, while the *Concord* had 130,000 pounds. This represented several times the weight in raw fish, and everyone was impressed. The Iceland experiment on the Westfjords had succeeded. Gloucester fishermen had found a new source of halibut, and the local vessel outfitters responded by sending an ever-expanding fleet to the distant grounds. In a little over a decade, 23 Gloucester firms would send 45 vessels to the Westfjords, making a total of 102 trips.

In undertaking this fishery, the skippers set up a station at the small Icelandic village of Thingeyri, on the south shore of Dyrafjordur, about 12

*In 1873, Captain John S. McQuinn of East Gloucester had made a brief trip to this area on the schooner *Membrino Chief*. He left Gloucester on May 23, arrived on the fishing grounds on June 9, and remained for five weeks. But the weather was stormy and he had only a limited opportunity to fish. "Captain McQuinn determined to abandon the enterprise, and on July 11th set sail for home, which he safely reached on the 13th of August." (Procter Brothers, 1882, p. 50; *Fishermen's Memorial Book*, 1873, p. 96.)

miles from the sea. The village had a fine wharf, a small hotel (the Niagra), 2 warehouses, a merchant's shop, 10 modest houses, and a ready source of fresh water. But there were only 18 residents, and once the halibuters arrived the village took on a decidedly American look.

The locals treated the Gloucester fishermen with respect and kindness. The merchant, Niles Christian Gram (who owned the wharf and all the buildings in the village), and his head clerk, Friedrich R. Wendell, were highly praised by the halibuters for meeting their needs. They managed the supplies landed by the Gloucester vessels, and went out of their way to meet the many and diverse requests of the fishermen—from ponies, to wheat, to alcoholic drinks. In 1886 the season opened on an optimistic note, as the fleet brought in a good stock of fletched halibut on its first baiting. They spent three weeks at sea and came back to Thingeyri with about 1,000 fish per vessel after about 10 days of excellent fishing. But then the weather turned. Snow fell in June and July, drift ice swept over the grounds, and by the end of the season the fleet averaged only 100,000 pounds per vessel.

The next year was similar. The first baiting was a disaster. Late April temperatures fell to as low as −15°F, and many of the men had frostbitten hands and faces. In May, heavy gales brought small icebergs and drift ice to the grounds, and one of the vessels, the *Arthur D. Story*, found herself anchored in the growing ice field. Captain Joseph Ryan had set his banks' anchor, hoping to ride out the blow, but the conditions became more perilous as northwesterly winds drove larger icebergs in toward the shore. One of the bergs crossed the *Story*'s bow, just as another passed within 15 feet of her quarter. As Captain Ryan looked out, he saw other bergs bearing down. He stood with a hatchet in hand, ready to cut the cable—better to be adrift among the bergs than sent to the bottom after one of them rammed his vessel. But then the wind changed, blowing from the shore, and the easterly breeze took the bergs harmlessly out to sea, passing to the lee of the *Story*.

While Captain Ryan may have escaped the full fury of the storm, within days the wind again turned to the northwest, bringing drift ice and small bergs back onto the grounds. For weeks on end, the *Story* and the rest of the fleet sat anchored in Thingeyri, the men filling their days with a variety of amusements. On board ship, they played cards and talked of their uncertain future. Ashore, they took pony rides into the surrounding valleys, walked the streets of the village, or partied at the Hotel Niagra. They drank, danced with local women, sang, and got into a fight or two, but they did little fishing. The 1888 season was proving to be the worst on record. The men failed to keep a consistent run on the halibut, and when they returned in September the fleet averaged less than 100,000 pounds per vessel. But then, just as quickly as the catch had dropped, everything turned around. Why this happened no one really knew; maybe it was weather, or maybe the fish were just more plentiful. Average catch levels rose to 160,000 pounds in 1888 and stayed there for years, and the Gloucester community responded by sending more vessels to the Westfjords. The peak came in 1894, with Gloucester outfitters committing 15 vessels to the Icelandic fishery.

With an expanded fleet, the Gloucester waterfront bustled with activity as skippers prepared their craft for the annual trip north to the Icelandic fjords. Vessels first went onto the marine railways, where workers scraped and painted hulls, checked for leaks, and overhauled rudders and steering mechanisms. They inspected the standing rigging, replacing all that was worn, and covered everything with a fresh coat of tar. At the sail lofts, workers examined each vessel's main- and foresails, adding patches and reinforcing stitching, as well as cutting and stitching new riding sails for the long season on the Icelandic grounds.

As the date for sailing drew near, the dorymen sprang into action. They stowed a staggering amount and variety of supplies, taking days to get everything below.* As the day of departure drew near, they stowed their personal gear, lashed the 12 dories on deck, and bent on the sails. With all in readiness, the skipper called for a tug to pull his vessel away from the wharf and out into the channel, or the "stream," as the locals called it. The trip was about to begin, and as the tug towed the schooner out of the harbor, the men stood on deck or climbed into the rigging, waving a fond farewell to those left behind.

For one such trip on the schooner *Concord* under Captain John Baptiste Duguo, of which he was part owner, Alex Bushie, mate of the vessel, kept a detailed log of the ship's progress. It is an unusual account.[32] The trip began with a tug towing the schooner away from John Pew & Son's wharf. It was March 26, and the *Concord* would have a fast, uneventful 16-day passage to Thingeyri. With a crew of nine, Captain Duguo planned to hire local Icelanders to round out the complement (men from the surrounding countryside who had joined with him in years past), a practice that had become commonplace within the fleet. As the *Concord* rounded Eastern Point, the men set the sails and went below for dinner. Many felt lonesome, talking longingly about the young women they had left behind; love was in the air.

Early in the trip the men did no more than the ordinary chores about ship—standing watch, taking the wheel, tending sails, and getting reacquainted

---

*For a five-month trip they lay in 10 tons of coal, 55 bushels of potatoes, 27 barrels of flour, 18 barrels of beef, 350 pounds of lard, 800 pounds of butter, 100 pounds of evaporated apples, 160 pounds of tea, 87 pounds of coffee, 130 quarts of peas, 400 quarts of beans, and so on.

)(*The schooner* Concord *lies alongside the John Pew wharf in 1894 as she is readied for the voyage to the Icelandic fishing grounds.*

with the men who would be their constant companions for the next six months. On the first day out, Bushie, Captain Duguo, and Frank Peterson engaged in a penny-ante game of cards. Bushie made $1.30, and as the men bunked in for the night, Bushie and the skipper paid 10 cents each for oranges that Peterson had brought on the trip.

On the third day out, the men began to prepare gear for the Icelandic fishing season. They checked lines and fixed dory thole pins. But it was still early in the trip, and on the following day, Bushie and the skipper played "a game of cribbage for a cent a hole." Bushie lost 3 cents, and when he went on watch, Frank Peterson took his place tackling the skipper. As the men arose on day four, the skipper fixed a regular schedule for working on the gear, fixing ganging hooks, and rigging trawl lines. The men had "lots of it to do, [for] the ganging hooks didn't last long [once on the fishing grounds]."

On April 1 the *Concord* spotted a large iceberg as well as a few smaller ones as she passed across the Grand Banks. A brief period of snow and hail followed. Over the next few days the crew sighted larger icebergs, but this did not prevent the men from catching one of a group of porpoises swimming around the vessel. Even though it was Good Friday, "a holy day on board of this ship," the men worked, skinning the porpoise. Bushie said, "The old man [Duguo] got all the oil he could and it didn't smell very nice. The liver and heart is very good eating, it's just the same as a hog's. The meat is too tough."

But Bushie was eager to be clear of the icebergs: "Cold chaps, I don't like this sailing amongst them worth a cent." On April 5, he got his wish, and the men redoubled their efforts "in rigging gear, putting beckets on the ground lines, and putting handles on fletching knives, and making dory plugs. The afternoon passed in the same way until 4 o'clock. . . . (They) swept out the cabin floor, (and) called it square for a day's work." The next day, Easter Sunday, April 6, began with a fried ham and egg breakfast—"good Gloucester eggs, provided especially for the occasion." It was a quiet day on board the Concord. "There was nothing done, [for] Sunday is naturally a very quiet day outside of working [the] ship, and telling stories." Bushie, a French Canadian Catholic from Arichat, said, "It would be better to be reading good books [or probably 'the' Good Book], but such is the life of a fisherman."

The next two days saw the men making rigging gear, some making skates (baited sets of trawl line set on a canvas bottom), others becketing trawl lines, and others tarring twine to ganging hooks. The men also began to look for land, and at daylight on April 11, they sighted Brede Bright Head on the southwest corner of the Westfjords, below Dyrafjordur. Two Icelandic fishermen came out to greet the ship. They had been dorymen on the vessel for several years, and the skipper paid them 108 krones due from the previous year's work. They would have earned only 20 krones working on one of the local sheep and potato farms, so jobs on the halibuters were in great demand.

The *Concord* anchored at Thingeyri at 3 o'clock in the afternoon. Men off-loaded supplies and stored them in Gram's warehouse. They then waited two days before going to the fishing grounds. "The old man shaved himself and went ashore and carried the mail (which would later be picked up by a passing ship heading west to North America), the rest of (the men) . . . got shaved and got ready to go ashore."

Heavy snow was falling outside the fjord, and the crew stowed gear and sat out the storm as the skipper detailed his plan for the season. The eight Icelanders who joined the crew would fish one man to a dory. Gloucestermen would fish two men to a dory (easing the burden of having only one man to do all the work), the men drawing lots for dory mates. As the next day opened cold and windy, the men filled the time sharpening ganging hooks and performing odd jobs about the ship. On April 17 the weather finally improved and the men worked on deck. They repaired the jib stay, checked the dories, and as Bushie noted, "if we should sleep here a week we would always find something to do." At 4 o'clock the following morning the *Concord* sailed out of Thingeyri for the first time in the season. After loading enough supplies for up to four weeks, the men fished on the first baiting along the coast of the Westfjords. On the first day, with the wind from the southeast, the *Concord* stopped at the small Haukadalur River, about three miles east, toward the sea, to fill the freshwater tank. When the *Concord* at last got to the headlands of Dyrafjordur, however, the weather proved too rich for their blood. It blew heavy and rough. They returned to Haukadalur, where five of the Icelandic fishermen spent the night ashore with their families.

For days the wind did not let up, and the men had nothing to do but play cards. On April 20 they finally made it 15 miles out from the coast, but a stiff breeze made it impossible to set trawl and the *Concord* went north with the Gloucester schooners *Senator Saulsbury* and *A. D. Story* to jig for bait. The snow continued to fall in heavy sheets and the three schooners sought safe anchorage in Isafjordur. Many of the men on other vessels took the opportunity to become "foolish drunk." But the men on the *Concord* felt tired just lying around doing nothing, and saw little pleasure to be gained in going ashore, for they could not understand a word spoken by the local people.

As the weather subsided on April 25, the *A. D. Story* was the first to leave. The others soon followed. They anchored 15 miles ESE of Isafjordur Bay in 90 fathoms of water, and there jigged for cod to use for bait as clumps of frozen ice floated nearby. The next day, as the sun rose at 3 o'clock in the morning, they went out to set trawl and later landed their first halibut— seven or eight fish, and still had enough bait to make up another 12 tubs of trawl. On April 27, because the weather looked questionable, Captain Duguo did not send out his dories, but the *J. W. Bray* did, and one of his dories got several loads of fish. Bushie anchored a mile from the *Bray* and set two tubs of trawl in each of five dories. The next day, seven of his dories hauled and the men "got a fine lot of halibut," 60 fish. But they also lost three times as many fish when the wind backed around to the east and they had to cut away the trawl lines. Now it was time to process the fish. Boards were placed along the railing, a rope tied to the tail of the giant fish, and it was hauled up and filleted. It was then washed clean, salted, and stored in one of the vessels' 12 compartments below deck—2 were traditionally left empty and 10 were filled with salt, and the newly salted fish went into one of the empty compartments. It was midnight by the time they finished cutting the fish and packing them in the salt pens. They were exhausted; the two other vessels had long since left the grounds.

The next day saw the *Concord* again sailing back to the grounds, jigging for bait on the way out. She anchored in 50 fathoms and set her gear. At 6 o'clock the men went to haul, and with a good pull, landed 100 fish. Some dories hauled twice, and once again it was midnight before all were aboard. They had breakfast and then dressed the fish.

On May 1 the vessel anchored at 11 A.M. and baited and set trawls. At 6:30 P.M., the men hauled, landing about 60 fish. The *Concord* then pulled her anchor about halfway up, moved to another location, and set three skates to a dory. The men then came back on board and had a nap. At 8 o'clock the next morning they had their second breakfast, and then "dressed the prior night's catch . . . (and) went to haul." They got 150 good fish. On May 3, the men baited at 6 A.M., after the skipper had moved the vessel. Six dories

✗ *A fisherman prepares to club a halibut as he hauls it onto the dory, c. 1921.*

set trawls about eight miles off the coast in shoal waters and caught 60 fish. The skipper repeated the pattern of moving, setting, and hauling for about a week. He returned to Dyrafjordur on May 9, ending the trip of the "first baiting."

Concord stayed on the Westfjords until early September, when she began the long sail back to Gloucester. On September 21 she reached Little Arichat, Cape Breton, and went in for supplies, having run out of almost everything. A week later she arrived at Cape Ann. She tied up at John Pew's wharf and off-loaded 120,000 pounds of flitched halibut (ready for smoking once back in Gloucester) and 40 barrels of fins—the large, thick fins of the halibut were for some a prized delicacy, and for others an ingredient in isinglass, a gelatinous substance used in brewing.

With such fine trips, the Gloucester outfitters continued to send a large fleet to Iceland for several years to come. The Gloucestermen viewed the fishery as unlimited, and in February 1891, the fleet left earlier than ever before. Outfitters hoped that by leaving earlier they would be able to return earlier in the summer. But the early sets in 1891 proved unproductive, although the fleet had good success later in the season. The first of the fleet to return, the *Senator Saulsbury*, landed 240,000 pounds of fletched halibut. In the following two seasons, 1892 and 1893, the schooner *Maggie F. McKenzie*—the largest vessel in the Gloucester fleet at 109.8 feet and 170.13 gross tons—brought in the largest fares: 254,000 pounds of fletched halibut in 1892. Captain Andrew McKenzie described the trip that year as one of hardship, but by its end prosperity crowned his efforts. In the following year, as the summer unfolded, the fleet had fine weather, but for the first time found fewer fish: It was the first sign of over-fishing. The *McKenzie* had one of the better fares, but even that came in at only 175,000 pounds—not a particularly impressive catch for so large a vessel.

By mid-March 1894, four vessels had already sailed from Gloucester, and the schooner *Alton (Alan) S. Marshall* had just become the fifth to leave. But she didn't have a normal trip, for while passing over the vastness of the Grand Banks she ran into a fierce storm, the storm raging unabated for two days. The wind "blew the top off the sea" and Captain Marshall could make no progress against the hurricane. Laying under bare poles, they drifted with the storm. On the third day, as the winds abated, they got up some canvas, only to run into another fierce storm. Two days later an enormous sea boarded the vessel, striking the fore-rigging. Her fore-rigging lay in the water for more than 15 minutes, and she began to fill with water. Captain Marshall thought his time had come, as he began to lose consciousness even as the vessel began to right herself as his men cut away the masts and rigging. The cargo had somehow shifted, and the skipper thought this simple act might have proved the difference between life and death. But two men had gone overboard: James Barry and Herman Olsson, the latter from Gullholmen, Sweden. Somehow they were able to grab onto the rigging; their lives were spared, and all they had to show for their ordeal were painful heads and injuries to their chests.

Even so, the *Marshall* was not yet out of danger. She tossed about helplessly in the turbulent sea, and the fate of the men still hung in the balance. Then on the morning of March 28, the steamer *Mohawk* spotted the distressed vessel and bore down with all speed. The men were saved, and as a final act, the skipper of the *Marshall* set the schooner afire, for afloat she represented an obstacle to passing vessels. Among the rescued men were five from Gullholmen—Nils Lund, Herman Olsson, Oscar Johnson, John Johnson, and Steve Olsson.

For the Gloucester vessels that made it to Dyrafjordur, the fish proved scarce early in the season. Many of the local people were sick. Typhoid and measles raged at Isafjordur. Although Gloucester crews somehow escaped the scourge, many of the local inhabitants died. In a letter home, Captain George Dixon of the schooner *Rigel* described the winds as blustery, with frequent snow squalls. By mid-May, snow still covered the mountains.* On May 17, a crewman of the *Talisman* described the weather as the worst he had ever seen in Iceland. It blew a gale of wind every second day. By early June, the weather had let up, but halibut remained scarce. Skippers had even tried fishing off the edge of the ice, but with little luck. By mid-month the *Rigel* and *Maggie E. McKenzie* prepared to leave Iceland. As soon as the weather moderated, they sailed to Greenland. Heavy westerlies had driven ice on the Banks, closing in the entire coast and making fishing impossible. But once off Greenland, their luck did not improve, and the *Rigel* eventually returned with only 70,000 pounds of fletched halibut and 37 barrels of fins.

But there was more to report than a final accounting of a broken trip and scanty fish in the hold. For on this trip, the *Rigel* had delivered the crew and passengers of the lost steamer *Miranda* to North Sydney, Cape Breton. The *Miranda* had struck a submerged rock off the coast of Greenland, which tore an enormous hole in the hull, and her pumps proved useless. The skipper sent several men on a 140-mile open ocean trek in search of help, and they found the *Rigel* at Neplast, Greenland. Captain Dixon took up the call. On arriving at Sukkertoppen, the men of the *Rigel* made room for the steamer passengers, removing 59 hogsheads of salt from the schooner's after-hold.

*Meanwhile the men of the *Rigel* had given up all hope for the safe arrival of a second Gloucester vessel, the *Ambrose H. Knight*. She had a crew of 16, 13 from Scandinavia, and all were lost.

This space connected with the skipper's cabin, and the steamer passengers stayed in this area. Sails covered the salt, with mattresses on top of the sails. As the *Rigel* left Greenland with the steamer passengers on board, all joined in singing "My Country 'Tis of Thee."

In thanks for this rescue, the passengers and crew of the *Miranda* raised several thousand dollars for Captain Dixon and the men of the *Rigel*. It was the difference between a losing and a successful trip.

Because of the poor 1894 catch, only eight vessels went to Iceland in 1895. On April 12 the *Rigel* arrived after an 18-day passage. She did not go into port, but immediately made a set and found good fishing. By mid-April, with poor weather again setting in, only the *Rigel* had made it out to the fishing grounds. The cook, Albert Willey, described the weather as severe and cold. Captain Andrew McKenzie of the *McKenzie* wrote that the rest of the fleet had taken few fish.

About the same time, Captain Dixon of the *Rigel* sent word that one of his dorymen, a young Swedish immigrant, had drowned. While fishing some 40 miles off the coast, Olof Johan Jacobsson and his mate Oscar Johnson (two Gullholmen natives) had begun to haul trawls, when at 11 o'clock in the morning a gale sprang up from the northeast. The water was shallow and the sea became rough. One by one the dories began to row back toward the *Rigel*. Jacobsson and Johnson joined the others in making for the safety of the vessel, but a wave broke over their dory, capsizing it. It flipped over and floated bottom up in the rolling sea. Jacobsson succeeded in grasping the plug strap of the dory; Johnson grabbed the becket strap in the bow.

The men on the *Rigel* saw the predicament of their two companions. Captain Dixon positioned members of the crew to keep an eagle eye on the dory as it drifted toward the anchored schooner. He believed they would be saved. But the two men found it difficult to hold on in the icy water. Time and again, the force of the wind and the rolling waves nearly wrenched their arms from their sockets as the dory suddenly lifted high in the air on the crest of a huge wave. Again and again they lost their hold of the becket strap and painter; but in a desperate battle for life they renewed their efforts, succeeding in reaching the dory and regaining their hold. The struggle continued for about an hour, even as they drew closer and closer to the anchored schooner. At length Jacobsson called out to his mate that he was utterly exhausted. He could hold on no longer. Johnson shouted back to his friend to keep up his courage; he could see the *Rigel*. The wind was blowing them down on their ship. In a few minutes they would surely be safe. But as they drew nearer and nearer to safety, the schooner suddenly broke away from her anchor. The cable burst like a pipe stem, and the two floundering men saw it happen. The wind filled the *Rigel*'s sails and the schooner drifted away from the men. Jacobsson called out desperately to his mate, "I am dying." Yet still he clung to life, thrusting his arm up to the elbow through the bight of the strap. But even as he continued to hold on, he became still. He had lost consciousness. His head hung down in the water and there was nothing Johnson could do. Back on the *Rigel*, Captain Dixon had climbed into the rigging. On deck, the helmsman regained control of the giant ship, and Dixon called for volunteers to launch a dory to rescue Jacobsson and Johnson. Three men stepped forward, and working through enormous swells and breaking waves, they reached the overturned dory and brought the two men on board. By then, Johnson, too, had lost consciousness, but he continued to breathe. Jacobsson had been in the water too long; his head had hung downward for half an hour, the sea washing over his limp body. Despite the heroic efforts of his friends to revive him, it was clear that he was gone. He had been dead when taken from the water.

The crew described Jacobsson as a sober young man of excellent character, and the men of the *Rigel* carried his body ashore for burial at the Lutheran Church at Dyrafjordur. The little cemetery sat in a place called Pleasant Valley, three miles distant from Thingeyri. Jacobsson made the trip to the churchyard in his dory, pulled by his mates. The Lutheran minister said the prayers of their shared faith, and his mates carried Jacobsson to the churchyard and lowered his body into the ground. There the minister prayed again, and the men of the *Rigel* sang hymns.

His mates held Jacobsson in high regard. Many had grown up with him in Gullholmen, including the Johnson brothers, Sam Andersson, and Steve Olsson. These men and the skipper chipped in to buy a four-foot-high marble monument for the grave site. It consisted of a base, plinth, shaft, and cap. They pulled it by dory for one and a half miles, and then carried it on stretchers the rest of the way to the churchyard, where it still stands today.*

In the first of his notes on the loss of Jacobsson, Captain Dixon also

---

*The monument cost about 100 kroners, or $27. This was the first monument ever erected to a drowned American in Iceland, albeit a man from Gullholmen. The inscription on the monument said: JOHN JACOBSEN, Schooner *Rigel*, Gloucester, Drowned April 29, 1895. The crew of the *Rigel* also had a picture taken of the stone for John's parents in Gullholmen—Jacob Kristiansson and Olivia Larsdotter. It would be passed down over the generations.

In a trip to Thingeyri in May 2003, the author of this story sought out this stone, the final resting place of his great-uncle. His guides were the Icelanders Johann Diego Arnorsson and Valdimar Gislason. Prior to starting the trip, Johann had a sad story to tell. It seems that the cemetery had long since gone into a state of disuse, and the small church next to the graveyard had long ago been dismantled. The cemetery sat in a small fenced-in area in the middle of a sheep pasture, and two years earlier all of the stones had been picked up prior to the re-leveling of the ground. Unfortunately, the local man hired for the task became deranged and drove over the stones with a tractor, destroying most of them. Johann did not know the fate of the Jacobsson stone but held out little hope that it would be among the half dozen to survive. But, when Johann, Valdimar, and the author arrived at the site, which sits in the midst of a vast green valley, survived it had, as had the stone of the merchant Gram.

reported on his early season success in finding a good school of fish. But as bad weather soon followed, he still had no more than 25,000 pounds. He also reported that on hearing reports of leprosy in Iceland, he made inquiries and learned that several families had the disease, and it had been passed on to their children. But Dixon deemed it unlikely that it would affect his men.

By summer's end, three other crewmen from Gloucester's Icelandic halibut fleet were lost, one a skipper. In July, Captain Duguo of the *Concord* lost two men in a storm on Breiðafjöður: an American and a native of Thingeyri. Then, on September 1, the crew of the *Arthur D. Story* returned to Gloucester with the body of their skipper, Captain Joseph Ryan. He had died in May, while out on the fishing grounds.

It seemed that while his men were out pulling in trawls, a strong breeze had sprung up. As Captain Ryan grabbed for the wheel, his heart gave out, and he died at the helm. The cook took the wheel and kept the vessel near the dories, signaling to the men to come aboard. They put into Thingeyri to land the body, and the "utmost kindness was shown by the Icelanders," who attended the funeral service and followed the remains to the Lutheran cemetery. There the crew arranged for temporary burial in a metal casket. Later in the summer, when the *Story* left for Gloucester, the English, French, and Norwegian vessels then in the harbor, as a mark of respect, displayed their colors at half-mast, for the men of the *Story* had retrieved Captain Ryan's body and placed it on deck in a tightly caulked box filled with salt. There it remained until the vessel arrived in Gloucester.

As the summer of 1895 drew to a close, many of the remaining vessels in the fleet again tried to salvage the trip by sailing to Greenland. But they had little success, and by season's end the fares for the combined Iceland–Greenland trips were low. Outfitters made nothing for their efforts.

The return of the last of these schooners to Gloucester effectively ended the Icelandic flitched-halibut fishery. Although in 1896 and 1897 a few vessels did go back, the fish were again absent. It was not a success.

WHILE ICELAND had seemed the halibut fishery of the future, some Gloucester outfitters had continued to send out a few schooners to other banks during the 1890s. A handful went to Georges and the Scotian Banks in search of fresh halibut,* while an even smaller number sailed north to the Funks for flitched halibut. Fishing on the Funks kept the flitched halibuters limping

---

*Among the fresh halibuters, the loss of the schooner *Falcon* and its crew was one of the strangest stories of the era. The *Falcon* was last seen in November 1896, after which she simply disappeared. Then, months later under the heading "News from the Dead," the local paper reported on her loss. It seemed that Captain Andrew Nelson had left Gloucester in the middle of November on a halibut trip to Georges. He had stocked for a four-week trip and told his wife to expect him home for Christmas. But when he had not returned by January the vessel was given up for lost. Unlike other vessels that disappeared, the *Falcon* left a message. In August 1897,

forward until the end of the decade. Among the leading captains were Joseph Bonia, Robert Porper,[†] Lemuel Spinney,[§] and Charles Young.

The small fleet of flitched halibuters faced the most severe of challenges, laying trawls along the edge of the great northern ice floes that surrounded the island of Newfoundland. Arctic ice made the grounds inaccessible before May, and the fishing had to close down before winter set in. In a May 1897 trip, Captain Lemuel Spinney on the *Gladiator* reported miles and miles of ice on the Burgeo Bank. He sailed for 75 miles and didn't know how much further it extended. So thick was the ice, the fleet did not dare sail at night. The *Eliza H. Parkhurst* became stuck in the ice as the skipper looked to set his trawl. There she remained for 20 days. On May 14, the *Lewis H. Giles* reported vast schools of ice on Burgeo, St. Paul's, and the Bay of St. George, Newfoundland (on the southwest coast of the island). Captain Young tried to go to Bacalieu Bank for a short trip, but the bank was one vast field of ice. Some vessels found ice to be a problem as late as July. Captain Joseph Bonia, commanding the schooner *Carrie Babson*, reported striking ice and losing

---

Captain Joseph Swim of the schooner *Mariner* spotted a floating bottle and a long oar brought it alongside and aboard. It contained whiskey, and when the whiskey was spilled on the deck, the crew busied themselves in having fun setting it on fire as it ran along the planking. The captain then discovered another bottle drifting near. Again a long oar went out, and the bottle was gradually hauled in. It was a plain bottle that had been in the water for a long time. A thick coating of sea grass covered the sides. Looking into it they found, not whiskey as they expected, but a carefully folded note. They broke the bottle and unwrapped the paper. On it, written in pale blue ink, was the following: *"We are sinking; sprung a leak. Two men washed away. Cable parted; rudder gone. All hands give up. May God have mercy on us. Who picks this up, make it known. Good bye. Sinking, schooner Falcon, on Georges Bank."*

There was no date, nor were the names of the crew listed. The bottle was found 25 miles from where the *Falcon* had been fishing. The *Falcon* had a crew of 12 and it is only right that we make known who they were, for they surely wished to be remembered:

Captain Andrew Nelson, 50, 101 Mt. Pleasant Ave., Gloucester, left a wife and four children.
Frank P. Bedmund, cook, Rocky Neck, Gloucester, left a wife and five children.
Jabez Smith, 50, Port Medway, N.S., left a wife and two children.
Thomas Murphy, 44, 78 East Main St., Gloucester, left a wife and three children.
Albert Barter, 40, Islejau, Me., single.
William Brown, 65, single.
Nelson Jacobson, 45, Sweden, single.
Cornelius Hutt, 30, single.
Patrick Ryan, 40, single.
Henry Thomas, no information.
Martin Mattison, 36, Norway, single.
William Brown, no information.

[†]Porper came to Gloucester from Manchester, Nova Scotia, in 1873 at the age of 14. He first sailed under Captain Henry Matheson on the *Mary E*, moved to the *Frederick Gerring Jr.* under Captain Edward Morris (who was highline of the 1874 halibut fleet), and then in 1880 assumed command of the *Gerring* itself. He lost his next command, the *Bessie M. Wells*, on Seal Island in 1885, and the crew spent seven days marooned on the island. It was only when Porper moved to the *Gladstone* that he cemented his reputation for landing record fare after record fare.

[§]Like Porper, Spinney came to Gloucester from Nova Scotia, from Argyle, in 1881 at the age of 15. Eleven years later he had his first command, the *Mabel W. Woolford*, followed by the *Gladiator*, a slightly larger vessel. A decade later, Captain Spinney became a Director of the Master Mariners' Association and by that time was recognized as one of the city's up-and-coming halibuting and dory trawling skippers. He made his lasting reputation as skipper of the *John Hays Hammond*, a vessel on which many of the men from Gullholmen sailed, including the Lund brothers and Steve Olsson.

part of his craft's head. The schooner *American* reported having "a novel way of renewing her supply of water, for whenever the butts were getting low, a fresh supply was secured from the icebergs, which were too plentiful." Yet, despite the weather, the flitched halibuters cruised to the edge of the ice floes and had a good fishery in the waning years of the 1890s. The number of craft was small, only about a dozen in 1898, but they brought in a consistent catch.

CHAPTER 14

Gloucester's Move into the Fresh-Haddock Fishery

Gloucester had been principally a salt-cod port, reaching its height in 1892, when 49.8 million pounds of salt cod came through the port.[*] At the same time, many Gloucester vessels landed fresh fish, mostly haddock, in Boston, where there was an active market economy operating out of T-Wharf. Through much of this period, Gloucester's primary fresh-fish species had been halibut, but at levels much lower than the fresh cod and haddock brought into Boston. T-Wharf merchants were largely responsible for the sale and distribution of fresh fish, while Gloucester's fleet was largely responsible for the catch. In any year, 50 or more Gloucester vessels brought fresh fish into the Boston market over the winter and into the spring. The catch was substantial, and for many Gloucester fishermen this was their primary focus.[†]

As the decade opened, some in the Gloucester community began to agitate for the fishery to move east to Cape Ann; but the call was not from the fishermen, it came instead from the newspapers and a few of the owners. Early in the decade there was no real urgency behind the proposal; the waterfront bristled with business from Rocky Neck to Harbor Cove. Skippers such as Clayton Morrissey[§] brought in huge fares of salt cod and were well paid

---

[*]Forty-five percent came from the Grand Banks, 35 percent from Georges, and the rest from the Scotian Banks and other nearby grounds.

[†]The Gloucester haddock fishery goes back to the beginning of the dory era. In 1866 the haddock fishermen moved to 40- to 60-ton craft, working some 25 or more miles off Cape Ann, on Jeffreys, Tillies, and Middle banks. Seven years later, the schooner *Eastern Queen* was to make the first dory haddocking trip to Georges, and within a few years Gloucester had a haddock fleet. By the winter of 1878–1879, with the disappearance of the mackerel, 150 craft were outfitted for the fresh-haddock fishery. They fished on Georges and La Have banks, leaving in early November on trips of 10 or more days at a time, bringing in catches of 75,000 to 90,000 pounds.

[§]Clayton Morrissey first commanded the *Effie M. Morrissey*, one of 18 vessels outfitted by the John F. Wonson Fish Company from its East Gloucester wharf. The *Effie M. Morrissey* was later sold to Captain Bob Bartlett, who sailed her on trips to the Arctic; she is still afloat today, sailing out of New Bedford. Clayton was born in Lower East Pubnico, Yarmouth County, Nova Scotia, in 1873. His forebears had followed the sea, and for years his father, Captain William Morrissey, skippered Gloucester schooners, coming south in the spring to command Gloucester vessels in the mid-year salt-bank fishery, ending the year in the Newfoundland herring trade. Clayton first worked with his father as a boy of 13. By age 16, he went out as a doryman on the Grand Bank; and in 1892, when he was only 19, he rose to command the *Effie M. Morrissey* when sickness forced his father to step down after the first baiting. Clayton's success was immediate: A driven salt-banker, while other skippers made two trips during the year, Clayton often made three. He skippered the *Morrissey* for two years, moving in 1896 to the *Procyon*, and in 1897 to the *Joseph Rowe*.

for their work. The salt fishermen began their season in early February, sailing to the far northern grounds, their holds filled with salt (about 400 hogsheads), their icehouses brimming with frozen herring, and the larder filled with enough food to sustain the men for three to four months.* Depending on the weather and the flow of ice, most returned from their first trip in June, landing 200,000- to 300,000-pound fares of salt cod. By mid-July, upwards of 10 million pounds sat out on Gloucester's fish flakes or soaked in the brine barrels. Salt cod was everywhere: A visiting journalist said that having "smelled the Gloucester fish smell you are ready to believe that five out of every five [cod] fish-balls are made in Gloucester. With that fragrance in your nostrils you are convinced not only that Gloucester is the most important fishing town on the Atlantic coast, but that it is the only one—anywhere." A whimsical summer boarder added that "every part of the cod is used except the smell."[33]

In Gloucester's view, there was no reason for the American consumer to stop buying salt cod. It was a pure, wholesome product. However, in 1893, as the national economy began to unravel, local outfitters sent fewer schooners to the northern grounds, and the cod catch dropped by 8 million pounds. They adopted a risk-avoidance strategy, preferring to tie up their capital in support of shorter trips to Georges for fresh haddock.

Then, in 1895, as the national economic picture worsened even further, Gloucester began to experience an unprecedented decline in its fleet and salt-cod landings. Now everyone, skipper and processor, could see an uncertain future. The salt-cod fishery seemed vulnerable; the catch was falling. The fresh-fish market was expanding, tastes had changed, and while it was still difficult to ship iced haddock over long distances, the community could see the need of moving the fresh-fish business from Boston to Gloucester. Gloucester vessels supplied the lion's share of the fresh fish to the Boston market—over 40 million pounds per year. Schooners sailed to Georges, set and pulled trawls until they had 15,000 to 40,000 pounds of iced haddock, hake, and cusk in the hold, and then piled on sail in a race for Boston; they hoped to catch the market when stocks were low and prices high. Fresh-fish prices fluctuated widely and all tried to catch the market when stocks were low and prices high. Of the many fresh-haddock fishermen, Arichat native William H. Thomas was the unquestioned leader of the fleet. A perennial highliner, he called Gloucester home but sold his fish in Boston.

---

*The schooners carried up to 22 men and went to sea with a great deal of food: 25 barrels of flour, 2 barrels of salt pork, 10 barrels of salt beef, 6 cases of canned fresh beef, 3 cases of condensed milk, 150 pounds of dried apples, 50 cans of tomatoes, 50 cans of corn, 1,100 pounds of sugar, 60 pounds of tea, 40 pounds of coffee, cases of eggs, prunes and other fruit, and so much more, including 6½ tons of coal and a cord of wood.

)(*Processing fish on the dock. In the foreground, an idle fish flake stands propped up against the wall.*

In 1896, the fresh-fish season opened full of promise, and although the early catch was light, demand remained strong and prices held at a most acceptable level. For Captain Thomas, the only event of note in early 1896 was a rogue wave that washed over the *Parker*, taking Samuel Landry to his death. Captain Thomas and his men raised $300 to help the Landry family.

The loss saddened Captain Thomas, but it was soon overshadowed by another much closer to him—the death of his 9-year-old daughter, Lottie. It was mid-July 1896, Captain Thomas was at sea, and his friend and business partner Captain John Chisholm was driving six children down Elm Street in a democratic wagon (a two-seat flatbed wagon, without a top, drawn by a single horse). At the same time, a Long Beach electric car was traveling westward up Main Street, heading for the exchange point at the Gloucester waiting station. The motorman rang his bell as the horse appeared at the corner of Main and Elm. He pulled on the lever to set his brake, but to no avail. The electric car entered the middle of the intersection and Captain Chisholm could not stop the democratic. The car struck the horse and the front wheel of the democratic, pushing the wagon sideways up the street. Lottie, who had been sitting on the front seat of the wagon, fell under the iron fender of the electric car. When the men got to her side, they found her throat cut, a long gash penetrating to the jawbone, and within the day she died. Captain Chisholm was prostrated by the accident, and confined to his home; the fault lay with him. Captain Thomas would only learn of the tragedy when he returned to Boston to unload his catch, days after the accident.

The Thomas family held Lottie's funeral from their home, after which Captain Thomas went right back to sea; life went on whether one lost a crewman or a daughter. The fresh fishermen went to sea to make a living, to support their families; they had no other option. Gloucester may have had a huge fleet, but each went to sea on its own. Thus it was that on July 27, the *Parker* returned to Boston with 40,000 pounds of cod, 30,000 pounds of hake, and 8,000 pounds of cusk. One month later, the *Parker* arrived at Boston from Middle Ground with a similarly fine fare, the crew bringing in the catch on four sets. Captain Thomas then went north to Rose Blanche and secured a fine baiting of squid. As a haddock fisherman, he took the vessel where he could find fish, and sailed until the end of the season in November, joining other craft in landing their fares at T-Wharf. There could be no question that the fresh fishery was thriving—the Gloucester fleet brought 40 million pounds of fresh fish into Boston in 1896.

It was an attractive business, and as the economy continued to sour, Gloucester was at last ready to try to capture it. For the fishermen, Boston prices fluctuated during the course of the year, and late winter prices were particularly troubling: They needed price stability. Skippers were willing to listen to plans to move the fishery from Boston to Gloucester. Higher, stable prices would be key. For the community, the only question was whether the city could pull itself together to carry off such a plan. The captains would have to be convinced that it would be to their advantage to land their fares in Gloucester. If it was to happen, Gloucester had to find a way to buy the 40 million pounds of fresh fish the fleet then landed at Boston, and the prices would have to be higher than those offered in Boston—difficult requirements at a time of national depression.

Gloucester's business community soon geared up to lead the fight. Merchants, bankers, and fish processors agreed to take stock in a new fresh-fish company; with control over the fleet they assumed they could capture the market and make a profit. There would be no competition, Gloucester would have a single firm, and that one firm would now have to do better than all of the T-Wharf buyers—a daunting challenge. The die was cast, and only time would tell whether the game could be won. The first two moves seemed encouraging; they raised $75,000 for the stock company, and almost all the vessels in Gloucester's fresh-fish fleet agreed, under the penalty of a fine, to land their catch in Gloucester. The community had come together. In a time of economic uncertainty, all were willing to participate in a new experiment; but the commitment was to a single new company, not one or more of the city's many established firms. There was to be a two-year trial, and the new Gloucester Fresh Fish Company was at the heart of the experiment. Its president, William H. Jordan, had 10 years' experience in outfitting vessels and processing fish. By the end of November, the firm had leased three wharves. Boxes were made, floors put down, icehouses built; scales were calibrated, knives sharpened, and the hoist tackle checked out: Gloucester was tooled up to land, process, ice, box, and send out fresh fish.

On Saturday December 4, teams brought ice from the Fernwood Lake Company, and workers stowed it for the start of business Monday. On Sunday, the streams of people visiting Low's wharf found three vessels waiting for the Monday opening—the *Ralph E. Eaton* under Captain Joseph McClelland, the *Horace B. Parker* under Captain Billy Thomas, and the *A. S. Caswell* under Captain Antonee Courant. At the Atlantic wharf, the *Minerva*, under Captain Charles Ellsworth, and the *Doretha M. Miller*, under Captain James Thompson, had waited since Saturday. They hoped to be the first to off-load at the new company on Monday morning. Their T-Wharf experience had taught them that the first in got the best price. On this day, the honor of being first went to the *Horace B. Parker*: She arrived early on Sunday and hauled in at the head of the wharf, ready to unload Monday morning.

On that first morning, and throughout the day, crowds thronged the buildings and wharves from one end to the other. They hampered workmen, but everyone was good-natured. As the day wore on the community began to celebrate. Flags flew about the city. Stores flew the national colors and nearly every vessel in port had its colors flying. In the evening a spontaneous parade wound through the city. People went wild with joy. "Fresh fish" became the rallying cry of the evening, and drew forth continuous cheers and hurrahs. For one evening, at least, fresh fish had become king in Gloucester. Of course, the real issue was jobs and wages: jobs for the shore crews and stable wages for the fishermen.

So ended the first day. It had been a good day, and in the weeks to come there would be many more. On December 8, 4 schooners with 75,000 pounds of fresh fish landed at Low's wharf; in Boston 12 vessels landed 125,000 pounds. However, 2 of the craft in Boston were Gloucester vessels that had not yet received notice of the new arrangement. They knew of the plans in Gloucester, but they did not know the Fresh Fish Company had opened for

)(*Newly landed fish have been taken out of barrels to be sorted on wharfside tables at the Reed & Gamage Company wharf in East Gloucester; in the foreground some already processed fish dry on a flake.*

business. On the following day, 7 fresh-fish trips landed at the wharves of the Gloucester Fresh Fish Company, totaling 180,000 pounds. In Boston, 10 craft landed 60,000 pounds, less than half the usual amount.

Boston dealers pretended to be indifferent to the success of the Gloucester Fresh Fish Company, but they knew they were in for a fight. Several Boston firms threatened to build a fleet of their own if Gloucester vessels refused to sell to them. At the time, Boston had only 42 fishing craft, including 14 small boats in the 5- to 11-ton range, and they would eventually make good on the threat.

Even as this action began, Gloucester skippers made it clear that they were driven by neither civic pride nor a desire to provide jobs for the men who worked Gloucester wharves. Rather, they entered the agreement to ensure a fair return for their catch. Money was key, and by late December, the *Provincetown Advocate* ominously noted that prices at Boston and Gloucester had been remarkably strong owing to the sharp competition. T-Wharf dealers were offering higher prices than Gloucester in their endeavor to induce vessels to continue to land at Boston. But in the second week of the company's operation, Gloucester held her own.

By the end of the first month, only one of the vessels committed to the Gloucester combine sold to a Boston firm, the schooner *Two Forty*. In the fourth week, Gloucester landings came in at 1,062,000 pounds, while Boston came in at 667,000 pounds. In the following week, the Gloucester Fresh Fish Company opened a branch called the John P. Locker Company on Commercial Wharf, Boston. to help keep up with the volume of trade. The move incensed the T-Wharf dealers, and more than any other action it ushered in a period of active confrontation. Boston was not about to see its generation-old business so easily lost to the dealers in Gloucester. Boston firms banned Locker from T-Wharf and extended a general embargo against all Gloucester firms buying fresh fish from the wharf.

But it was not the Boston merchants who would cause the eventual failure of Gloucester's fresh-fish venture; rather it was the community's belief that the business could be carried by just one company. Skippers believed the company would pay a reasonable price even in the face of increasing supplies. They were naïve, and their faith was shaken to the core on the first fish glut day of the new year. Seven vessels landed 400,000 pounds of fresh fish in Gloucester. In Boston, nine vessels landed 253,000 pounds. In both ports, the price for haddock dropped to 80 cents per hundredweight—a losing price for the skippers. The Fresh Fish Company let the market dictate the prices. There would be no premium to the skippers for landing their catch in Gloucester. It seemed good business practice, but the skippers felt betrayed. If the company would not be loyal to them, why should they do anything other than act in their own self-interest? Gloucester's grand plan began to unravel, and as it did Gloucester's one chance to surpass Boston in the all-important fresh fishery of the twentieth century died with it. The fishing community was used to working in a "company town," with local interests controlling all aspects of the business: Local businessmen set prices, owned the wharves and processing plants, owned many of the schooners, and controlled the national distribution of their product. But now the decisions of individual skippers and a competitive T-Wharf market changed the future. Captain John Greenlaw of the schooner *Annie Greenlaw* sold his next catch in Boston since the T-Wharf dealers out priced the Gloucester firm. In open violation of the agreement, the *Greenlaw* was liable for a fine of $200; yet, the Fresh Fish Company stepped back from imposing it. Captain Greenlaw saw it as an empty gesture, and he would have gladly paid the fine. He said it would have been more than offset by the higher prices offered in Boston. First one and then a trickle of other skippers took their catches to Boston, and the Gloucester Fresh Fish Company failed to achieve a commanding position in the trade. Boston had held her own.

By October more fresh fish came into Boston than Gloucester—5.5 million pounds versus 3.6 million pounds. It was not the way it was supposed to be. What had begun as a project grounded in civic pride and economic necessity, now seemed governed by issues of price, embargo, and personal bickering.

Within less than a week, on January 11, 1898, the depth of the problem came home to the community when the *Horace B. Parker*, under Captain William H. Thomas, landed fresh fish at T-Wharf. She had been one of the many Gloucester craft under agreement to land all her fares at the Gloucester Fresh Fish Company; Captain John Chisholm, her part owner, had helped spearhead the formation of the Fresh Fish Company and had pledged his fleet (some 10 vessels) to the effort. But now the crown jewel of that fleet had landed her fish in Boston. Something had to be done: They could meet the competition head-on, or retaliate against the skippers. Retaliation became their response; the company sued John F. Wonson & Company and Robert M. Brown Company, the next two firms with vessels landing in Boston, for violating the agreement. Fines were $200 for each violation, and the Wonson Company posted a bond to secure immediate release of their vessels. So much for civic pride.

If the Gloucester Fresh Fish Company thought their new aggressive stance would alter the behavior of the skippers, they soon found out just how independent these men could be. In January, fresh-fish landings at Gloucester came in at 2.5 million pounds, against 3.3 million pounds in Boston. By year's end, both Boston and Gloucester had received substantial landings of fresh fish—53.4 million pounds in Gloucester, exclusive of halibut, and 55.4 million pounds in Boston, inclusive of halibut. The *Horace B. Parker* had the distinction of being highliner of the haddocking fleet. Captain William H. Thomas stocked $25,441, with each crewman receiving the remarkably high share of $760. Thomas sold all of his winter haddock and other fresh fish in Boston; he had abandoned the Gloucester Fresh Fish Company.

In one sense, the two-year effort had reversed the dismal situation of 1897. Gloucester's total fish landings grew from 75 million pounds in 1897, to 104 million pounds in 1898, and to 121 million pounds in 1899. But as the new year opened, landings of fresh fish in Gloucester fell even further behind those at T-Wharf: For the week ending March 18, 2.6 million pounds came into Boston, against only 628,000 pounds in Gloucester. On April 2, the weekly total in Gloucester was 892,000 pounds, while in Boston it was 1.1 million pounds.

In an April letter, one stockholder of the Gloucester Fresh Fish Company, a man who had invested $500, stated the obvious: "Boston proved too much for us and many of the owners and skippers were prevailed on to carry their trips to that port instead of marketing them here at home."

As for the company's owners, in the fall they made a commitment to continue the business, but no one said how it would run. They hoped to secure vessel owner signatures on a new agreement, but could provide no rational reason for why a vessel owner would sign. In their view, "Gloucester should market every possible pound of her catch of fish at the home port." The company felt they had been handicapped in getting enough fish to meet the demand and naïvely hoped the vessel owners would decide to return to Gloucester. However, by year's end, at least one-third of the 60 million pounds of fresh fish landed at Boston had come in Gloucester vessels. Boston would be the fresh-fish capital of America. Gloucester would have a role to play, but the T-Wharf dealers had survived and their business would flourish in the years to come.

Looking back, Gloucester business interests had committed themselves to strategies that had worked in the past: They focused on a common threat and organized to create a monopoly. But in adopting this approach, they failed to account for the response of the T-Wharf buyers as their livelihood was challenged, and they considered themselves blindsided by the response of Gloucester fishermen when the company failed to sustain reasonable price levels. One can only wonder what might have happened had two or more of the city's strong processing firms joined the fray, or if the price floors had been maintained in that first year, or if the company had been better capitalized. The men of the Gloucester fleet were willing to give the new venture a chance, but the company failed the fishermen. They faced a volatile market, with wide price fluctuations. If the company had kept its promise, the fishermen would have kept theirs, but instead they were forced to choose between community interests and their own livelihood. It was as simple as that. The Fresh Fish Company was undercapitalized, it could not respond when the market fell—even if it had wanted to—and one must conclude that this half-hearted effort had little or no chance of success. The failure would have ramifications in the years to come.

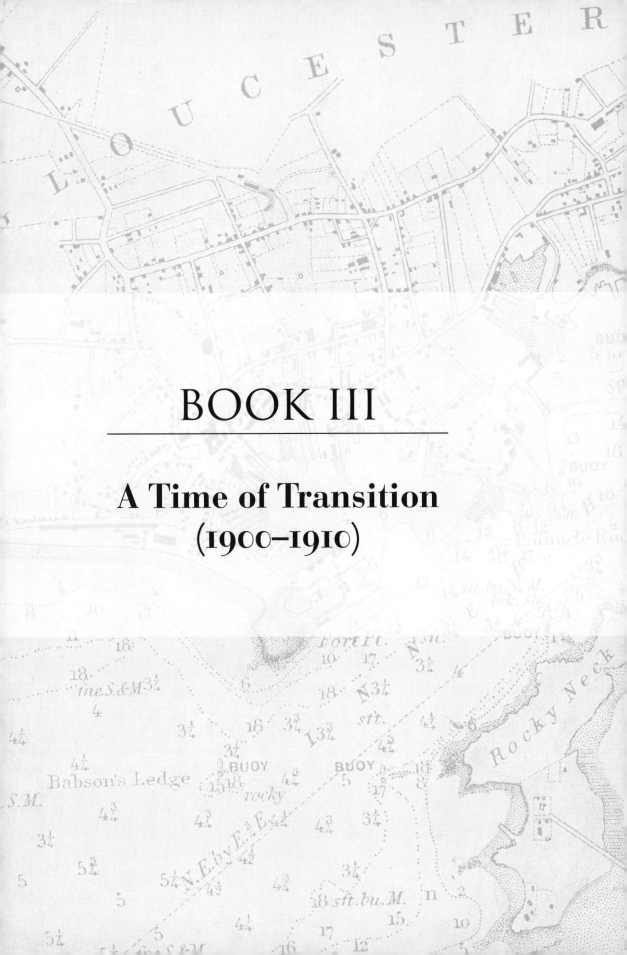

BOOK III

A Time of Transition
(1900–1910)

In 1900 something unexpected happened: The mackerel catch rose fourfold to 16.1 million pounds, followed in 1901 by an almost equally impressive catch of 13.6 million pounds. People were ecstatic. A sense of optimism swept along the waterfront, and the fishing community soon forgot the troubling days of the late 1890s. Outfitters ordered new vessels, gasoline-powered auxiliary schooners entered the fleet, and fishermen reverted to their old individualistic ways. Collective action became a relic of the past as the people of Gloucester refocused on old concerns—making a living, getting ahead, and worrying about the drinking habits of the immigrant fishermen.

An almost completed schooner at the James yard in Essex is prepared for the move to Gloucester in the early 1900s; another is being planked up on the shore.

Optimism abounded once more, and a fanciful 1902 account from the local newspaper captured the people's pride in the achievement of their fleet and the magnitude of the local fisheries. The writer asked readers to imagine the fleet's annual catch put up in 500-pound boxes, as was generally the case with fresh fish. He then said to imagine all the boxed catch ready to go to market on the same day. In his calculation there would be 292,750 boxes of fish ready to move from the packing houses to the depot. Calling 10 boxes a wagonload and bringing along the two-horse jiggers (or wagons) then used to carry the fish from the wharves to the trains, he said that 29,275 jiggers would be needed to bring the fish to the depot.

He next asked his readers to imagine themselves looking out from the window of the Board of Trade office on Pleasant Street, where prices were set for the cod and halibut landed in Gloucester. Glancing up the street one would soon see a

> stream of jiggers coming up and around the post office out of Main Street, from the eastward. Here they come, passing by the window at the rate of five a minute. It is just 8 o'clock of a Monday morning as the first one heaves in sight and you bite the end of a cigar and smoke it. Then you get hungry and go home for dinner. When you come back, they are still coming, and no sign of the end. You sit down and watch, then at evening go home to bed—but the jiggers keep moving just the same. You repeat your performance of Monday on Tuesday and on Wednesday, and Thursday you do the same. Remember there is no halt to the string of jiggers; no breakfast, dinner, or supper hours, no time to sleep, but a steady march 24 hours out of 24. Friday morning you once more slip into a seat and again you smoke and wonder. Finally, just at 10 o'clock, the last jigger goes by and you have a chance to cross over to the post office and get your orders from the west which came on the 9:28 train.

As another upbeat wag said at the time, there were enough fish to "provide dinners for every man, woman, and child in Massachusetts for over two months and there would still be fragments enough for the family cats and other domestic pets." The future looked bright. Outfitters confidently replaced lost vessels with newly commissioned schooners—about 25 new schooners each year. Gloucester may have failed in its attempt to take on the entrenched brokers at T-Wharf, but the community felt comfortable in reverting to the successful fishing pattern of days gone by. Salt cod, mackerel, and halibut were flowing through Gloucester, even if much of the fresh fish was heading to Boston.

CHAPTER 15

Stabilization of the Fleet and the Coming of the Auxiliaries

At the start of the 1890s, Gloucester had 383 schooners; by the end of the decade there were only 254. Those lost at sea had not been replaced, the shipyards had gone quiet, and owners sold to parties beyond Cape Ann. As many as 2,000 fishermen had lost their jobs, the fresh-fish experiment had failed, and there seemed little reason to hope. Then the mackerel returned, and early in the first decade of the new century, each year saw the commissioning of about 20 traditional sailing schooners, plus two or more of a new class of auxiliary-powered vessels.

In 1900, the first auxiliary schooner, *Helen Miller Gould*, was commissioned by Captain Solomon Jacobs; her arrival set in motion a process that would transform the fleet in less than 20 years. The new vessels had two types of propulsion: a full set of traditional fore-and-aft sails, and a gasoline-powered propeller-driven engine.

The *Gould* was a large, traditional schooner, with a 108-foot keel, a 25-foot 4-inch beam, and a gross displacement of 149.64 tons. She had conventional sails, with an 80-foot mainmast and 75-foot foremast, and a 130-horsepower auxiliary gasoline engine—although, when launched, she went out with a much smaller 35-horsepower engine.

Outwardly nothing about the *Gould* suggested that she was anything other than a conventionally designed sailing schooner. Her designer had merely dropped the gasoline engine into a space in the stern, known as the run of the vessel, normally used for refuse storage, and placed the gasoline tanks beneath the floorboards. The *Gould* was truly a sailing schooner with an auxiliary gasoline engine for use in specific situations.

From the beginning, the auxiliary-powered schooners made money on the mackerel grounds. The skipper of a seining vessel now quickly approached the fish, set his nets, and encircled the mackerel from the traditional seine-boat and dory, then steamed back to port regardless of the wind conditions. It was a decisive advantage. Within the year, other skippers joined Captain Sol in placing orders for auxiliaries. But the early powered vessels had seri-

)(*The auxiliary schooner* Oregon *glides past Duncan's Point wharves in Gloucester's inner harbor, pre-1920. The Fitz Henry Lane house is at far left rear.*

ous problems, including fires, explosions, and breakdowns. It would take a decade for the designers and the fishermen to work out the kinks and ensure that fishermen could go safely to sea in an auxiliary-powered schooner. As for Captain Jacobs, while he had moved alone into uncharted waters, the rewards for his pioneering effort would be great. To fully understand his decision it is informative to go back to 1899, one year before he commissioned the *Gould*. Mackerel were in scarce supply off the North American coast, and Captain Jacobs sailed his seiner, the *Ethel B. Jacobs*, to Ireland's southwest coast in search of fish. Never satisfied to do the ordinary, Jacobs could be counted on to do the daring, the unexpected. He left early in the summer, and by September the local paper was heralding his success. It was said that he had taken 270 barrels of mackerel at a single set of the seine and had already shipped 800 barrels (representing 160,000 pounds of fish) back to Gloucester.

But the first reports were illusory; there would be no mass exodus of Gloucester mackerelers to Ireland. Early in October, Captain Jacobs returned unexpectedly, coming by transatlantic steamer to care for his sick wife. He spoke of a season of unfulfilled promise. Yes, the mackerel had been plentiful, but they schooled close to shore, in the rocks, and for a summer's work he had caught only 353 barrels. None of the fish had yet been sent back to Gloucester, and he said it would have been better had he never sailed to Ireland.

But this was not the end of it, for the story soon took a tragic turn. Word reached Gloucester that the *Jacobs* had been lost with all her gear and the full cargo of mackerel taken over the summer. She had gone aground on the rocks, in a thick fog, on Abbey Island, Darrynane, the waves lifting her high up onto their crests and tossing her down with a maddening crunching of her timbers. As she broke up, the men scrambled onto the rocks and into their dories, even as the incoming tide washed her clear of the rocks and into the roiling sea.

Nevertheless, Captain Jacobs decided to remain in seining—he was not concerned with the scarcity of mackerel off the North American coast. He used the insurance money to commission the *Helen Miller Gould*, named after a patriotic New York philanthropist and daughter of one of America's richest capitalists. The *Gould*'s cost, about $20,000, exceeded that of any other Gloucester schooner ever launched, and yet Captain Jacobs never questioned his decision. He believed that auxiliary power would give him a great advantage once on the grounds. On a windless day, when other skippers were sitting becalmed, he would be able to engage the engine and bring in a catch. It was not a complicated vision, and it resonated with other skippers, but in 1900 only Captain Jacobs acted on his instincts. It was another of his daring decisions. No one else would even dream of stepping forward to invest so enormous a sum in the still-moribund mackerel fishery; it was his genius and good fortune to be so often ahead of others in the community.

The launch of the *Gould* at Gloucester's Bishop shipyard in March 1900 was a true community event. More than 3,000 people came out to cheer; they stood on open wharves, lumber piles, barrels, shed roofs, and buildings. Sixty people found themselves standing on the rickety roof of an old oil storage shed and would not listen to the foreman's pleas to come down, for it surely could not stand the weight. Soon the sound of hammers and axes signaled that the launching was imminent, and as the last blocks were knocked away, the *Helen Miller Gould* started on the short journey down the ways. In an instant she was floating on the smooth waters of the cove, the cheers of the people filling the air. And as the people on top of the shed pushed and shoved to get a better view, the old roof gave way. All 60 of them came crashing onto the oil barrels below, but their injuries were limited to a few scratches and bruises. On the *Gould*, no one noticed: All eyes were on Miss Jacobs, the skipper's daughter, as she broke a bottle of wine over the side of the new auxiliary.

Two weeks later, on April 11, the *Gould* made a trial trip to Boston, leaving harbor under motor power, even as the crew put up the sails. Off Ten Pound Island, with the canvas trimmed flat, she moved south and then west in a light southwest breeze. On the next day, following her successful trial run, she sailed south on her maiden seining trip, and a mere twelve days later Captain Jacobs landed 200 barrels of medium mackerel at New York's Fulton Market. He had the first mackerel catch of the season. As he entered New York Harbor, men on the nearby wharves were said to have looked on in wonder as the *Gould* lost no headway even as the crew lowered the canvas. As the little engine kicked in, the propeller started to churn and the *Gould* glided to the dock. Captain Jacobs was proud of his new schooner; she had shown her mettle on the grounds, the men having made hauls in the dead calm of night when other schooners could only sit and watch.

When Jacobs finally returned to Gloucester on June 12, he expressed pleasure with the sailing qualities and economic prowess of his new vessel. He remained in port a few extra days to have the originally specified larger gasoline engine installed, and he then headed out to the Cape Shore.

Captain Jacobs was now running neck and neck with Captain Charles Harty of the schooner *Marguerite Haskins* for highline of the southern fleet. The *Gould* had stocked $10,997—the largest on record in spring and early summer southern seining fishery. But the record held for only a matter of days, for Captain Harty also had a fine southern season. His craft stocked $12,104, and with this report there was a buzz about the waterfront—the mackerel had returned. Skippers were setting new records, and by the end of the fall season the *Gould* shattered all seining records for the port. Her final stock came in at $40,664. No one else was even close, and with her success many other skippers got the message. Five soon placed orders for new auxiliary schooners, and by mid-December the Essex shipyards were humming. Nine vessels sat in various stages of construction, some with auxiliary power, but many more without. The return of the mackerel and the success of the *Gould* had revitalized the people of Gloucester.

Over the winter, as the men at the Essex yards worked on the next generation of Gloucester schooners, the *Gould* sailed north to the Bay of Islands for herring. She returned in early January with the first cargo of the winter season, 850 barrels of frozen herring and 150 barrels of salt herring, making the return trip of about 1,000 miles in only 97½ hours. It was the end of the first year's work of a Gloucester auxiliary, and what a year it had been. Gloucester would never be the same.

The following spring, Captain Jacobs began once again to fit out the *Gould* for the trip south, to seine for mackerel off the coast of Virginia and Maryland. The smell of boiling tar filled the air as the men coated their nets. The fleet had expanded, and by year's end, close to 200 schooners would go out after mackerel, almost double the number sent out in 1900. In early

)X(*A schooner is planked up—sides and decks—while thick snow lies on the ground. It is hard to imagine such a fine vessel being built under such conditions today.*

March, the second auxiliary, Captain Charles H. Harty's *Mary E. Harty,* was launched. Five hundred people came to the Bishop shipyard to see her go; they looked on as she ingloriously hung up on the bottom of the cove. But she was soon free of the mud, rescued by the tug *Eveleth.*

On her trial trip, the *Harty* sped effortlessly about Massachusetts Bay, her engine meeting all expectations for speed and endurance. As March ended, a third gasoline auxiliary schooner, the *Victor,* came off the ways at the James & Tarr yard in Essex. Even as a tug towed her to Gloucester to be fit out, the *Gould* was sailing south to open the southern mackerel season. On April 9, Captain Jacobs landed 35 barrels of large fresh mackerel at Fortress Monroe, Virginia, and one week later the *Victor* arrived at Newport after a successful 10-hour test run using her gasoline engine. The crew had gone to the engine room once an hour to check on the bearings and oil cups, but the engine had worked perfectly.

By late May, with the southern season drawing to a close, the *Gould* returned to Gloucester as highline of the southern fleet, having stocked $9,600. But although the fishing had been fine, the engine had given Cap-

)((*A new schooner is framed up in Essex, 1905.*

tain Jacobs trouble throughout the season, and he now made plans to replace the engine later in the year. For now, he had the existing one repaired and fished locally. On June 20, Captain Jacobs came into Boston with the largest mackerel fare of the season: 200 barrels of salt mackerel and 400 barrels of fresh mackerel. The fish were taken two days earlier, 60 miles south of the No Man's Land. Many of the fleet were there and got fish, but the *Gould* was the leader. As the summer unfolded, however, the engine continued to act up, and the *Gould* was laid up for repairs during a portion of the best part of the season.

By early fall she was back to sea, sailing to the Maine coast and the Cape Shore, Captain Jacobs making port at North Sydney, Cape Breton. The *Gould*'s stay was brief, and within a matter of days the crew set the sails and powered up the engine to make for the offshore grounds. But as the *Gould* slowly pulled into the harbor, gasoline had begun to leak unseen from the fittings connecting the tank to the engine. It puddled on the floor near the engine, and for a reason never explained, caught on fire. Jacobs's first inkling of the problem was when black smoke began to pour out of the cabin and engine room. The fire was spreading, and as the flames grew, the men below could not get on top of it. The engine room became an inferno, and as the men streamed on deck they felt lucky to have escaped the spreading flames.

They quickly made for the dories and withdrew to a safe distance, watching as rising flames ate away at the schooner. Fire leaped high into the night sky, burning sails flapping in the wind and dense black smoke curling skyward.

Within hours the *Gould* was gone. In the light of dawn only a mass of charred embers remained. It had been a short career, yet praise poured forth from the seafaring community. *Motor Review* magazine came closest to the point, saying that the success of the *Gould* had proved "beyond question the value of auxiliary power in the fishing fleet." Yet with the *Gould*'s loss to fire, and with the loss of several other auxiliaries over the next few years because of fuel leaks and downtime due to faulty engines, orders waned. Gasoline engines were temperamental; not all the problems of modifying them for marine use had been worked out, particularly propeller shaft installations. Poor alignment, shaft vibrations, and wobble in some of the early installations destroyed otherwise perfectly good engines in short order. Only the mackerelers and swordfishermen had an obvious need for auxiliary-powered craft, needing to close quickly on fast-moving fish, and even those skippers tempered their enthusiasm as expensive auxiliaries went up in flames. Yet, as each auxiliary entered the fleet it succeeded as a fisherman, and the designers slowly became more adept in fitting the engines, laying out the drive trains, and installing the propellers. In the next decade, with the arrival of safer, less expensive marine diesel-powered engines, auxiliaries became commonplace.[34] For the next 20 years, dual-powered, conventional schooners would comprise the lion's share of the Gloucester fleet.

As for Captain Jacobs, he received a $17,000 insurance payment for the *Gould* and within three months had commissioned a new vessel for the upcoming mackerel season. But it would not be another gasoline auxiliary or a conventional sailing schooner. For his new craft, Captain Jacobs all but abandoned sails as a means of propulsion, going instead to a steam-powered vessel—the harbinger of the fully powered vessels that would make up the Gloucester fleet many years later. Once again, Jacobs was ahead of his time. As before, however, while other skippers looked on with interest, the design was not then copied. One can only assume that the novelty of the experiment and the excessive cost of a steam-powered vessel, plus the available option of commissioning an auxiliary, kept others on the sidelines.

On March 17, 1902, Captain Jacobs launched his new coal-fired steamer, the *Alice M. Jacobs*. She looked like nothing that had come before. She was big—141 feet long, with two pole masts, no spars—and was said to resemble an oceangoing tug. Her sails consisted of only trysails, brailed (connected by ropes) to the masts. She had no need for spars, booms, gaffs, topmast, or bowsprit. For propulsion, she had two Alma steam boilers and a 450-horsepower engine. One boiler was used when on the grounds; two came on line to help her reach her top speed of 11 to 12 knots when she went back to market. The bunkers of the *Jacobs* held 100 tons of coal and 500 gallons of water, the steam from the boilers breeching skyward through a single smokestack rising 25 feet over the deck.

By the end of May, Captain Jacobs was not highline of the southern fleet. Two other skippers sat before him as head of the fleet, both auxiliaries. In second place was the new gasoline auxiliary schooner *Saladin*, under Captain Flar McKown, while highline honors went to Captain Thaddeus Morgan on the gasoline auxiliary *Constellation*. The auxiliaries were continuing to prove their worth, and when a reporter asked a skipper what he thought of them, he replied that "next year he would go 'hand' in one rather than be skipper of a sailing seiner." But they continued to be touchy craft, the *Saladin* having to sail back to Gloucester, her crankshaft broken.

By late September, Captain Jacobs said he was well pleased with the new steamer. In four months at sea, she had used only 60 tons of coal. By early October she came near breaking the record for a single season's mackerel stock, passing $41,000. But that would be it; Captain Jacobs converted the *Alice M. Jacobs* for the Newfoundland herring fishery.*

Nevertheless, while auxiliaries proved attractive to some mackerelers and swordfishermen, there was no early cry to place gasoline engines in Gloucester's dory-based schooners. The dorymen "loved their beautiful sailing schooners and seemed . . . in no great rush to install power."[35] The economics of their fishery gained little from a move to auxiliaries. An auxiliary required an additional crewman, the engineer, whose sole responsibility was the care of the engine. Thus, an auxiliary engine had to increase productivity enough to offset the additional expense of the engine (including fuel and maintenance) and an extra crew share. Until 1910, the Gloucester fleet sent out an average of only six auxiliaries a year. Sailing schooners remained the

---

*The next two years saw Captain Jacobs achieve new records as highline of the mackerel fleet. As the winter of 1903 approached, he again steamed north to engage in the herring fishery. But then disaster struck. The *Alice M. Jacobs* went down at Cape Ray, Newfoundland—the captain and his crew coming away safely. As the *Jacobs* had left North Sydney the skipper wired home "that he expected a good passage across to Newfoundland." But a gale whipped up in the Gulf of St. Lawrence, with blinding snow and a thermometer that fell to −13°F. The *Jacobs* had both boilers going, putting out a full head of steam. But without warning, the propeller simply fell off the shaft, and the vessel floundered helplessly in the heavy sea. Soon, the *Jacobs* came ashore on a rocky reef at Gerrands Island, off Cape Ray—with breakers all around her. The men huddled on deck, launching a dory, only to see the waves dash it to pieces. They set out a second dory, but it met the same fate. By then, the men's clothing had frozen into solid sheets of ice, faces and hands had become frostbitten, and their fate hung in the balance. It was at this point that the constant beating of the waves actually came to their rescue. For the *Jacobs* heeled over, creating a safe lee side to the wind, and the men succeeded in putting a dory over the side. They pulled the boat to and from the shore with a line, getting the men off two at a time. They made shore with only the clothes on their backs—Captain Jacobs had left behind the $2,000 in cash that he had on board to buy supplies. As the men watched from shore, the steamer was beaten to pieces on the rocky coast.

vessel of choice, and the first decade of the new century saw outfitters commissioning many more traditional schooners than auxiliaries. Yet, there was always room for incremental improvement, and early in the decade, designer Thomas McManus brought forward a new type of sailing schooner. Nicknamed the knockabout—after a small, bowsprit-less recreational daysailer that yachtsmen used to knock about Massachusetts Bay—the type slowly gained acceptance across the fleet. In a knockabout schooner, men worked the headsails from the relative safety of the deck, rather than from the more dangerous bowsprit shrouds and stays.[36] It was the fishermen themselves who coined the term "widowmaker" when working the headsails from the long pole bowsprits; men were routinely swept away by the heavy seas, or lost when the rigging on which they stood failed.*

The frame for the first knockabout, the *Helen B. Thomas*, was laid at the Oxner & Story yard in January 1902, and she was a strange-looking vessel. She had a lengthened bow to take the place of the bowsprit, although her middle and stern were along conventional lines. Her trial trip from Gloucester to Boston occasioned deep interest up and down the coast, and went off without a hitch. The second knockabout, however, would not be commissioned until 1905, when the Oxner & Story yard built the *Shepherd King* for Captain Jacob O. Brigham. It was not until 1907, when the Essex and Gloucester yards had several knockabouts under construction, that the type finally took hold. In the earliest knockabouts, the bow did indeed look odd, the lack of a bowsprit made possible simply by elongating the bow. The additional hull length provided no economic advantage and was not popular with owners because shipyards charged for their vessels by the foot. Thus, the building costs were more but the returns were not, and owners could not be expected to put the interests of the men before their own. Only after ways were found to alter the rig so as to maintain speed and handling characteristics on a par with bowsprit-equipped vessels in knockabouts with normal bow length and profiles did these vessels begin to enjoy wide acceptance. The Thomas McManus–designed *Pontiac* was the first of the refined type. Launched in 1906, she no longer had an extended bow, and her entrance was long and sharp.

---

*Thomas McManus, a Boston designer, had noted the many small pleasure craft in the Cape Ann area that sailed without a bowsprit, and believed the arrangement would work on large banks' schooners. These small, keel sloops had outside ballast and no bowsprits. They handled well, even in rough water, and it was their robust performance in all weather that gave them the name knockabout. But having an idea, even a good idea, was not enough. McManus needed a client willing to commission such a vessel, and Captain William Thomas of Boston was that man.

CHAPTER 16

An End of Innocence

There was no denying the importance of the return of the mackerel: It brought stabilization to the fleet and growth in jobs. Yet, this new prosperity also brought the old forces of rivalry to the fore. Where unity had been the key to success in the depression years of the late 1890s, division and raw personal ambition became the hallmarks of the early years of the twentieth century. While the late 1890s had seen the divergent interests of owners, managers, fishermen, skippers, fish processors, and the important leaders of the community put on hold in the interests of survival, now these groups felt free to pursue their own agendas. For the first time in the city's history, fish-processing workers talked of forming a union, and the issue of temperance—which may have seemed frivolous at a time when putting bread on the table was a pressing priority—once again reared its head. Such trends were not unique to Gloucester. Indeed, what happened there was, in many respects, less intense than the union organizing and temperance movements elsewhere in the country.

GLOUCESTER FIRMS were once again prospering, living off the reinvigorated mackerel fishery and the continued success of the salt-bankers. Along the waterfront, a growing workforce unloaded and processed the fish. This was as it should be; it was the future all had hoped for in the 1890s. But there was an unforeseen side to the prosperity: The wages of the dock and loft workers had not grown, and they began to ask why.

In the national depression of the mid- to late 1890s, Gloucester's business community had come to see stable wages as the natural course of labor. Edward J. Livingstone, a principal at the Fresh Fish Company, made the point plainly, saying, "They do not permit [the workers] to bank any [money for the future]." But it was the job of his and other fish processing firms to provide the jobs for today; to sustain the community for today, not for tomorrow. If the business community provided the jobs, they saw no contradiction in the fact that the wages of the workers did not allow them to grow a surplus.

Men such as Livingstone saw no need to apologize for low wages. His remarks were simply an acknowledgment that the managers had not been extravagant with the owners' money. Yet, while it might have been acceptable to workers in the late 1890s, when jobs were scarce, it was unacceptable in the more robust days of the early years of the new century—workers wanted a more equitable return on their labor. Firms were prospering, and workers looked for a fairer share of the rewards. And when the firms failed to act, the workers at last turned to a new vehicle for collective change, the American Federation of Labor (AFL). On January 13, 1902, 220 fish skinners and cutters came together as "brothers" under the umbrella of the national union organizing effort. They met in Sinclair Hall, on Duncan Street, at what the AFL said was the most enthusiastic gathering of its kind ever held in Gloucester. They raised their voices together and created a new union: the Gloucester Fish Skinners, Cutters and Handlers Union, Number 9582 of the AFL. They claimed 400 members.

The union drafted a proposal asking for higher wages, shorter hours, and a union shop. In response, the fish packers and handlers appointed a committee to consider the union demands, and the early response seemed positive. Packers and handlers conceded that the workers had legitimate grievances and voted to make a limited number of concessions to their employees. But the firms would not agree to all that the union had demanded.

Five weeks later, after three joint meetings and little progress, the packers and handlers formed an official organization to represent the firms in Gloucester. All agreed to abide by the action of the majority, and in their first joint action the firms formally voted against the union demand to employ "none but union men." In response, the union bargaining committee vowed not to confer with the bosses until they agreed to the demand, setting August 1 as the strike date. The union committee said the firms had also failed to agree to other demands: that the members be given certain holidays off, and an increase in pay to 27½ cents an hour (up from the current rate of 25 cents an hour). For the workers, the issue of wages took center stage, and on July 30 the union voted to strike. They claimed now to represent 700 workers, but that figure seems doubtful, as only 240 members cast a vote: 223 to strike and 17 not to strike.

The call for a strike came at a time when the new union was ill-prepared for a job action. They had few dollars in the strike fund, and, as would soon become clear, the union was unable to offer help to the families of striking workers. Nevertheless, on August 1 union members put down their knives and left their benches. The strike was on.

※ *By the beginning of the twentieth century, Gloucester's wharf space was almost completely taken over by fish flakes on which salt cod was dried in the sun and open air. Within 12 years of this picture being taken, the arrival of Birdseye's freezing techniques would revolutionize Gloucester's fish processing, c. 1912.*

Only four of Gloucester's many fish processing firms mounted gangs of nonunion men to spread fish on flakes and take out fares from arriving vessels, and all firms closed their skinning lofts. On the second day of the strike nothing changed. Yet, it became clear that the union had not succeeded in fully closing down the firms. Some skinning lofts that had been closed on day one opened on day two, and many wharves remained operational.

Then the union made a mistake. Showing either their weakness or lack of sophistication, they decided against posting members outside each of the fish-processing firms. Instead, the union said it would do its "missionary work" among the men as they met them on the streets. But why missionary work, and why so weak a response on day two of the strike? The most that could be said for the union was that it talked a good game. One member told a newspaper reporter, "I want you to understand that you cannot starve a fish skinner. They are used to it. My stomach can laugh even when it is empty."

Thus the union put forward the impression that it was confident of winning, but so were the owners. By day four, several firms added to their workforce, and to all intents the strike had been broken. Day five saw more of the same, with strikers remaining quiet and orderly on the sidelines, even as firms continued to operate. The firms filled most of their orders using processed fish already on hand, and given a spate of poor weather, they said they had no need to put out cod for airing.

One week into the strike saw 40 men (largely skinners) and 11 women (the latter having long worked as packers and bone pullers) at work at Slade Gorton & Company, and the manager said he had all the men on the wharf he could handle. In fact, the men had taken out a 120,000-pound fare from the schooner *Maggie and Hattie* with no outside help. It was an unusual strike. The union set no picket lines, firms continued to operate, and many strikers were engaged in other activities. Some stayed home, while others went berrying, sailed on the bays of Cape Ann, played baseball at the Cut, or swam at one of the Cape's many fine beaches. But soon the strikers began to feel the economic effect of being out of work. The union had promised $5 a week for families, but they could not deliver, and most of the men had little or no savings. In the past, men in need would have turned to the firms for support, but now they found a changed world. Many workers had been accustomed to living in an unending state of debt to their employers, but now employers refused to help. Worse still, when the men went on strike, many found that the firms had deducted from their pay the full amount the men owed in store bills. Some men had gone home with almost nothing in their pockets.

Workers were vulnerable, and the union itself was in a weak position. On day 11 of the strike, the two sides finally sat together before the State Board of Arbitration. The packers' committee said it would not budge, and some firms said that if the union did not vote to abandon the strike they knew of men among the strikers who would return to work anyway. The next day three union men returned to Cunningham & Thompson's loft.

Four days later the strikers held a peaceful march; 264 men parading past the city's lofts and wharves. It was said to be the first sizable union march in Gloucester's history, but it was all show and had no effect. Even then the packers had begun to bring in nonunion men to take the jobs of the strikers, guaranteeing them permanent positions.

On August 20, after 19 days, the union voted to end the strike on the original terms offered by the firms. The vote was almost unanimous. But the next day, four firms would not guarantee to take back all the men, and the union voted to go back to all the firms except those four. The firms would have none of it, however. Not one union man would be taken back by any firm as long as the union discriminated against the four firms. The union had no choice; they caved, with most of the strikers going back to work on August 22.

Members soon left the union in droves, and two years later the Gloucester Fish Skinners, Cutters and Handlers Union, Number 9582, ceased to exist. It had not delivered for the workers: not on wages, not on benefits, and not on working conditions. Workers had been led to believe that their collective actions would make a difference, but in the end it was the association of fish processors that prevailed. More troublesome, however, was that the common goals that had once united the fishing community had been shattered. It was only a matter of time before workers and owners again came into conflict, and, when they did, the fishermen would sit at the center of the storm.

WITH THE EMPLOYEES of the fish-processing plants back at work, the community turned its attention to the "plight" of those who worked on the vessels—the dorymen, the cooks, and the skippers. But here the concern was not economic, nor was it related to the role these men played in sustaining the community. The good people of the community, including Protestant and Catholic temperance leaders, focused on social issues, with many self-selected reformers taking the fishermen to task for drinking, drunkenness, rowdyism, and visiting prostitutes. Such men were said to bring trouble upon themselves; they neglected their families and brought the community into disrepute. Gloucester had a growing summer population of affluent Americans.

Gloucester Harbor, April 1905, view from Banner Hill, overlooking Smith Cove and the inner harbor, toward the downtown area.

The local paper ran regular "society" notes about the comings and goings of the rich and famous who called Cape Ann home over the summer season, and somehow it just seemed improper to have drunken fishermen walking about the town. There was also a sense of danger, as the fishermen engaged in fights and blocked the streets.

Gloucester was one of the few communities in Massachusetts to permit alcohol to be sold in saloons, and the reformers thought it was time for change. In fact, in their view, change was long overdue, and the Cape Ann reformers came together to rein in these most independent of men. Such cohesive action would have been inconceivable in the 1890s, but with the new sense of economic security, a call to change the image of this rough-and-tumble fishing port could now be sustained politically. Clergy and temperance advocates spoke about the need for change. If they could only convince the voters to stop the sale of alcohol in saloons, fishermen's lives would improve, Gloucester would be a better place in which to live, and the summer visitors would not be disturbed.

At the same time, temperance advocates knew their campaign would not be easy. Most fishermen resented the interference of others. Drinking was an accepted part of life, and the votes of fishermen at the ballot box had almost always been sufficient to keep Gloucester in the license column. But the reformers thought they had a chance to win and redoubled their efforts, mounting a series of campaigns that easily exceeded those of prior decades. Newspapers printed graphic stories of families traumatized by drunkenness. They began with page-one accounts of fishermen gone wild, followed by stories of how they and others had been brought down by an addiction to alcohol.

Alcohol abuse of the poor thus took on a wider social character. In one typical story, the local paper told of two men who had been gaffing for fish from a dory at the John Pew & Son's dock and hooked something below the surface—the body of John Ryan, a local fisherman. He had been last seen on Saturday evening and, the paper said, "having imbibed a little too freely," he started to his vessel at Perkins's wharf and fell overboard.

The message was clear. A life had been needlessly lost, and accounts of troubled dorymen would sit at the heart of the campaign to close Gloucester saloons. Of these stories, none was more poignant than that of the Gullholmen doryman John Lund, and the paper repeated aspects of his story for months on end.

John Lund was a good, seasoned fisherman, but he made poor choices in his private life. Beginning in March, the daily newspaper told the story that spoke loud and clear to reformers. There was a drunken doryman, an injured wife, and destitute children, and the trouble all seemed to begin during the prior fall when Ida Lund left her husband, John, following months of quarreling and drunkenness. She moved to Main Street with her eight-year-old daughter, Jennie, became the manager of a fishermen's boardinghouse, and did her best to move on with her life. But it was not to be. After Ida left, John continued to drink heavily, and in March he bought a 32-caliber Smith and Wesson revolver and a box of cartridges, and went to his wife's home, where he found her with their daughter and two boarders, Stephen Olsson and Albion Backman, two other Gullholmen fishermen (Backman having come over in 1894). Also present was Eliza Thompson, a neighbor.*

The paper then published a sensational account of what followed. John was said to be drunk, and when he asked Ida to return to him, she refused. He asked her for something to drink and eat. Again she refused. John left, and when he came back, gun in hand, he rushed at Ida saying, "You didn't think I had the _____ (nerve) to kill you, but I have, I will show you." He fired a single shot, but the bullet went wide of its mark. John then chased Ida, but their daughter grabbed him, crying, "Papa, papa, don't kill mama!" Olsson and Backman struck Lund's arm as he fired the next shot. The bullet again missed its mark, and the two men helped Ida into the backyard. But John followed, firing two more shots at his fleeing wife. One hit Ida in the left shoulder, and she ran into the house of Mrs. Thompson, crying for help. Seeing what he had done, John tossed the revolver into the yard of a neighbor and started toward his home.

Police Officer Herman R. Joyce found John near the store of John Pew & Son (at the foot of Union Hill) and placed him under arrest. At the station, Joyce said Lund acted like a man who was "not bright," being perceptibly under the influence of alcohol. Lund was in an almost hysterical condition. He claimed that "both himself and his wife had been somewhat to blame for the trouble, and but for the fact that he was under the influence of liquor he would never have committed the crime." But he gave no explanation as to why he had bought the gun.

The next account spoke of Mrs. Lund going to the Addison Gilbert Hospital, where the doctors used an X-ray (a technology less than 10 years old) to look for the bullet. They could not find it because it had passed through

---

*The lives of John Lund and the other immigrant fishermen from Gullholmen had changed over the prior decade, demonstrating the complex nature of the changing lives of the dorymen. Lund was now separated from his wife. Lund's two brothers made return visits to Gullholmen to take brides. Olof Johan Jacobsson had drowned. Sam Andersson also returned to Gullholmen, married Anna Jakobina Khritiansson (Ina), one of the sisters of Olof Johan Jacobsson, and never returned to Gloucester. Meanwhile, in Gloucester, Steve Olsson continued to live in a Lund boardinghouse, going to sea with the Lund brothers. As for the four remaining Gullholmen immigrants who came to Gloucester in 1893, their lives diverged from those of Olsson and the Lunds. They now served on different vessels and lived in other boardinghouses.

her shoulder.* Over the next few months the paper retold the tale in a series of follow-up stories on page one. At first, they expressed concern for Mrs. Lund, although it soon became clear that she had come through the ordeal. The paper then moved to keep its readers abreast of John's alcohol use, bail, trial, and imprisonment. Finally, on June 7 a jury found him guilty of assault with a dangerous weapon, and the judge sentenced him to three years in the House of Correction. The paper emphasized what Ida Lund had recounted that day, weeping during the examination. She had left her husband "because of his drinking." To the righteous reformers in the community, the key to this sorry affair was found in Ida Lund's final words: "He was kindhearted but drink caused the trouble. When he was sober, he was all right." The social and political implications of the newspaper account could not have been clearer: He was a good doryman, and when at sea he was sober, an excellent halibuter, a good companion. But when he came back to port, there were the saloons. Alcohol was the culprit, and many in the community came forward to insist the town rigidly enforce the saloon laws, hoping the December vote would be for no license.

By late May, the police had gotten the message and began a series of regular weekend raids at fishermen's boardinghouses—but not at the hotels and guesthouses of the rich summer visitors. The Lund case helped stir up a hornet's nest. Some raids found liquor; some found nothing. It did not matter. On page one the newspaper reported the names of all the working-class people who had been raided. The message was clear. Gloucester's papers and the local police were responding to the calls of the reformers. Police remained out on the warpath, making weekend raids of fishermen's boardinghouses throughout the summer. If you had money or social standing, you were left alone; if you were poor, the police were at your door.

The pace slackened in the fall, but the issue did not die. November saw a series of no license rallies before the December community-wide vote. At one rally at the Baptist Church, a minister from Boston spoke about the allure of the saloon and the weakness of the fishermen. "Sin is made safe by the liquor business and I am opposed to anything that clothes sin and vice

---

*As Lund awaited his next court date, his wife improved and was discharged from the hospital. But then there was to be a rather unusual twist to this story. On May 18 she appeared as a defendant in Police Court on a warrant sworn for illegal liquor selling. Carl Andrew Parkman, the first witness on behalf of the government, stated that on the day of the shooting affair, he visited Mrs. Lund twice at her premises on Union Hill in company with his friend Steve Olsson. "On the first visit he and Olsson had a drink of whiskey apiece, and Mrs. Lund . . . (gave) each a cigar. They paid for their drinks but not for the cigars. On the second visit about 6 P.M., they each had a bottle of beer and paid for it." Mrs. Lund testified just the opposite of Parkman. The men had visited but she received no pay for the liquids furnished. She was in the habit of furnishing beer to her friends and those of her husband, of which Parkman was one. The judge did not think the evidence sufficient for conviction and ordered her discharge, but the newspaper had now done its best to tie all of these dorymen, as well as the boardinghouse manager Ida Lund, to the taint of alcohol.

in garments of respectability. If a man wants to sin, sin, but let no man's sin and vice be hidden by respect. I know of nothing that makes a drunkard of a man so quickly as the practice of treating [to a drink], when every man in the crowd wants to treat his neighbor."*

But the message did not prevail. The local fishermen were home in December, it was a good time for them to vote, and they came out in great numbers. The referendum was lost in a one-sided vote of 2,156 "Yes" to 1,143 "No," and Gloucester remained one of the few North Shore communities to vote for license. Needless to say, the vote did not end the debate. The issue never went away. In December of each subsequent year the parties contested the issue at the ballot box, but Gloucester was a working fishing port, and the skippers and dorymen almost always prevailed.

As for the Lunds and the other Gullholmen immigrants, after 1900 they faded from public view. By mid-decade, the Lund brothers and Steve Olsson were living at Vincent Street. John had been released from prison, Ida had divorced him, and he joined his brothers and Steve Olsson on halibuters. Once again they were simple dorymen, and in a surviving daybook from this period, one gets a glimpse of the mundane nature of their lives.[37] Steve Olsson and Nils Lund served together on several vessels, including craft skippered by Captain Frank Stream, and the account book records the supplies they bought for their time at sea: boots, nippers, sails, and hats. But by mid-decade the Gloucester fisheries had again begun to change.

---

*This was a time in which the clergy were quite public in "looking out for the welfare of the hard working seaman," and stories such as that of John Lund fed into this process. Religious leaders had a distinct vision of the community, and saloons were not a part of their worldview. This was a time in which individuals who engaged in unusual behaviors were closely watched. Few people were divorced, and divorce hearings were front-page news. In 1909, of Gloucester's 26,184 inhabitants, 14,553 were single, 9,700 married, 1,869 widowed, and only 62 divorced. At the same time, however, Gloucester may not have been as religious a community as the clergy believed. In a 1909 survey by the "religious work committee" of the Prospect Street Methodist Church, a substantial segment of males in the community were found to belong to no church. "Only slightly more than one-third of all the people in Gloucester and Rockport were associated with the (35) churches, both Protestant and Catholic, in any way; but making allowance for infants, the committee considered that at least half of the people of suitable age . . . [were] not . . . associated with the churches in any way." The committee was astounded to find that there were at least 5,000 over 18 years of age who were not under any organized religious influence whatever. Upwards of 2,000 more boys between 12 and 15 were in the same condition. But this was really no surprise, for Gloucester had become a community of working people, of captains and dorymen, and their concerns focused on this world, on the mundane issues of life. With so large a number of unaffiliated male voters, the preachers of the community faced a daunting task in sustaining their moral crusade.

CHAPTER 17

The Changing Fleet and
the Changing Gloucester Fisheries

In the early years of the first decade of the 1900s, outfitters replaced lost vessels with new craft from the Essex and Gloucester shipyards. The total catch hovered at 105 million pounds per year, and when a seasoned master such as Captain Joe Mesquita lost the schooner *Mary P. Mesquita* off Chatham, he quickly commissioned a new schooner to take her place.* Similarly, when Captain Robert Porper lost the schooner *Anglo Saxon*, Cunningham & Thompson contracted with James & Tarr in Essex to build a new vessel for his use—the *Cavalier*.†

As the decade moved on, however, catch levels dropped by an average of 15 million pounds per year, halibut falling to about 4 million pounds and mackerel to less than 2 million pounds. As the catch declined, so too did the fleet. For the first time in almost 100 years, Gloucester's fleet had fewer than 200 schooners and only 300 total craft in 1907.§ For anyone in Gloucester at the time, the annual changes would have been hard to detect. Each year some outfitters added new vessels, while others sold existing vessels to parties outside the port. But the fleet did drop in size, and the fisheries were once again in decline.

---

*The community had initially feared for the crew of the *Mesquita*. Debris washed up on the beach—three dories, oars, kid boards, thwarts, a dory mast, and a rolled-up dory sail. People walking the beach also found a foghorn, a bottle of kerosene, companionway slides, hatches, fish pens, and some pine decking. But no bodies came ashore. Four days later a telegram made its way to Gloucester from England announcing that with one exception, the men were safe. They had been picked up by the Cunard steamer *Saxonian* after she had rammed them.

†The *Anglo Saxon* went aground on Noody Island, N.S., where she was pounded heavily by the sea. She was refloated and Captain Porper and five of the crew made for Yarmouth. But the pumps could not keep up with the water pouring in through the opened seams and the *Anglo Saxon* was abandoned, sinking in 30 fathoms of water. The men rowed 7 miles to land. Captain Porper was said to have felt the loss of his vessel "keenly," being unable to account for how she got so far off course. She had just been overhauled, and all he could imagine was that the new compass was off, as the course he set would have given land "a wide berth."

§At the same time, the fleet landed over 5,000 trips in Gloucester during the year. The most visited fishing grounds were the nearby shores frequented by the market fleet (location unspecified), 2,611 trips; followed by landings from Georges, 831 trips; South Channel, 778 trips; Middle Bank, 512 trips; Nantucket Shoals, 510 trips; Jeffreys, 354 trips; and the Cape Shore, 168 trips. Less frequented were the major northern banks, including Quero, 83 trips; La Have, 82 trips; the Grand Banks, 74 trips; and Bacalieu, 31 trips.

Even more striking than the loss of craft was the drop in the number of large outfitting firms. While 11 firms had outfitted 10 or more craft in 1904, only 5 sent out that number in 1908. The big change came in 1906, when 4 of Gloucester's most respected salt-cod processing firms came together to form a new commercial powerhouse—Gorton-Pew Fisheries. The so-called big four were Slade Gorton & Company, John Pew & Son, David B. Smith & Company, and Reed & Gamage. Together they had a capitalization of $2,500,000. John Pew & Son, founded in 1849 when the railroad came to Cape Ann, ran the oldest fish business in Gloucester.* At the time of the merger, it owned a fleet of 15 vessels. David B. Smith & Company was the largest wholesale producer of fish on the East Coast and owned the largest fleet, 27 schooners. Both Slade Gorton and Reed & Gamage were renowned as packers of fish, and in the words of a local newspaper reporter, "wherever fish is sold or eaten, these names—Pew, Smith, and the other firms coming together—are bywords."

Their coming together created a firm that, in the long run, would prove central to the viability of the Gloucester fisheries. But in the short run, the effect was to cut down on the number of firms and the demand for schooners.

A quick look at what the Smith Company brought to the merger shows the scope of the resources that would now be available to this single firm, including vessels, land, leases, stock, horses, wagons, trademarks, machinery, cash, and much more. The Smith Company had made cash advances of $5,125 to the crews of 19 vessels then at sea, and the value of the firm's goods and property approached $300,000, which included $67,210 for fish stocks on hand, $90,000 for wharves and buildings, and $116,500 for vessels. Many interesting fishing items were included in the total, including 18 brass dory compasses, 6,900 pounds of mooring rope, and 118 herring nets. Doryman-related items included 4½ dozen pairs of overalls, 18 sou'westers, 333 pairs of nippers, 199 pairs of Goodyear rubber boots, 24 pounds of tobacco, and 6 dozen corncob pipes.

The principals of the firm said they had come together to seek new ideas for furthering the salt-cod business. They hoped to find new markets for salt-cod products, and in so doing ensure better prices and greater profits. They also held out the hope that the merger would result in better conditions for those who caught and worked the fish, and this view won favor across the business community. One small competitor said, "Not only is it a good thing for Gloucester, but it is a grand good thing for us smaller concerns. It will help keep up prices and that means everything to us."

---

*This firm still operates out of Gloucester today as Gorton's of Gloucester. Its ownership has moved out of the city to Minneapolis, the Netherlands, and now Japan, but its roots remain in the community.

The merged firm had 39 craft under its control, with Gorton-Pew the sole owner of record for most of them. Their book value was $218,954, or $5,614 per vessel. In addition, over the next few years, when prices seemed right, Gorton-Pew bought schooners from local owners, and by so doing kept them in port.* In one such instance, in October 1907, H. & J. S. Steele (previously George Steele Company) offered 5 vessels for sale, and Gorton-Pew bought them at an average price of $4,125 per vessel. By 1908, Gorton-Pew controlled 51 fishing vessels; only four other Gloucester firms owned 10 or more vessels. Cunningham & Thompson had 16 vessels, Orlando Merchant 12, Davis Brothers 11, and Sylvannus Smith 11—101 of Gloucester's 169 schooners were owned by just five firms.

Yet, even Gorton-Pew, with all of its resources, could not prevent the decline in Gloucester's cod landings, and as the decade drew to a close, Gloucester's fleet declined. The cod catch dropped from 50 million pounds per year early in the decade to 35 million pounds per year by decade's end. America's taste for salted cod was waning as fresh and frozen fish first entered and then dominated the market. Even so, Gorton-Pew worked to update its product, and with product innovation and good marketing, there would still be a demand for salted fish. In the years to come, Gorton-Pew would market existing and new salt-cod products, and eventually canned codfish would become a staple across the country.

The type of vessel that brought the fish to port also changed in these years. As the new century was ushered in, a fleet of 50 or more salt-bankers sailed on two extended trips during the year, leaving on their first trip in March and their second trip in July. But as the decade drew to a close, fewer salt-bankers went north in this grueling fishery. Being away from home for months on end was hard on families, and with a more settled population in Gloucester there was a declining interest in joining the crews. For Captain Clayton Morrissey, his two 1905 salt-bank trips on the schooner *Elector* each took more than three months to complete. In 1906, an early season trip by Captain Joe Bonia on the schooner *Maxine Elliot* took over four months to complete, with little to show for the effort. Poor weather and constant ice floes greeted Bonia on the grounds, and his meager catch did not even cover the cost of outfitting the vessel. But even for those who made two good salt-

---

*In addition to adding to its fleet, Gorton-Pew expanded other holdings. The firm purchased the salt-importation business of George Perkins & Son, giving them virtual command of the salt importation at Gloucester. They also purchased the Merchant Box and Cooperage Company. Benjamin A. Smith, manager of the Gorton-Pew fleet, led the effort to acquire this company to provide Gorton's with a supply of boxes for their fish trade. Nor was this acquisition the end of Gorton-Pew's 1907 buying spree, for in June they purchased the Gloucester fish cutting and packing concern of Shute & Merchant. Shute & Merchant was located at the head of the harbor on Parker Street. One more firm had been lost.

)(*Two fishermen retrieve cod from a ground line—the sea is calm and there is neither rain nor snow falling; it would not always have been so.*

bank trips in 1906, each of the dorymen received a share of only about $450. It was acceptable money, but hardly great, and lower than the shares of many of the fresh fishermen. Why would a good fisherman want to be gone for so long, for so little?

Thus, with each passing year, the number of salt-bankers declined, as did the number of craft willing to go out on a second salt-bank trip. Fishermen on Cape Ann had more profitable choices that did not involve months away from home. In fact, as time went on, it was only the seasonal migrants from Nova Scotia who kept the salt-bankers at sea—several hundred men came south in February.

Yet, even as the salt-bank fleet declined, Gloucester continued to receive fares of salt cod. They came in on what the locals called shackers, but at lower levels that were more in keeping with the reduced demand for salt fish. On a shacking trip, the crew of the schooner salted the cod and hake brought in on early sets and iced the fish caught in later sets. The salted fish had to be carefully dressed, with clean seawater used to remove bloodstains, and the fish were fully packed in salt to ensure whiteness. The shackers' fresh fish had to be properly gilled and cleaned before going into ice, for they hoped to receive a premium price for them.

Shackers usually went out on trips of a few weeks' to a month's duration, sailing from May into September, fishing from Georges to Cape North.

)(*Once on board the schooner, fish were quickly cleaned and treated for the sometimes long voyage home.*

The many large schooners of the fleet used their full capacities, while also taking advantage of the expanding demand for fresh fish. Under this plan, boats went to sea for shorter periods of time; they made more trips during the year, and the men received excellent shares for their work, while also spending more time with their families. They preferred to sell their fresh fish in Boston, then sail to Gloucester to sell their salt fish to firms such as Gorton-Pew. But summer prices and demand could be low in Boston, so in the summer much of the fresh cod was brought to Gloucester for the splitters. Thus, when the demand for fresh fish waned in Boston, the splitting firms in Gloucester took the catch at lower prices, cleaning the fish (splitting them open, removing the head, stomach, and backbone), and then processing the fish in salt (and later in ice) for sale.*

Two other of Gloucester's four great fisheries, halibut and mackerel, also declined in this period, and the city and fleet faced an unprecedented period of scarcity. If the early years of the first decade had brought back a sense of

*Of all the fish landed in Gloucester, 58 percent came in fresh and 42 percent salted, while in Boston, 99 percent came in fresh.

excitement, later in the decade everyone looked for even the smallest suggestion of the fish returning. They looked in vain.

As the catch fell, some halibut skippers continued to bring in fine fares, while other equally talented men had little to show for their efforts. When Captain Robert Porper came in with a good fourth trip, many along the waterfront offered congratulations. But Porper would have none of it. He shook off the praise, saying that heavy northerly winds and difficult tides had kept down the sets and challenged all who engaged in the fishery. He was unwilling to take any special praise for his luck. For these men, the traditional halibut grounds off Nova Scotia and Newfoundland were producing fewer and fewer fish. But as halibut brought a premium price, such trips were still warranted.

Captain Joseph Cusick, on the schooner *Dreadnaught*, sailed to latitude 54.4°N in the Labrador Sea, said to be the farthest north that any American vessel ever caught halibut. Even the highline of the fresh-halibut fleet, Captain Lemuel Spinney on the schooner *Dictator*, was forced to fish on the dreaded Sable Island grounds—an area that even under the best of circumstances required keen judgment and constant vigilance if the vessel was to arrive home in one piece. But no matter the challenge, the men went out, for halibut commanded a price four to five times higher than other fresh fish. It enjoyed great popularity as an unrivaled white, firm, delicate fish. Halibut were in demand, and a dedicated group of skippers and dorymen pursued them.

Among the small fleet of flitched halibuters, Captain Joseph Bonia continued to have success early in the decade, sailing to the northern Bacalieu Bank. But, as halibut became ever more scarce, Captain Bonia in the schooner *Senator Gardner* and Captain John Marshall in the schooner *Tacoma* decided to try their luck again on the once-productive Icelandic halibut grounds. Maybe the fish had come back. The *Gardner* sailed in late March,* but she was laid up for days by high winds, snow, and icebergs on her southern passage across the Grand Banks. Finally, on April 16, Captain Bonia reached Dyrafjordur, only to hear the locals say the fish had yet to strike in. Bonia off-loaded his supplies and sailed outside the fjord to make his first set, but the dorymen, including his nephew Charles Daley, found few fish. A combination of encroaching ice and a developing storm drove the ship back to port; the men enjoyed a concert ashore.

They were soon joined by the *Tacoma*, and the two vessels waited for the weather to clear, for it was blowing a steady gale and the whole island was

---

*Much of this account is from the logbook report of Michael Wise of the *Senator Gardner*. It was reproduced in the local daily newspaper.

covered with snow. When the schooners next went out, the men set four to
five skates of trawl per dory, meeting with limited success as they brought in
10 dory loads of fish. But the wind again kicked up and Bonia lost a full set
of trawl line. Two days later, they lost three more skates of trawl. Nevertheless, Bonia ordered his men to bait up, and sailed to the northwest in search
of a patch of good water. He set three skates to a dory, but again the weather
closed in, and Bonia made for the safety of the nearby fjord before hauling in
the trawl. Another five skates of trawl line were lost. By late May, Bonia had
had enough, and he abandoned the Iceland experiment as a failure, setting
a course for the Labrador Sea to try his luck on the more familiar Funks and
Bacalieu banks. If nothing else, the halibuters were persistent.

On June 23, when north of the Arctic Circle, Bonia's men set two skates
of trawl per dory, but got only five fish. He then sailed south, but for days
made minimal headway through ice floes off Lady Franklin's Island. On the
Fourth of July, the ice finally abated, and the *Gardner* made land at Komaki-
owick. But the village was deserted, and Bonia continued southward. Finally,
in early August, at 58.33°N his men caught 10 dory loads of fish, and over
the next week or so they brought in a catch. On September 12 the *Gardner*
made Gloucester with 100,200 pounds of flitches.

Bonia had proved three things: The Icelandic flitched halibut fishery had
passed into history, and the Bacalieu and other more northern grounds continued to be difficult and unpredictable places to fish; and bad weather could
prevail over the entire season, with ice everywhere and scarce fish. Nevertheless, over the later years of the decade, the small fresh-halibut fleet brought
in a consistent catch of around 4 million pounds per year and received top
price. But it was a trade fraught with danger.

In February 1907, Captain Joseph Bonia tried his luck on an extended
northern halibut trip to Green, Grand, and Burgeo banks on the *Harry A.
Nickerson*. But he had little success. In looking back on the trip, one sage on
the waterfront said the *Nickerson* "went out about in time to catch the last
of the winter gales and all that came . . . [in] the spring and she . . . [had]
been used up more than roughly." As for Captain Bonia, he described the
March weather on Green Bank as "the limit of anything he had ever run up
against." A storm came from the southwest and lasted 30 hours. He put the
craft under bare poles, laid out a string of cable as a drogue sea anchor, and
then asked the men to bend on a piece of the riding sail in the teeth of the
gale. After three hours of trying, a monster sea came at them and the men
dashed for the rigging and cabin. The sea rose to its full height just before
reaching the *Nickerson*, catching the vessel squarely under the quarter. She
shivered from stem to stern, and the sea forced the counter under until the

)(*A bald-headed (no bowsprit) schooner leaves the inner harbor, c. 1920. Men work on the deck and on her dories as she passes two steamers hauled out on Burnham's Railway.*

vessel was hoved down, her mastheads laying in the water. Men were torn from the main rigging and swept into the sea. Somehow they survived, and in Captain Bonia's words: "The Lord got them back, that's all there was to it. I don't see how they ever did it, but they swam and grasped a rope when they could find one and hauled themselves back as the vessel righted. It was a miracle, that's all you can call it."

In March the following year, Captain John Stream of the *Kineo* brought news of the loss of Captain Augustus Swinson on Georges while working a sail—the first of several losses to be reported within a matter of days. Swinson was working the sail on the *Kineo* when a gust of wind swelled it out, knocking him into the sea, and on hitting the water he disappeared beneath the waves. Then came word of the loss of three men when they became separated from another of the fresh halibuters, the *Waldo L. Stream.* According to Captain Frank Stream, the fog set in when his dories were spread all over the grounds, and one dory did not make it back to the ship. Through the night, Stream burned a torch on the masthead and sounded the horn. In the morning, with the fog holding sway, four dorymen took a backup foghorn, ran out a short distance into the fog, and sounded the horn with the hope of striking the missing dory. They saw no immediate signs of the missing men, but when they rowed to windward they found them. Now they, too, were lost in the fog. They then anchored, arranging the men three to a dory, and waited for the fog to clear. But it held on until the men on one of the dories decided to try to reach the schooner. They hauled anchor and rowed away, one man waving back to his comrades as the wall of fog closed in behind them. Two days later, when the fog lifted for a short time, the men in the second dory made out the beacon fire. But the fog quickly shut in and they were forced to wait until morning, when the fog cleared, before rowing to safety. The other dory was never seen again; they had rowed away from the schooner and perished.

At almost the same time that Captain Stream told his tale, there came word of another strange story, although this one had a happier ending. The key players were the captain of the halibut schooner *Monitor*, John McKay, and his brother, William McKay, a doryman on the *Cavalier* under Captain Robert Porper. In this account, Captain McKay drove the *Monitor* hard on the long trip from Gloucester to the southern edge of the Grand Bank, and as he approached his favorite halibut spot, he saw another vessel jogging around. It was the *Cavalier*, and Captain McKay began to run down along the line of dories in order to speak with Billy. Suddenly one of his crew shouted, "Skipper, skipper, there's a dory bottom up just ahead." As they neared it, McKay saw two men struggling for their lives. Instantly, he had a dory in the water, and two pairs of brawny arms rowed as they had never rowed before, for as they left the side of the vessel McKay yelled out, "For God's sake boys, drive her; that's Billy on the stern there and he's pretty near all up." In fact, his rescuers came upon the dory just before Billy's grip failed. They hauled in the two men and brought them back to the *Monitor*, warm blankets taking the place of wet clothes. The McKay brothers soon held a family reunion.

Though life could indeed be strange for a Gloucester halibuter, as the decade drew to a close, the small fleet followed its normal schedule. In 1909 the highline, based on combined stock of halibut and non-halibut trips, was Captain Lem Spinney on the *John Hays Hammond*.* He stocked $30,352, bringing in $21,473 from seven halibut trips—a fine stock. Among the straight halibuters who did not spend part of the year in another fishery, two skippers stood out: Captain Daniel McDonald on the *Mooween* stocked $28,600, with each of his men drawing a most respectable $700 share, while Captain Bob Porper on the *Cavalier* stocked $24,500. Among the Georges

---

*Lem Spinney launched the schooner *John Hays Hammond* in 1907, naming her after the richest man in town, and she had a remarkable record in the halibut and haddock fisheries. On the day prior to her maiden trip, Mr. Hammond presented Captain Spinney with a complete set of bunting and a first-class chronometer. On the next morning, Captain Spinney looped his schooner back and forth off the Hammonds' summer home (in the outer harbor, off West Gloucester), so that Mr. Hammond could see her under sail.

fresh halibuters, special note was made of the Swedish skipper Charles Colson on the *Selma*, who stocked $22,498. He was a young man, one of the city's up-and-coming halibuters.*

THE MACKEREL FISHERY did not hold up as the decade unfolded, however. Instead, it went from a robust fishery in 1901, to a declining fishery in 1906, to a failed fishery in 1910. The dividing point was 1906: In March of that year, as 70 skippers busily prepared their vessels for the trip south, Captain Wallace Parsons was so convinced that the fishery would succeed that he enthusiastically walked 90 miles from his snowbound parents' home at Bay of St. George, Newfoundland, to board a steamer for the trip to Gloucester. However, once he and others reached Hatteras, North Carolina, they found it too rough to put out a seine-boat. It was April 1 before Captain Sol Jacobs brought in the first catch of the year, and that was a mere 3,000 tinker mackerel. The next craft heard from was the schooner *Cynthia*, skippered by Jeff Thomas, but she made the news only for running aground on Romer Shoal in lower New York Harbor. Multiple tugs, a derrick lighter, and a corps of divers refloated the vessel, and once her rudder post was repaired she rejoined the southern fleet.

Then there came a report that seemed to suggest that the fleet's luck had changed. The fleet ran into a 10-mile-long raft of mackerel off Cape Henlopen, Delaware. The catch was good, but it was to be the one big sighting of the southern season. By mid-May, as craft made their way back to Gloucester, skipper after skipper reported a poor southern season. Then, as the weeks passed, first the Cape Shore and then Georges produced little or no fish. Schooners on the Cape Shore brought in an anemic catch at best, while by early July schooling mackerel had not made their usual appearance on Georges.[†] A few mackerel had been seen on the Rips between Georges Bank and Nantucket Shoals, but they were spread out and unpredictable.

---

*Charles Colson came from the headland across from Gullholmen, and men such as Steve Olsson and the Lunds served on his ships for over 20 years. He was born in Lusichea, Sweden, in 1874. There was a natural affinity between him and the men from Gullholmen, and Steve Olsson later became a friend and neighbor. Charles entered the United States around 1895 when he jumped ship in Gloucester, and lived in one of the Scandinavian boardinghouses. After his marriage and the birth of his only child, Nina (Colson) Groves, in 1908, he moved to 139 Mt. Pleasant Avenue, East Gloucester. When a neighboring house (143 Mt. Pleasant Avenue) became available 10 years later, he suggested to Steve Olsson that he might want to purchase it. Steve did, and from 1918 onward the two men lived almost next door to one another.

†In fact, the 1906 Cape Shore catch consisted of only a little over 4,000 barrels, compared to 15,000 barrels in 1905. Among the vessels on Georges was the schooner *Esperanto*, from the fleet of the Orlando Merchant Company. She made her maiden trip on July 7, and while few noticed her at the time, she later became one of Gloucester's greatest racing schooners.

One mackereler described the conditions on Georges as being "as dry as a prohibition town on Sunday."

By year's end, the season had proved a bitter disappointment to fishermen and buyers alike, and the 1907 southern fleet dropped from 70 to 58 craft. But then the unexpected happened—the mackerel catch improved, and 1907 was a better year. It was to be the high point of the decade. In each of the succeeding years the mackerel fishery became progressively more distressed; more men pulled out, and as they did the Gloucester fleet declined. The 1909 and 1910 seasons opened poorly and never improved. Apathy reigned, vessel owners got deeper and deeper in debt, and many abandoned the fishery. There was no longer a spring rush to be the first to leave, not even when reports came in of mackerel schooling to the south. In March 1910, Captain Thaddeus Morgan set the tone when he sent word that he would not be leaving his Nova Scotian home to engage in the southern fishery. In fact, 19 craft had abandoned the mackerel fleet over the past year. By year's end, the 1910 fleet landed the smallest catch on record: 3,395 barrels of salt mackerel (2,380 landed in Gloucester) and 19,882 barrels of fresh mackerel (490 barrels landed in Gloucester). The highline honor went to the *Oriole* with a stock of $9,000, but its owner, like so many others, abandoned the fishery in disgust. There was no longer any money to be made in mackereling.

With the disappearance of the fish, the U.S. Fish Commission sent out the schooner *Grampus* to search for answers, and many skippers weighed in with their own opinions. Captain Charles A. Dyer of Portland spoke for many when he suggested closing the season until July, giving the mackerel time to throw their spawn off the coast. A different viewpoint was advanced by Captain A. Everett Bassett of Harwich, who said fishermen had to adjust their schedule to that of the fish. In his view, mackerel had their feeding grounds, and if a man could tell when they went to these grounds the fish could be caught as before. In Gloucester, Captain Jacobs said he believed the bonitas in southern waters had driven the fish north at great speeds, making the mackerel impossible to catch.

As for the *Grampus*, on her trip from Cape Hatteras to Labrador she netted mackerel on the southern grounds in May. It proved that some fish had been there even as the skippers of the southern fleet considered leaving to refit for the Cape Shore, but because the fish did not rise to the surface, the fleet had no hope of making a fare. In early June, the *Grampus* saw schools of mackerel 25 miles southwest of Block Island, but that was it. Over the rest of the summer she cruised from Long Island to the Cape Shore, and into the Gulf of St. Lawrence, but netted few fish. The mackerel had disappeared.*

*By 1911, only 151 schooners remained in Gloucester's fleet, down from 255 schooners a decade earlier.

CHAPTER 18

Gloucester's Thriving Fresh-Fish Fleet

As the first decade of the twentieth century moved to a close, Gloucester's fresh fishermen landed banner fares of haddock, cod, pollock, and other ground fish. At the height of the season, vessels sat two and three abreast at Boston's T-Wharf as dorymen and dockworkers hustled iced fish into waiting carts. In the dead of winter, as the Lenten season began, 60 or more vessels crowded in, landing up to 2 million pounds of fresh fish over a two-day period. Lent was a time of spiritual preparation, where Catholics and other Christians abstained from meat and turned to fish. As Lent ended and demand fell, prices would begin to fall, and as the cod reappeared on Georges and the Scotian Banks, the larger vessels in the offshore fleet refit for shacking. In another month, with the reappearance of swordfish off the New England coast, many of the smaller schooners would be refitted for that fishery. The summer was thus a period of mixed fishing: Some schooners continued to trawl for haddock and pollock, while others went out after cod and still others after swordfish. In September, as swordfish and cod became scarce, most of the fresh-fish fleet fit out once more for the ground fisheries, skippers and dorymen spending the rest of the year trawling for haddock and pollock.

Such was the time-honored annual cycle of life on the grounds, governed by the appearance of the fish and fluctuations in the market. Gloucester's fresh fisheries were quite productive. The annual catch came in between 60 and 80 million pounds, providing one-third to one-half of the fresh fish moved through T-Wharf. But not all the fresh catch went into Boston. About one-third came to Gloucester, where it was purchased by either one of the city's large splitting firms or one of the few remaining small fresh-fish firms. When prices were high, skippers sold to the Boston dealers. They landed their fish at T-Wharf, refit in Gloucester, and quickly went back to sea. When the Boston market was glutted and the prices fell, skippers sold to the Gloucester splitters. Together, these two markets sustained Gloucester's shore and offshore fresh-fish fleets, for even when the Boston markets were slow, skippers had an outlet for their fish—good money was to be made.

In the days between Christmas and New Year's, the skippers of the larger schooners began to fit out for winter dory trawling on Georges, Brown's, and even the more distant Scotian Banks. Some of Gloucester's finest masters manned these vessels, including the Nova Scotian expatriates Jeff and Billy Thomas, Clayton Morrissey, and Lemuel Spinney; and the Azoreans Joe Mesquita, Joseph Silveria, Frank Cooney, and W. P. Goulart.* In going to sea in the dead of winter, these men had to fight bitter cold, high seas, and strong gales. They often returned with less than a full catch when their supplies ran out, having spent days looking for an opportunity to set their trawls. But in January 1907, an unusual spate of mild weather led to banner trips; there were reports of dorymen fishing in their shirtsleeves. The schooner *Catherine G. Howard* even reached the Grand Banks, where the dorymen struck fish wherever they set, and the skipper returned with a full load of iced fish.

But the fine weather did not last; toward the end of January winter gales again closed in, and vessels returned with short fares, lost gear, and damaged sails. Captain Joe Mesquita thought it providential that he had made it back to port, for while sitting at anchor on the grounds, gale force winds swept in and tugged at his cable. The schooner *Mesquita* began to jump high in the air, only to come crashing back down on an unforgiving sea. The anchor had to be raised to save the vessel, and as the men worked the windlass, a giant wave swept over the deck, leaving the anchor and cable hanging dangerously under the vessel. It swung wildly from side to side, putting the vessel out of control, pulling her head under the waves. Only Captain Mesquita's quick action in cutting the cable saved the vessel and his men.

February proved to be a disastrous month on the grounds. Men felt fortunate to set trawl on one or two days a week, and when they came in they often had little to show for hard weeks on the grounds. Late in the month, as the weather continued to challenge the men at sea, the local paper lightheartedly reported that the storm held no special terror for the Master Mariners' Association, as the skippers had come together for their annual winter banquet. As the guests arrived at City Hall they were greeted by falling snow and howling wind, although once inside everything was delightfully warm and cheerful. It was a different story at sea, and by its end the storm had

---

*Among them the Thomas brothers continued to be central players in this fishery. As the 1902 season ended, Captain Billy Thomas and his partners had sold the schooner *Horace B. Parker*, ending the skipper's 14-year association with the vessel. She had been a consistent highliner, with Captain Billy bringing in an average annual stock of $21,000. Over the 1904–1905 winter haddock season, while in command of his new schooner, the *Elmer Gray*, Captain Thomas made one of the best records in the history of this fishery, landing big fare after big fare. However, by year's end, his younger brother Jeff supplanted him as highliner of the fresh-haddock fleet. Captain Billy took command of the *Thomas S. Gorton* after her launching on August 14, 1905, at the James & Tarr yard of Essex. The schooner carried a crew of 21 men, was 106.6 feet at the waterline, and weighed 140.2 gross tons. She was owned by Slade Gorton & Company.

taken both lives and vessels. Five men from the schooner *Lady Antrim* were swept to their deaths, fragments of their schooner and two of their bodies washed ashore on Marblehead Neck. The schooner *Golden Rod* also went ashore but her crew was saved; and on the schooner *Agnes*, Captain James H. Goodwin was lost when a huge wave swept over the stern, washing him into the bubbling deep.

Once the storm had passed, the remainder of the late winter and spring seasons passed uneventfully. The offshore fleet made regular three- to four-week runs to the grounds and had a fine Lenten season. But by May, as fresh-fish prices fell and cod made their annual reappearance on the Georges and Scotian banks, many skippers began to refit for a summer of shacking.

By May 1, the first three offshore craft had converted to shacking. Another 10 joined by month's end, and still others came on line in June and July. Craft were gone for up to three weeks at a time, and in early season trips they typically returned with about 100,000 pounds of salt cod and 50,000 pounds of fresh. A prudent skipper, facing rough seas and days of inactivity on the grounds, salted his early catch and iced only those fish that his men caught in the week prior to leaving for home. Buyers had little interest in "old," deteriorating iced fish.

But in 1907, the spring fleet made surprisingly quick trips to the Peak and Scatteri, and skippers returned with up to 200,000 pounds of fresh, iced cod. Over the summer, as the 59 vessels in the offshore fleet made 103 shacking trips, large quantities of bait squid were found on the grounds, and with no need to make port for a second or third baiting, almost all of the fish came in fresh.

For the shackers, September was a transition period. Only 26 made trips to the grounds. By late month almost all had gone back into the fresh-haddock fishery, where they averaged about 60,000 pounds per trip: 30,000 pounds in haddock, the rest in other ground fish. Demand was strong; fresh haddock sold for about $3.25 per hundredweight, and the men were making good money. But it was not long before fall gales overspread the grounds, and setting trawl became difficult. Supplies ran low, trips were cut short, and small fares became the order of the day. Yet, even though all understood the challenges, no one was prepared for the news that the Portuguese schooner *Clara G. Silva* had lost 10 of her men when a fierce gale swept in as the men sat setting their trawl.

According to the skipper, a man named Silva, the gale came up as he was setting out the last 2 of his 14 single-man dories. The dories were all over the fishing grounds, and seeing the danger the skipper blew the horn to call them in. But he said the men did not respond; they continued to set their

)(*Fish are off-loaded onto the wharf as a group of fishermen look on, c. 1910.*

trawl, oblivious to the immense storm that was even then closing in over the grounds. Silva said only he saw the peril, and he brought his vessel about, intending to go down the line to pick up the men. But as he did a squall burst over the *Silva*. Driving rain and snow cut down his vision, and he had no option but to tack away from the men; they disappeared into the gloom.

Meanwhile, on the schooner *Frances P. Mesquita*, Captain Joe Mesquita faced the same squall. But to Mesquita the danger of the approaching blow seemed obvious. He did not send out his men, and fearing the worst, he dispatched two men aloft to look for stray men or overturned dories from other boats.

Soon one of the lookouts sighted a dory to the lee, bottom up, with no man in sight. A little farther off, the same lookout saw a second dory. The doryman was waving frantically with one hand while bailing with the other. At least one of the men of the Silva had survived, and as Captain Mesquita swung to his rescue, the second lookout spied another occupied dory to windward. This man, too, waved for help, and as he was the nearer of the two, he was the first to be rescued.

Having retrieved the two men, Captain Mesquita signaled Captain Silva to swing over to receive them. When he did, Mesquita told of having passed two overturned dories, and Silva acknowledged that he had passed four, all with no sign of life. But the story did not end here, for many in the community came to ask why Captain Silva had not followed the example of Captain Mesquita. Why had no other vessel lost a man? Why did Captain Mesquita keep his men on deck, while Captain Silva set all of his dories? The idea that Silva's men refused to come in made no sense; the people on Portagee Hill had lost confidence in Captain Silva. He had left men on the grounds on two other occasions in the prior six months. Other skippers had to bring them in, and for many it was an unbelievable coincidence. They questioned the skipper's judgment and his commitment to his men. Surely this was not the best Gloucester had to offer.

As people gathered on the waterfront, they said Silva had sent out his dories without careful consideration. Some even questioned what really happened when the squall broke over the dories. How could Captain Silva have personally retrieved only a single man? Why had the men failed to respond to a signal to return? What made them stay on their trawls? It made no sense. People talked, and as they did Captain Silva retreated to his cellar, afraid of what people were saying and what they might do to him. They called him "rat." A story even circulated saying he could have picked up one of his dories but left the dory and its man astern without taking him aboard.* It was a horrible thought, but such were the rumors spreading through the community.

In a time of sadness, many in Gloucester's Portuguese community came together in faith, reiterating their trust in a just and forgiving God. They paid tribute to the 10 lost men at a solemn high mass of remembrance at Our Lady of Good Voyage Church, and with this simple act of closure moved on with their lives. Within days, Captain Joe Mesquita announced that he would stay ashore for the rest of the year. By mid-November, despite his reputation, Captain Silva had a new crew. In tough times, even a "rat" can find a crew, and the *Clara G. Silva* sailed back to the fishing grounds.† But the weather did not improve and the catch fell. December saw some of the fiercest gales of the year, and a few of Gloucester's finest offshore vessels did not make it home for Christmas. But most did come in, following the lead of Captain Billy Thomas by completing their last trip of the year well before the Christmas holiday.

---

*The Gloucester doryman Pat Kelley said his father was fishing "in that storm and the story went through the fleet that he [Captain Silva] picked up one of his dories [and doryman] but then put him astern without taking him aboard. When he saw all the other dories going he let him [the dory and the doryman] go too." (Jeff and Gordon Thomas Oral History Collection.)

†Gloucester skippers came up from the ranks of the dorymen; they were talented, ambitious, and capable of managing a crew when on the grounds. To succeed they had to handle the vessel, find the fish, and maintain the loyalty of the men. A skipper who too quickly left dorymen on the grounds was to be avoided; such a skipper would never go to sea with the best of crews.

Like so many other Gloucester fishermen, as a family man, Captain Billy Thomas was now able to be a loving husband and devoted father. It was a simple role to play, and Billy liked nothing better than to sit in the family den with his wife as she darned socks and mended clothes, with their children dashing about the skipper.[38] Billy loved his children, and as Christmas drew near he joined them in selecting and decorating the Christmas tree, piling on balls, garland, and wax candles. He could be counted on to ask his wife and others to point out who in their circle of acquaintances might need help at this special time of year. He could not stand to see others in need when he and his wife had so much. Invariably Billy gave one of his children one or two $10 bills to buy something special for those in need.

Christmas was a magical time for the fishermen of Gloucester, and the city opened its heart to the visiting fishermen and the less fortunate in the community. The Salvation Army distributed baskets of goodies to the city's poor. Schoolchildren supplied the food and interested citizens provided a variety of other useful articles, including a toy for each child in the family. Even the children of the poorest families had to have a toy at Christmas.

For the many dorymen whose families lived in far-distant lands, the Fishermen's Institute held a standing-room-only celebration at the McClure Chapel. There were prayers, Bible stories, and Christmas hymns, and each man received a small, practical gift for use at sea. They also partook of a fisherman's "mug-up"—buns, coffee, cheese, oranges, apples, peanuts, and candy. Some participated in other activities later in the day, such as dances, games, and, of course, having a drink with other fishermen of the fleet.

Meanwhile, for men still at sea, there would be no hymns, no gifts, no warm family gatherings. Rather, on Christmas Day they sat in their craft being tossed about by the fiercest gale of the winter. Two men lost their lives, and while all of the schooners came in, the returning men told of narrow escapes as the snow closed in on dorymen working their trawls. Nevertheless, with the last of the fleet in port, the city looked with pride on the success of its offshore fishery. The typical vessel had landed upwards of 700,000 pounds of fresh fish during the year. Men had money in their pockets and were looking forward to going back to sea before the first of the year.

There were, however, smaller vessels in the fresh-haddock fleet—the shore craft. They also went out after haddock in the fall and winter, but their skippers more closely monitored the weather, for with their size they went out for a day or, at most, a few days at a time. The shore fleet included both traditional, 40- to 80-foot schooners that trawled on Georges and the Gulf of Maine, and small motorized boats that fished within the sight of land. The schooners sold their fish in Boston, while the small motorized boats sold theirs in Gloucester. But regardless of the size of the craft or where the fish were sold, it was a profitable, growing fishery. Dorymen from other fisheries were moving into the shore fleet in increasing numbers, and it was at this time that Steve Olsson and Charles Daley made the change from long-haul halibuting to shore fishing. Steve signed on as a doryman in 1908, joining Captain Manual Sears on the Portuguese schooner *Emily Sears*. He was in his late 30s, and by moving into the shore fishery he had time to court, and soon marry, a young woman from Gullholmen, Hilda Christiansson.* As for Charles Daley, who was 10 years younger than Steve, he left long-haul halibuting to assume command of the 74.6-foot schooner *Almeida*, a Gorton-Pew shore craft. It was his first command, and his uncle Captain Joe Bonia was instrumental in the appointment.

Charles's assignment to command a Gorton-Pew–owned shore craft was a relatively rare occurrence. Gloucester's Portuguese community owned about half of the schooners in the shore fleet, while newly arriving Sicilian fishermen owned almost all of the small motorized dories.

The schooners in the shore fleet fished from early January to late December. They went after haddock in the winter; swordfish, cod, and pollock in late

---

*Hilda was seven years Steve's junior, and when Steve left for America in 1893, he saw her as only a child. They had barely known one another, and Steve had made no special trip home to seek his bride. In fact, Hilda first settled not in Gloucester, but in Boston, and it took several years for the two to establish a relationship. Yet, from the beginning they knew of one another's presence in America, for many threads tied them together. Hilda's older brother had come to America with Steve, but had drowned in the Iceland halibut fishery. When Steve's friend Sam Andersson returned to Gullholmen he married one of Hilda's sisters. Another of Hilda's sisters, Bertha, who had settled in Gloucester, married Steve's close friend, the doryman Macker Simmons; and one of her brothers, John Jakobsson, a friend of Steve's in Gullholmen, also moved to Gloucester. They all lived in close proximity to one another on Main Street and Harbor Terrace. It was through Hilda's two siblings that Steve and Hilda met, became friends, and slowly fell in love—the Swedes came together when the men were in port, they talked in Swedish of their old and new lives, and nature did the rest. When Hilda briefly returned to Sweden in 1906, Steve wrote to remind her of his love and wish for her return. Hilda responded with a touching love poem, saying: "From the heart a greeting I send thee. It is a burden to long for so long time. I hear the longing for years. When can I see you? How rapidly does time fly together with friends. As a memory one day disappears. If the beloved is far away the day becomes so long. Already the morning is greeted, my solitude let he hear. Hoping that everything goes so well. Please write some time soon." Once back in Gloucester, Hilda moved in with her sister Bertha, and Steve took pleasure in spending time with them; they would be his closest confidants for the remainder of his life. But Steve and Hilda waited to marry, saving for the day when they could have their own apartment. Finally, on the last day of December 1908, Hilda and Steve made their vows before the Rev. W. H. Rider, D.D. They were Gloucester's 149th, and last, marriage of the year. By mid-1909, Steve and Hilda Olsson had moved into an apartment at 65 Taylor Street—the same building in which Bertha and Macker (Sven) Simmons resided. They were to live in Gloucester's Portagee Hill section until 1911, while Steve sailed on a shore schooner under Manuel Sears, his neighbor. Here Hilda and Steve had their first child, Gerda. Steve left a record of his shore trips over a 20-month period. The account begins in November 1909 (less than a year after his marriage) and ends in May 1911. Over this period, he made 35 to 40 trips a year. He went to sea every month, and in 1910 he made 34 trips. His per-trip earnings increased dramatically over this three-year period—$10.06 a trip in 1909, $17.68 in 1910, and $22.93 in 1911. Steve made good wages, and he and Hilda saved for the future.

※ *While ground fish are off-loaded from one dory, previously landed fish are sorted into pens on deck.*

spring and summer; and haddock and pollock in the fall.* On a good winter's day, upwards of 50 shore craft landed fish at T-Wharf, but in early 1909, lingering snowstorms kept much of the fleet in port, as skippers looked ever skyward for a chance to go out and set their trawls.† Finally, as the weather cleared, the boats dashed out to Jeffreys, made one or two quick sets, and raced for Boston, where haddock sold for over $4 a hundredweight. Within the month, as Lent began, shore craft surrounded T-Wharf. Dealers had long backlogs of orders, and fishermen scheduled their trips to make the market. Fish moved quickly from the holds of the vessels to waiting handcarts and from there into the horse-drawn wagons and motorized vehicles of the express companies.

*Come spring, Gloucester sent a number of its medium-sized shore craft into the Georges, Rips fishery. Here schooners drifted across the Georges Shoals as men dropped handlines with baited hooks and lead weights to the seabed below. They used cockles for bait. In 1907, 45 craft participated in this fishery, and they sold their salted and fresh cod in both Boston and Gloucester. They brought in a fine-quality, hand-caught fish, and received top dollar. Among the craft going out in 1907 were the *Mina Swim, Olympia, Etta Mildred, Thalia, Good Luck,* and *James A. Garfield.* They averaged 80 feet in length and 75 gross tons.

†One skipper, Joe Mesquita, made good use of this time, when a large assembly of his friends gathered at his Prospect Street home to throw him an old-fashioned surprise birthday party. There was a bountiful supper, including Captain Joe's homemade wine. Well past midnight, they joined in song, dance, and talking. Some stayed the night, their children sleeping upstairs, as the adults danced downstairs till dawn. They then had a big breakfast and set off for home. (Mary and Francis Mesquita, the Jeff and Gordon Thomas Oral History Collection.)

But Lent soon passed, and while the men continued to bring in a good catch, fresh-fish prices began to fall. It was time for the men to make a change, and in late April skippers began to refit for the mid-year fisheries. Some remained in dory trawling. They rechecked their lines, hooks, and dories, and soon set off in search of haddock and pollock. Others moved into swordfishing, first off-loading their trawl gear and then bringing on harpoons, harpoon points, lines, and buoys.

For the Gloucester shore fleet of smaller schooners, swordfishing was a time-honored summer alternative. They dominated the fishery. In late May, as the fish returned to the New England coast, skippers such as Charles Daley made the conversion. They off-loaded their trawl gear and then brought on the swordfishing supplies needed for an extended trip to the grounds. Shore craft would now be gone for weeks, rather than days, at a time.

Once at sea, life on a swordfisherman often involved days of monotony, as vessels sat in the nothingness of the fog or waited on a quiet sea for the swordfish to appear. Men played cards, fished with handlines, and mended their gear, even as the lookout strained to see the first sign of the giant fish. Then, on a fine early season day, he would at last spy a commotion across the water. The swordfish were feeding, their two-edged swords flashing in the sun as they filled the water with tiny, fishy fragments from the small menhaden on which they fed.

The men moved to their stations. The helmsman slowly closed in on the fish. Men stood ready at their dories for the chase that would soon follow; the skipper stood in the pulpit at the end of the bowsprit with a lance in hand. The pulpit's iron railing provided the necessary support when the skipper harpooned the fish. The sputter of the small auxiliary engine settled down to a steady putt-putt as the vessel drew ever closer to the prey. Then, as the bowsprit glided over the fish, the skipper hurled the lance. Its detachable metal point cut deep into the fish, holding fast behind the gills, and the fight was on. Over the side went the line and buoy attached to the point. Soon the line was uncoiling over the waves, rotating as it trailed behind the fleeing fish. Its drag soon began to slow the wounded giant, permitting the one or two men in the trailing dory to close the distance. They pulled on their oars, and when at last they drew aside their prey, they grabbed the line and wrestled the swordfish into their boat. It was no easy task—swordfish of 250 pounds or more were common, while some weighed upwards of 600 pounds. Once the fish was on board, the men returned to the small schooner, where they wrestled the fish onto the deck, where it was beheaded, dressed, lowered into the hold, and stored in ice. A 10- to 11-foot swordfish would weigh about

250 pounds; its sword would comprise about 3½ feet of its length, and the dressed weight once in the hold would be about 200 pounds.

The swordfish season had officially opened, and on May 29, the *Almeida* and *George H. Lubee* where among the first for the grounds. But it took almost a month before the first of the fleet had come into T-Wharf, landing 37 fish for 18 cents per pound—this translated into a stock of over $1,200 and "a handsome check" of $80 for every man on the crew. The next day, 13 swordfishermen came in with varying sized fares, and finally on June 28, after 31 days at sea, the *Almeida* returned with 29 swordfish. The season was in full swing, but Boston's demand had its limits. On July 14, prices fell to 7½ cents per pound. Now, as vessels brought in trip after trip, there came reports of near calls on the grounds, as vessels came close to going down in thick fog on Georges' "liner alley." The story was always the same: Men heard the blast of a steamer's horn, they peered into the dense fog, and soon the high bow of a transatlantic vessel shot out of the fog and bore down on the men and their schooner. But over the summer, no vessels were lost. In each case, the steamer passed within feet of the schooner, and the men reported that they felt lucky to be alive. There were even accounts of men using their oars to push off a steamer as it passed.

Other accounts came in of lone dorymen tackling a determined sword-fish. In each case, a man was said to have been at work landing the giant fish when something unexpected happened. Perhaps the fish dove under the dory or it came rushing straight toward the man. But however it began, the fish invariably tore a piece out of the dory's bottom or side and poked a hole into the man's foot or torso. Then, with the dory leaking, the man tended to his wound, landed the fish, and patched up the dory. The result was always the same: The man felt lucky to be alive.

So, the swordfish season was at its height, and stories of hardship filled the air. By the end of the first week in August the season had peaked, the catch began to drop, and the prices began to rise. During August, most, but not all, had continued their hunt for swordfish, and by early September the fleet had slowly begun to refit, one vessel at a time, for the fall fishery in had-dock and pollock.*

By late September, the entire shore fleet had refit for the ground fishery, going after the old standbys of haddock, cod, hake, and pollock. Weather per-mitting, they made good fares, but poor weather and threatening seas kept craft in port for days on end. When they did get out, a changing sea or an

---

*For the two-month summer season, the Boston swordfish catch came to 1.6 million pounds, with a value of $143,000, or an average return of 8.9 cents per pound.

)(*Stacked dories, lashed down ready for the schooner to set sail, 1914.*

approaching storm often sent them scurrying for a safe anchorage. On the schooner *Almeida*, Captain Daley sailed on 15 fall and early winter trips, but he brought in fish on only 11 of the trips.

In late October, the *Almeida* joined 14 other vessels in sailing for haddock, while most of the other shore craft went after pollock. But an easterly closed in, and only 7 vessels landed fares in Boston. A few days later, with all of the shore craft out on the grounds, Boston saw one of the biggest fish days of the fall. Sixty-three craft landed 1.5 million pounds of fresh fish. A week later, with nearly the entire shore fleet—51 vessels—sailing to the grounds, Captain Daley was the first to return. His was a broken trip, for Captain Daley had lost a man, his St. Joseph's cousin Ambrose (Andy) Fagan. It was a strange loss, for there had been only the slightest wisp of a breeze rippling across the water when Charles set Fagan's dory on the fishing grounds. Fagan's was the first dory set, and as Captain Daley set the fourth dory, one of the men shouted out that he did not see Fagan. His dory had come up into the wind, but no one was in it. The buoy line had been thrown over the side, but the trawl line was still on board. Fagan was nowhere to be seen, and the common belief was that he must have reached for the buoy and slipped, fallen overboard, and drowned.

Following the loss, Captain Daley remained in port for a week, not going back out until November 19. But on November 15, the other vessels in the shore fleet made big landings. Fifty craft surrounded T-Wharf, rafted up seven to eight deep at its end. However, it was the last great fish day of 1909. On November 23 came the first of a series of coastal storms that bedeviled the shore fleet for the rest of the year.

When the weather did let up for a few hours, it just as suddenly closed in before the shore craft could get out much beyond Eastern Point. Finally, on Monday, December 13, 5 offshore craft and 31 shore craft made landings at T-Wharf. They did not soon go back to sea, for yet another gale swept by the coast. Those in port remained fast to their moorings, and over the next week or so, nothing changed. If schooners got out, their skippers did no more than take a look around and come back to port. The few who tried to stay at anchor off Eastern Point, waiting for a break in the weather, had nothing to show for the effort.

The shore fleet squeezed in a pre-holiday trip or two, with all craft coming home for Christmas. The fleet shut down on December 23, and the people of Gloucester celebrated the holiday in the traditional way, with children rising early to see what Santa had left in their stockings. Protestant churches held Christmas concerts, Episcopal and Catholic churches celebrated Holy Communion, trade stopped, and the wharves went quiet.

Gloucester also showed its heart this Christmas to a unique group of old, poor dorymen who could no longer go to sea. In good weather, they made a trip or two to the summer grounds, but in winter only the hardiest of dorymen went to sea. What money the old folks had made on the summer trips had long since run out, and they now had but meager resources for food, lodging, and coal. As the winter approached, they appeared in court as vagrants, hoping to be sent to the House of Correction at Ipswich. Judge York thought it "one of the saddest features of his court administration." But this Christmas season, after Judge York's account of their plight appeared in the local newspaper, a Boston philanthropist, Dr. John Dixwell, came forward with $2,000 to start a fund to return them to the community. It was followed by donations of $25 to $100 from others, including John J. Pew, President William Howard Taft, and someone who wished to remain anonymous. With the growing fund, Judge York made one of the biggest and most "noble" Christmas presents he ever gave in his life, signing a petition to release 12 aged fishermen from the Ipswich House of Correction. He personally took a carriage to the prison to tell the men of their good fortune. The deputy assigned to his court had found boardinghouses in which the men could come and go as they wished. He gave priority to the houses maintained by widows of fishermen who had been lost at sea. The release of the men came on Christmas Eve, and one of them, a man who had put in over 40 years on the grounds, "wept for joy."

For a glorious moment, the spirit of Christmas descended. Twelve old fishermen had come home. They enjoyed a hearty Christmas dinner, had new warm clothes, and if they so desired, on the afternoon of Christmas Day, they were able to join other fishermen watching a movie at the Olympia Theater, or they could go to an afternoon entertainment at the Union Hill Theater entitled "The College Girl."

On Christmas night itself, however, with the worst snowstorm in a decade seizing Cape Ann, all stayed in. By daybreak the snow was coming down in earnest, continuing into the next night. It cleared around midnight, the moonlight penetrating through the clouds making everything clear and bright.

Thus did the shore fleet's season come to an end. Many had had a good year, including the schooners *Maud F. Silva*, *Mary DeCosta*, and *Belbina P. Domingoes*. The leading schooner in the shore fleet, the *Mary E. Cooney* under Captain Frank Cooney, had stocked $35,000, and each of her dory-men received the fine share of $1,192.

BUT THE PICTURE is incomplete, for it leaves out the smaller vessels—the motorized dory trawlers and gill-netters, known as the "mosquito fleet." Both were relatively new to Gloucester, but their importance grew as time went on.

The motorized dory fishermen were almost all owned by Sicilian immigrants, coming to Gloucester from Boston. Originally arriving in New York from the fishing communities of Palermo, Terrasini (very near Palermo), Sciacca, Massaita, and Trapani, they found Gloucester an ideal site for their small-boat fishery. At first, they came only to fish, staying a day or two and then returning to their North End homes. But as they found the fish to be abundant off Cape Ann and local buyers stepped forward to take their catch, they began to move their families into the boardinghouses of the Fort (at the western edge of the inner harbor). Life in the community was hard at first. Few spoke English, the local fishermen offered little help, and many of the "locals" disdainfully called these men in their small dories "Ginnys." But none of it mattered; they knew how to trawl. They owned their dories, and with their sons and siblings they worked as a family in this self-sustaining fishery. Their year began in February and ended in November, the men setting trawl for flounder, cod, and haddock, and nets for herring and, at a somewhat later date, mackerel. They fished the shore waters of Massachusetts Bay from Thatcher's Island to Plymouth. If a storm kicked up, they darted for the safety of Gloucester Harbor, but only the worst weather kept them in port.

Each dory had two sets of trawl, the men going to sea with one, while the second remained behind to be checked and repaired by the women of the family. The men rose between midnight and one in the morning, and if the weather seemed OK they baited up and were off to the nearby sheltered grounds by daybreak. There they set the trawl and by 11 o'clock were bringing in a catch. If the weather turned, wind-driven seas would come near to swamping the boats, and a man intent on saving his trawl could go overboard with a sudden lurch of the boat, his mate struggling to pull him aboard as the turning screw of the motor kept the dory next to him. But this would have been an unusual day, and on most mornings, in about an hour they had their catch in the dory and began to motor back to Harbor Cove. There, a Gloucester dealer bought the fish (in the early years it was said that some of the old-timers took advantage of the new immigrants). Some of the fish were processed locally and some were boxed and shipped to T-Wharf, but the Italian fishermen were paid in either case, averaging $3 to $8 for a day's work by the two men in the boat.

Their catch was small, but as more dories were added to the fleet, it increased. In less than five years, the fleet boasted 30 motorized dories, giving work to more than 100 men, with a catch of several million pounds.* Everyone soon called these small boats "guinea trawlers," but the Italians took no offense to the term, for that is what they, too, came to call them (when used by others as a slur, the usual interpretation was that the person was nonwhite).[39]

In these early years, as the men fished out of their small boats, there was to be little integration with the diverse groups that made up the established schooner fishermen of Cape Ann. They stood alone, and in so doing found that few helped or welcomed them into the community. In response to their sense of exclusion, the Italians in the Fort came together to form a tightly knit community. It has been said that Sicilian mothers were reluctant to let the first generation of children out of the Fort. The children went to school, the boys and men sold their fish, but when all was said and done, daily life revolved around what happened in the Fort.

---

*In 1910, there were 263 persons in Gloucester who had been born in Italy. By way of comparison to other foreign-born, Gloucester had 3,806 from Canada, 635 from the Atlantic islands, 547 from Finland (largely in the granite business), 426 from Sweden, 386 from Newfoundland, 148 from Portugal, 95 from Norway, and 73 from Denmark.

)(*Fish are pitchforked from a schooner to sorting bins on the wharf, c. 1910.*

IN 1909 YET another shore fishery, gill-netting, came to Gloucester. Four- to five-ton vessels, crewed by four men and powered by small gasoline engines, laid long nets (62 fathoms in length) in shallow waters near shore. Lead weights on the bottom edges of the nets brought them to the seafloor, and buoys on the top edges held the nets upright. This created a catching fence in the shallows, and mature fish became entangled at the gills in the six-inch mesh nets while attempting to pass through. Younger fish passed through freely. Each boat carried 7 to 12 nets and housed 4 to a box.

In fair weather, boats went out early in the morning and hauled the nets set the previous day. The end of the net was attached to a power winch, and as it was recovered the crew stood at the ready with gaffs to strip the fish from the mesh. The wet nets were stowed on the craft and dry nets reset on the seabed below. The boats returned to port that afternoon, leaving the new nets "to fish" until the next morning. Once in port, the fish were delivered to the fish-processing houses, and the nets were wound around large wheels to dry in the air.

The gill-netting fleet grew quickly to some 60 small vessels, employing more than 600 men in the winter and spring seasons. Like all vessels in the shore fleet, success depended on the weather. In foggy or windy weather, few gill-netters ventured outside the harbor to set or pick up their nets, but on a good day, they could land 50,000 pounds of fish in Gloucester.

The fresh fisheries were prosperous. In 1909 to 1910, the catch of fresh and salted fish brought into Gloucester leveled off at about 88 million pounds per year, with another 35 to 40 million pounds going into Boston. Larger vessels in the shore and offshore fleets would bring more and more of their fish into Boston, while the smaller craft sold their fish in Gloucester.

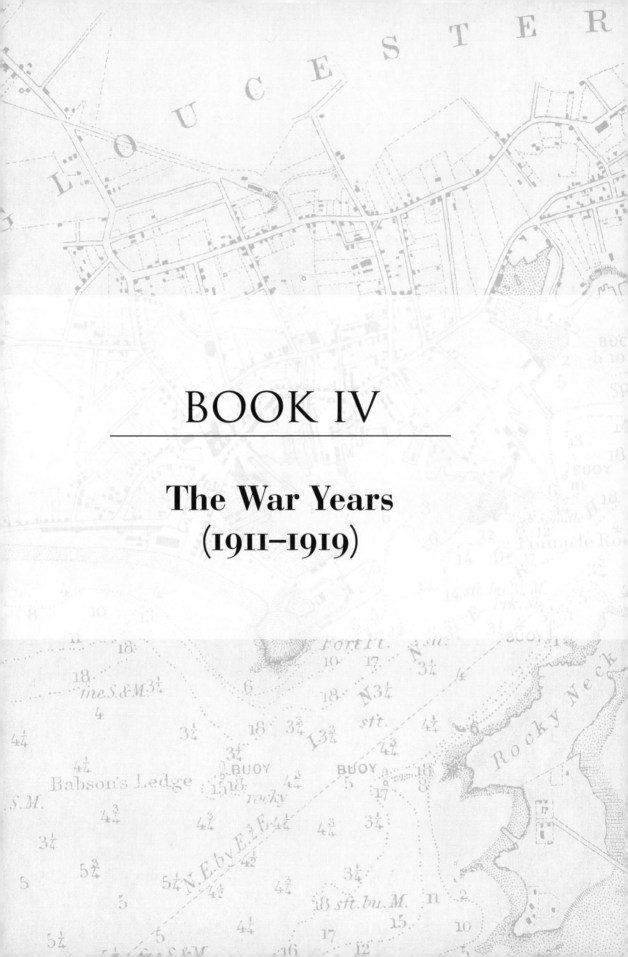

BOOK IV

The War Years
(1911–1919)

The years leading up to the Great War were a time of uncertainty. Mackerel had largely disappeared, salt cod was in decline, landings in Gloucester slowly fell, and duty-free Canadian fish entered the Gloucester local marketplace. The old ways had unraveled, and the people of Gloucester knew that they could no longer depend on the fleet to carry the community, let alone sustain itself. Over-fishing, the loss of much of the fresh-fish trade to the Boston dealers, and the arrival of cheaper Canadian fish were all parts of the problem. At the most basic level, the ecological consequences of new technologies seemed finally to have caught up with the local fisheries. One of Gloucester's oldest fishermen, Captain Sylvannus Smith, reminisced about days long past when the local waters still teemed with fish. He remembered a time when dory loads of halibut could be caught within sight of land; when schooling halibut were common on Georges. But all that had changed, and in his view, it was the practices of men—with their seines and trawls—that had caused the halibut and mackerel to disappear.

Whether Captain Smith was correct or not was of little consequence. Gloucester was bleeding vessels, and as they left the fleet, so too did the dorymen. Between 1901 and 1913, Gloucester's fleet dropped from 353 vessels to 251, and the catch landed in Gloucester fell from 106.4 million pounds to 69.9 million pounds.

The Great War would ultimately save the Gloucester fisheries, but in the years before 1914 no one foresaw the conflict that would come—a conflict that would lead to an unprecedented demand for food, with wheatless Mondays, meatless Tuesdays, and porkless Thursdays and Saturdays. The rationing would lead to a surge in Gloucester's fish landings: from 69.9

million pounds in 1913 to 144.2 million pounds in 1918, including 27 million pounds of salted fish that came south on Canadian vessels. The war would change everything, while it lasted. Gloucester feared for the safety of her wooden schooners as German U-boats prowled the North Atlantic, hyperinflation challenged the old economic order, and the great Spanish flu epidemic of 1918 took both young and old victims.

CHAPTER 19

Cheap, Duty-Free Fish Enters Gloucester

In 1905, Newfoundland's Premier, Robert Bond, had reopened the free-fish reciprocity debate. He asked the British government for help in protecting the Newfoundland fisheries, but Westminster proved an unwilling partner, and so Bond introduced local legislation to close off the fishing grounds to Americans. He asserted, with no basis in fact, that the treaty of 1818 between the United States and the United Kingdom gave Americans no right to fish in Newfoundland's bays, harbors, and creeks. The British once again refused to support Bond. The politicians were unwilling to engage in so blatant a confrontation with the Americans, particularly with U.S. Secretary of State Elihu Root reaffirming the right of U.S. vessels to use purse-seiners in treaty waters.

Nevertheless, the contest was on. A few American schooners were seized, many more were harassed, and a few even faced trials for fishing in Newfoundland waters. But it was all a charade, and before long the countries agreed to send the matter to an international tribunal at The Hague. There it sat for several years, awaiting a ruling on the points in contention. When the jurists finally issued their ruling in 1910, it largely favored the positions advanced by the Americans. The only substantive issue lost was a point on headland distance: American craft could no longer fish in the Bay of Chaleur. Otherwise, U.S. access was secure and intact: Any rules Newfoundland might issue had to apply equally to American and Newfoundland craft.

Even though the issue of access had been resolved, the larger question of duty-free fish did not go away. U.S. Presidents Taft (Republican) and Wilson (Democrat) both came out for reciprocity. President Taft entered office in 1909 at a time of national recession, hearing calls for lower prices on the necessities of everyday life. Entry of duty-free Canadian fish, largely salt cod but also including mackerel, seemed right around the corner. In Gloucester, Captain Sylvannus Smith spoke for many when he said that such trade would hasten the day when the local fleet would lay alongside the wharves, the skinning and cutting establishments would be closed, and laborers would be thrown out of work. In reality, the Taft administration's first salvo in tariff

X *Fish are sorted and weighed on a Gloucester wharf in 1906, as fishermen look on from a schooner tied up alongside.*

reform, the Payne-Aldrich Act of 1909, neither threatened Gloucestermen nor benefited the working families of America. The act reduced tariffs on 655 commodities but left unchanged tariffs on such necessities as dairy products, beans, flour, potatoes, eggs, and most fish. Gloucester was relieved, but the country as a whole was appalled, and the people took it out on the Republicans when they went to the polls in November 1910—the Democrats gained 56 seats in the House and 10 in the Senate. President Taft got the message. He called a tariff-reform conference in Ottawa, and a far-ranging trade reciprocity agreement was soon ready for ratification. If approved, fish tariffs would be a thing of the past, and duty-free fish would flow into the United States.

The people of Gloucester were taken aback, but for the first time in over half a century anti- and pro-tariff sentiments divided the community. The Master Mariners' Association, representing the skippers of the fleet, remained staunchly pro-tariff. They depicted the duty-free legislation as a threatening menace, and had a flying squadron of their members—Gloucester's proud, independent skippers—ready to swing into action on a call from Congressman Gardner. Around the harbor, skippers lowered their flags to half-mast as a sign of their grief at the impending sacrifice of the Gloucester fisheries.

But the presence of a pro free-trade group in Gloucester and in other ports belied the claim of universal grief. Lehman Pickert of the L. Pickert Fish Company, a Boston mackerel importer, made the case for all to hear: "The fish dealers in Boston as in Gloucester, in distinction to the owners of fishing vessels, favored a reduction in the tariff on fish and believed it would benefit the industry." The schism between skippers and processors that had become so public in the great fresh-fish debacle of the 1890s now became more of a chasm. What might benefit Gorton-Pew and the other processors might harm the skippers and the dorymen, and there was no bringing the two sides together.

On February 1, the meeting of the Master Mariners' flying squadron with President Taft came to naught. As reported in the local paper, Taft said that "the working out of the agreement would prove that they were wrong, and that Gloucester would not be hurt after all. 'You are seeing ghosts,' added the President, laughingly." It was painfully obvious that the fishermen of Gloucester no longer carried any weight on the national scene: The national recession in 1907–1908 had altered the political landscape—it would not be the 1880s all over again. On February 14, 1911, the U.S. House of Representatives voted 221 to 92 in favor of the agreement, and although the Senate vote was several months away, the result was not in question. As the people of Gloucester awaited the final ruling, Gorton-Pew Fisheries, the city's largest employer, took out options on properties in Lunenburg, Nova Scotia. The firm was prepared to prosper under the altered economic environment.

On July 22, in a special session of Congress, the reciprocity legislation passed the Senate by 53 to 27, with even Massachusetts Senator Henry Cabot Lodge voting for the measure. President Taft signed the legislation on July 26, and Canadian Liberal Premier Laurier dissolved Parliament, announcing a two-month pro-reciprocity campaign prior to a national election. His opponent Robert Borden and the Conservative Party quickly stepped forward to oppose the legislation, and the campaign soon became a referendum on future free trade with the Americans.

As the Canadians debated the issue, Benjamin A. Smith, vessel manager for Gorton-Pew Fisheries, made a two-week trip to Newfoundland, Cape Breton, and Prince Edward Island to gather knowledge of the northern fisheries at close range. On several of his stops, he stated that Gorton-Pew intended to adjust its business to meet the changed economic and commercial conditions should reciprocity pass. In fact, Smith bought processing plants at Rose Blanche and the Bay of Islands. A new day seemed at hand, and Gorton-Pew was planning to pursue a general fish business in Canada. They would take in and cure salt fish in Canada, and then move it to Gloucester for cutting. The fate of the Gloucester fleet and the city's many independent skippers was not a crucial issue in their deliberations.

On September 22, the much-anticipated Canadian election was held, with unexpected results: Canada had said "No" to reciprocity. The 15-year reign of Wilfred Laurier's Liberal government ended. The anti-reciprocity Conservative Party now controlled Parliament, and they turned to Great Britain for trade. The Americans were seen to be arrogant and expansionist, and Great Britain seemed a less threatening trade partner.

The news surprised and relieved the Gloucester fishermen. Their fleet was still economically viable; duty-free salt fish from Nova Scotia would not overwhelm them. But for the average American, there was shock. This was not what was supposed to happen, and in 1912 President Taft was voted out, replaced by the Democrat, Woodrow Wilson: Both the Presidency and Congress were now under Democratic control. The Democrats remained opposed to protective tariffs, and in October 1913, the United States unilaterally lowered tariffs on the necessities of life, including raw and processed fish from Canada.*

Three days later, the Newfoundland schooner *Palatia* arrived in Gloucester. She landed 275,000 pounds of duty-free salt cod at the Cunningham & Thompson Company. The final nails had been driven into the coffin of Gloucester's salt-cod fishery. The cost of constructing and outfitting a salt-banker in Canada was significantly less than in Gloucester, and all that was left of Gloucester's once-domineering salt-cod fleet was a handful of older vessels. The fishery had blossomed after the Civil War, had begun to decline in the 1890s with America's new preference for fresh fish, and now it all but died. There was still some demand, but Gorton-Pew and the few other Gloucester processors now bought most of their salt fish from Canadian sources—they were simply unwilling to pay a premium for salt cod brought in on American vessels. The entry of duty-free fish would also further swing the balance of power in the local fisheries economy squarely to the processors. The local businesses would make their profits on importation, processing, and marketing, rather than on the catching of fish.

Gloucester's one remaining fleet of any substance was the fresh haddock fleet, which landed fish in Boston, yet even its future was uncertain. And then, in faraway Sarajevo, Archduke Franz Ferdinand of Austria-Hungary was assassinated, and the whole world stood on the brink of change.

---

*A House committee recommended free entry of green fish (split and salted but not dried), with only a trivial duty on boneless or manufactured products. The bill also called for the removal of all tariffs on articles of food and clothing, reductions in the rates on all necessities of life, and a new income tax that touched the pockets of only those Americans whose net income exceeded $4,000—which applied to few in Gloucester. The Sixteenth Amendment to the Constitution, passed in 1913, permitted income to be taxed: After exemptions and deductions, however, less than 1 percent paid a tax, and that at 1 percent for income after exemptions, plus a 1 to 6 percent surtax.

CHAPTER 20

Gloucester on the Verge of War

With the fisheries in flux, skippers, cooks, and dorymen took different paths in the prewar years. Many with roots in Nova Scotia and Newfoundland headed north to join the booming Canadian salt-cod trade. Often they took command of craft that had once sailed out of Gloucester but now carried British registry.

Charles Daley and Joseph Bonia would head north to sail on Canadian and Newfoundland vessels, delivering Maritime and Newfoundland fish to the Gloucester processing firms. The American market was open, and Canadian cod was eagerly purchased by agents from Gorton-Pew, Cunningham & Thompson, and Davis Brothers.

Charles Daley skippered his last Gloucester vessel in 1911, taking out the old Gorton-Pew schooner *William E. Morrissey*. He went shacking over the spring and summer and ended the year on the Newfoundland west coast, gathering herring. In late November he left Gloucester, filled out his crew in Newfoundland, and set his herring nets in the Bay of Islands and Bonne Bay. A month later, he left Bonne Bay with a partial load of herring, planning to top it off on his way home. But as he reached the Gulf, a severe gale swept in from the west.

As the storm hit, the *Morrissey* sat on the lee shore at Bell Burne. Drift ice was closing in, and there was no hope of making the open waters of the Gulf. The crew's only chance was to ride out the gale. Captain Daley hitched himself to the wheel, and there he remained throughout the night, trying to bring his vessel through the storm. But she was soon badly iced up, and Daley's hands froze to the wheel. Late in the evening the *Morrissey* came ashore, driving far up the beach, at Table Point, 15 miles below Port Saunders (about 85 miles north of the Bay of Islands). The men walked ashore by running along the bowsprit and jumping safely to the ground below. It was two o'clock in the morning, and with no houses in view, they spent the night in the nearby woods. The mercury hovered at 0°F.

As dawn broke over the desolate scene, a local man stumbled upon the wreck and took the fishermen to his home three-quarters of a mile away. The nearest railway was 112 miles distant, and some of the crew began the long walk for help. But neither the frostbitten captain nor the cook was up to it. The woman of the house had already wrapped Captain Daley's frozen body in blankets and towels and soaked him with heated cod-liver oil to help fight the frostbite. He would come through the ordeal with his hands, feet, fingers, and toes all intact.

Even though Captain Daley recovered, Gorton-Pew Fisheries did not reassign him: The fleet was shrinking, Charles was a young captain, and he had lost a vessel. He returned to Newfoundland, where he joined the expanding schooner fleet sailing out of St. John's. By 1916, he had become master of the Newfoundland schooner *Flirtatious*, a vessel previously owned by Gorton-Pew (where she was known as the *Flirt*). She was now a transport vessel, and Charles sailed her south carrying Newfoundland salt cod and herring to Gorton-Pew's Gloucester processing plants.

Around the same time that Daley lost the *Morrissey*, his uncle, Captain Joseph Bonia, was also beginning the move from master of a Gorton-Pew salt-banker to eventual master of a Newfoundland merchantman. One of the great skippers of his day, Bonia had been a leader of the flitched-halibut and salt-bank fisheries for as long as anyone could remember. But in 1912 he made his last trips as Gorton-Pew master, completing two herring runs on the *Fannie A. Smith*.

The first run had been uneventful, and Bonia returned to Gloucester in early December, off-loaded 1,425 barrels of herring, and returned north for another load. The second run, however, was a different story. After securing a load and turning his vessel south for the return trip, Bonia ran into an extended period of bad weather. Spray from the frigid waters of the Gulf of St. Lawrence froze on the *Smith*'s deck and rigging, pulling her head deeper and deeper into the raging sea. In danger of sinking, Bonia pulled into Port au Basque so that his men could chip away the ice in safe haven. Then he sailed south at a leisurely pace, pulling into Nova Scotian ports at the first hint of a gale or head wind. It was a safe and proper farewell cruise for the master of a Gloucester schooner.

On his return, Bonia wasted no time waiting for others to decide his future, and set out by train for Seattle to look into the economics of the West Coast halibut fisheries. Since Captain Sol Jacobs's initial trip several decades earlier, several companies had established vibrant halibut fisheries in the Northwest, taking advantage of new railroads and a national demand for halibut. Why he made the trip was never explained, but once there, Bonia wired back that he liked what he saw and asked his agents in Boston to sign up experienced halibuters to come west. There were, he said, jobs to be had in the Pacific Northwest. In a matter of days, 61 halibuters responded to Bonia's call—five from Gloucester and the rest from Boston. One of the Gloucester men was Bonia's brother Maurice.

But then came a troubling telegram from the union representing the West Coast halibut fishermen. Bonia was bringing the men west only to break an ongoing strike against the vessel-outfitting division of the New England Fish Company, a company that had started in Gloucester in 1868 and had since become a major force in the Northwest fisheries. All other West Coast firms had settled. The East Coast fishermen should not be deceived; there was no scarcity of fishermen in the Pacific Northwest. But the message had little effect on the unemployed eastern halibuters. These were troubling times, and 54 of the 61 men who had been recruited soon left on the four-and-one-half-day transcontinental train ride. They believed Bonia, not the message from the union.

They soon found themselves in a world of trouble. A large contingent of local union men prevented them from entering the vessels of the New England Fish Company. Within the hour, the East Coast men were paraded to union headquarters, where they joined up; now they too were out on strike against the New England Fish Company. Feeling deceived and out of money, these men had nothing good to say about Bonia.

On April 22, Captain Bonia returned to Gloucester, saying he planned to remain for several months while making a trip or two with Captain Michael Wise on the steam otter trawler *Crest*. However, he wanted to set the record straight on what had happened on the West Coast and attempted to place a positive, face-saving spin on his actions. He said that all the East Coast men had been distributed among the halibut steamers, presumably other than those owned by the New England Fish Company, and that they were well-satisfied with their jobs. But the words rang hollow; this respected Gloucester skipper had deceived the community. Why he did it was never explained, but the Master Mariners' soon sent a representative west, Captain Lemuel Spinney, to investigate conditions and ascertain the prospects for employment, since they no longer trusted the information supplied by Captain Bonia.

Within days, William Griffiths, one of the Gloucester fishermen who had gone west, wrote to his family that Gloucester was now good enough for him, and many others of the party with whom he had made the venture were of like mind. As soon as they could amass enough funds, Griffiths believed they would drift back, one by one. He was using his nest egg to cover the cost of the train fare from Seattle to Gloucester.

As for Captain Bonia, he left Gloucester for New York to take command of the steamer *Heroine*, making landings at Fulton Market, New York.

Whatever had happened on the West Coast was now behind him, and when the demand was right, he would now bring a fare into Boston. Captain Bonia had moved on with his life. In the years to come, Captain Bonia sailed vessels out of his native Newfoundland, including one that had once called Gloucester home, the *John Hays Hammond*.

Though Bonia, one of the great skippers of the era, had left, many other skippers continued to man Gloucester vessels, doing their best to survive in the troubling prewar environment.

Captain Robert Porper retained his status as a halibut killer par excellence, bringing in big stocks on the *Cavalier*, while dealing with all the Atlantic could throw at him. He liked nothing better than to start the new year on the southern edge of the Grand Bank, where he often found great stocks of fresh halibut, but in 1912 the decision came at a cost. A severe gale took down the *Cavalier*'s mainmast and foremast, leaving her at the mercy of the winds. For 11 days the crew displayed the universal distress signal—the Stars and Stripes nailed upside down on the stump of one of the spars—hoping to be seen. But *Cavalier* was alone on the fishing ground, and the prevailing winds drove her more than 300 miles down the coast. When she was finally spotted, the revenue cutter *Androscoggin* towed her into Gloucester looking more like a derelict than a fine banks' schooner. But she was not beyond repair and was soon back at sea enjoying more successes. One harrowing experience could not deter an old salt like Porper.

Two of the other hustling skippers of the era, the Thomas brothers, also stayed the course. Jeff moved from the *Cynthia* to the *Sylvania*,* while Billy remained on the *Thomas S. Gorton*. They made fresh-haddock trips in winter, shacking trips in the middle of the year, and returned to the fresh fishery at the end of the season. Each year saw them competing for honors as highliner of the fresh-haddock fleet. And then there was Captain Lemuel Spinney, who mentored many young skippers who would go on to have exemplary records on Gloucester vessels during the war years and beyond; among them were Christopher Gibbs, Fred Thompson, Archie McLeod, and Charles Colson. As 1910 started, Captain Gibbs had taken out the *John Hays Hammond*, returning to Gloucester on January 9 after running into the hardest kind of weather. Fitting out again, he sailed in mid-month, becoming the first halibuter to complete four trips in the winter season.

---

*Captain Jeff's last trip on the *Cynthia* was in March 1913, after which he moved to the faster schooner *Sylvania*, a 105.6-foot-long, 136-gross-ton vessel. Captain Thomas had a one-quarter interest in the craft.

When Gibbs left the *Hammond* in 1911, Captain Fred Thompson*
became her master and headed for Quero Bank, continuing a relationship
with Captain Spinney that had begun on Thompson's first command, the
Spinney-owned schooner *Dictator*. But his first trip on the *Hammond* proved
dangerous. For 19 days he battled winds that blew from every point of the
compass and was forced to heave-to for 24 hours to weather a hurricane.
Later that year, Thompson shifted to another of Captain Spinney's schooners, the newly launched *Governor Foss*, of which he was part owner.

By 1914, Thompson had stocked $130,000 on the *Foss*. The local paper
described her as a "regular money machine." When Thompson moved to the
Foss, Spinney moved back to the *Hammond*, bringing in successful stocks on
one trip after another. Then, in 1913 the *Hammond* went out under Captain
Archie McLeod,† the youngest skipper in the halibut fleet; Spinney stayed
ashore. McLeod brought in big trip after big trip, following Spinney's pattern
of making one or two trips a month. Over the next quarter-century, McLeod
became one of Gloucester's great halibut fishermen.

Spinney had a knack for identifying hustling young skippers—none
more so than Charles Colson, a Swede from the small mainland village of
Lyskiel, across the bay from Gullholmen. In 1891, at the age of 19, Charles
jumped ship in Gloucester while serving on a Scandinavian vessel. He had
crossed the Atlantic to fish off North America, and now, with only $1.25 in
his pocket, he started a new life. During his career, Colson would become
known as one of the finest halibuters ever to sail around Eastern Point. He
bore the affectionate nickname "Blossom" on account of his swollen red
face, and as a skipper, he took pride in going to sea with a largely Scandinavian crew, including several of the Gullholmen dorymen. Captain Colson
stayed true to his roots.

By 1913 he was in his prime, one of Gloucester's finest halibuters.
He had first skippered the *Selma* in 1907, and now commissioned a new

---

*He came from the small fishing port of Haugesund, Norway. In Gloucester he was to skipper such craft as the
schooners *Dictator*, *John Hays Hammond*, *Governor Foss*, *Alden* (named for his son), *Florence* (named for his
daughter), *Elsie*, *Pioneer*, and *Avalon*, and the steam beam trawlers *Seal* and *Sea Gull*. His last trip was in 1928.

†Captain Archie (Archibald) McLeod (MacLeod) lived for 84 years and was one of Gloucester's greatest master
mariners, highlining in the halibut and haddock fisheries. He was born in St. Peter's, Cape Breton, on May 11,
1883. He arrived in Gloucester at age 16 and immediately went into fishing. He was a skipper at 26, and over
his career commanded the *Electric Flash*, *Agnes*, *John Hays Hammond*, *Catherine*, *Hortense*, *Louise R. Silva*,
Georgianna, *Bay State*, *Gertrude L. Thebaud*, *Dawn*, *Arthur D. Story*, and *Marjorie Parker*. His most famous craft,
the *Catherine*, was built by Arthur D. Story at Essex in 1915, and he sailed her for 18 years. The *Catherine* was
wrecked on Bald Rock Shoal at the entrance of Canso Harbor on December 31, 1933, when bound for port,
heavily iced with frozen spray. After he retired as a captain, Archie McLeod went as hand and was watchman on
vessels in port. For his last two years, in failing health, he was a patient at the Huntress Public Medical Center
in Gloucester.

schooner of his own, the knockabout *Natalie Hammond.** He would command her for the rest of his life. She was designed by Thomas McManus and bore the name of the 10-year-old daughter of Mr. and Mrs. John Hays Hammond, the wealthiest family in the community, and the man for whom Lemuel Spinney had named his craft. A newspaper account speculated that "in selecting a name of the new vessel, Capt. Lemuel E. Spinney, who with Capt. Colson is one of the owners, evidently believes distinguished names for vessels are omens for success."

She was launched on December 16, 1913, but her maiden voyage was put on hold until after the Master Mariners' Annual Concert and Ball, held on January 8, 1914. Finally, on January 12, the captain was ready to go halibuting. The *Natalie Hammond* left Gloucester in a breeze of wind, with a crew that included the dorymen Steve Olsson and John Lund. Fresh halibut were going for the fine price of 18 cents per pound for white and 14 cents per pound for gray. Captain Colson was now eager to get to the fishing grounds.

Once outside the harbor, the winds picked up, and the *Natalie Hammond* soon disappeared below the horizon off Eastern Point. On the following day the weather turned nasty. The temperature fell to −13°F, as a northeast gale swept up the coast. At Eastern Point Light, the wind reached 71 miles per hour. On land, the gale howled and whistled through the night, threatening to sweep houses from their foundations. Streets were deserted, and before dawn the superintendent of schools had the "no school" signal sounded.

The next day the wind continued to howl, and the vessels at sea became iced up. The *Natalie Hammond* had been struck hard, but there were no early reports of how she had fared. The first vessel heard from was the *Avalon*, under Captain Daniel McColsh; she arrived in Boston with seawater frozen in a mass over the entire vessel. Then the 39-foot Boston schooner *Two Brothers* reached Nantucket with her decks buried under two feet of ice.

The first vessel to reach Gloucester was the steam otter trawler, *Heroine*, under Captain Joe Bonia. He came in for coal, having gone through one of the toughest experiences of the winter. Several of his crew were badly frostbitten, and ice encased everything on the vessel, including the hull and masts. Bonia had run out of coal and his men had to feed wood fittings from the cabin and forecastle into the boiler to keep the vessel going.

Bonia reported that on Tuesday the *Heroine* had barely missed colliding with a schooner laying to under a close-reefed foresail, which he assumed must have been the *Natalie Hammond*. He said the weather had been so

---

*She was an impressive vessel at 110.67 tons gross, 110 feet overall, and 102.7 feet at the waterline, with a 22.9-foot beam, a 11.1-foot draft, and a 60-horsepower oil engine (129-horsepower diesel engine later).

※ *Captain Charles "Blossom" Colson at the wheel of the* Natalie Hammond, *pre-1920.*

thick that he did not see the schooner until his steamer rose to the crest of one of the big waves and he found the schooner dead ahead. Bonia threw the wheel over and just missed her.

But no one could say what had happened to the *Hammond*, and there were rising concerns for her safety. On February 11, 30 days after she left, word came that she had been sighted on the Gully on January 21—she had weathered the gale. The wives of the crew could now take a collective breath. Captain Colson, Steve Olsson, the Lund brothers, and the other 22 men in the crew had survived, and yet, there had been other gales since January 21, and no direct report from the *Natalie Hammond*. Finally, on February 20, the *Times Fish Bureau Column* wrote the words all wanted to see: "Charlie Colson is at Portland." The maiden voyage of the *Hammond* was completed.

Once back in Gloucester, Captain Colson reported that the *Hammond* had come through the harsh winter weather with little or no damage. She had shown herself capable of weathering the worst of gales, and Colson, like others, would continue to fish out of Gloucester. With luck, they might still get by.

CHAPTER 21

The World Changes: War Comes to Europe

For Gloucester, 1914 was a time of uncertainty. The fresh fishermen continued to have success and Gorton-Pew managed to keep its fiscal head above water, but by year's end another 18 vessels had left the fleet. Salt cod was coming south on Canadian vessels, and ominously, on March 30, Boston's T-Wharf was replaced with a new, much larger fish pier. Located in South Boston, the earth-filled Commonwealth Pier was immense: 1,200 feet long, 300 feet wide, with roads, railroad tracks, processing buildings, and a cold-storage facility. Slips were dredged to a 23-foot depth at mean low tide, and there was room for 50 vessels to unload simultaneously: It was the nation's largest fish mart. On the day it opened, more than 1.5 million pounds of fish were processed, and Captain Joseph Bonia in the *Heroine* was first to tie up to the new wharf. For many it seemed that the writing was finally on the wall. The Commonwealth Pier would surely attract even more of Gloucester's fresh haddock and halibut trade, just as most of her salt cod would now come south in Provincial vessels.

In June, Archduke Franz Ferdinand, first in line to the Austro-Hungarian throne, was assassinated, and the world faced a new day. In Gloucester, the first report of the assassination appeared on the left side of page 8 of the local paper, next to an article entitled "Seiners Salting Lots of Tinkers." On page one, the lead article told of eight men of the schooner *Rex* having gone astray on the Grand Bank. For the next few days, at least, the people of Gloucester continued as if nothing had happened. Several halibuters and shackers, including the *Natalie Hammond*, refit for trips to the Banks. Each day saw craft pulling out of the harbor and sailing for brief summer trips to Georges and the Scotian Banks. However, within the month, even as the *Hammond* unloaded fresh halibut from her recent trip, people were beginning to learn that the future had been altered. A full-fledged European war seemed imminent: On July 31, the front page of the *Gloucester Daily Times* reported that the German Kaiser had put his whole country under martial law. It was rumored that Russia had already invaded Austria. Meanwhile, in England the political parties were said to have united in the face of a common danger, and 30,000 armed Canadians were reportedly ready to help the mother country. Events followed at a rapid pace. Germany declared war on Russia and France. Germany invaded neutral Belgium, and on that same day, in response to the invasion, Great Britain declared war on Germany.

Gloucester newspaper accounts were immediate and alarming. During the first days of August, readers learned that German troops had poured into Luxembourg and France, Russian troops had invaded Prussia, and a big naval battle had taken place in the North Sea. At the same time, Washington made it clear that America would not soon join the fray. The country had only a small standing army and President Wilson rattled no sabers.

As events unfolded, the *Gloucester Daily Times* asked prominent members of the community to comment on the possible impact of the war on Gloucester and the local economy. Thomas J. Carroll, manager of Gorton-Pew Fisheries, forecast increased fish prices, an influx of Nova Scotian fishermen to the United States to sail on American craft, and the possible presence of German naval craft on this side of the Atlantic. He believed that Provincial fish would be prevented from going to the West Indies and other English possessions and would thus come flooding into the United States. Charles E. Fisher, president of the Gloucester Safe Deposit and Trust Company, believed that Gloucester businesses would benefit from the war. The only trouble he foresaw was difficulty in buying raw materials, although he did not consider it a serious issue. Unlike Carroll, he thought it unlikely that they would see German raiders on the Grand Banks. In his words, "Germany would have more weighty matters nearer home to command the attention of the fleet."

By mid-August, vessel outfitters expressed concern that their business would feel the impact of future price increases. Under the rules for dividing the stock from a fishing trip, the price of food was subtracted before the rest of the money was divided among the owner, captain, and crew. In the view of one outfitter, vessels making shacking, halibuting, or salt trips of three or four weeks would be the first to notice the increase in prices. Yet, by late summer, the increases had been limited to a few items, including flour and sugar. The food bill of the typical vessel had yet to go up, and concern for the future was still mere speculation. Nothing had changed.

By year's end, those who had forecasted better days were proved right. Fresh cod landings rose by 1 million pounds, salt cod went up by 300,000 pounds, and haddock rose by 500,000 pounds. Gloucester's shackers and fresh fishermen brought in banner catches from the Scotian, Georges, and Grand banks.

X *The schooner sits at anchor, the dories have been launched, and the men await the skipper's call to head out to set their trawl.*

September saw Captain William Thomas on the *Thomas S. Gorton* land 115,000 pounds of fresh cod, while Captain Clayton Morrissey on the *Arethusa* landed 100,000 pounds of fresh fish, 20,000 pounds of salt cod, and 2,000 pounds of fresh halibut. The *Natalie Hammond* also had a fine season. Captain Colson landed halibut in Portland and returned to Gloucester with small fares of fresh fish for the splitters. But it was the massive increase in salt cod coming in on British-registered Canadian and Newfoundland schooners that really stood out. In early September, the schooner *Saratoga* came in from Quebec with 350,000 pounds of salt cod for Gorton-Pew Fisheries, followed by the *Blanc Sablon* from Labrador with 567,000 pounds, and the *Arcadia* from Caraquet, New Brunswick, with 400,000 pounds—the latter two fares going to Cunningham & Thompson. Meanwhile, the one-time Gloucester schooner *Flirt*, now owned by Newfoundland interests and under the command of Captain Charles Daley, delivered 222,000 pounds of salt cod and 3,000 pounds of flitched halibut to Gorton-Pew Fisheries. Daley then went back out, only to return in 7 weeks with 1,375 barrels of salt bulk herring and 52 barrels of pickled herring. In early November, a Norwegian steamer, the *Bauta*, arrived with 1.5 million pounds of Icelandic salt cod for the Gorton-Pew Fisheries and Cunningham & Thompson Company. She was the first Norwegian steamer to deliver to Gloucester, and it was the largest amount of fish ever brought into Gloucester on a single vessel.

But the Gloucester fleet was also doing well, and the processing firms could support increased landings by local and foreign vessels. The war had pushed demand skyward.

On December 22, as the season drew to a close, the local paper led with a headline that spoke to the fishing families: "Fleet Coming Home for Xmas"—the annual cycle of life on the Banks was about to come to a close. Among the skippers off-loading their pre-Christmas catch in Boston were Jeff Thomas on the *Sylvania*, Fred Thompson on the *Governor Foss*, and Charles Colson on the *Natalie Hammond*. Some of the schooners' family men were excused, and they rushed over to the North Station to catch the Gloucester train. Others remained with their vessels, helping the shore gangs finish unloading the catch, after which they too sailed with the skipper toward home on Cape Ann.

But the weather took a turn for the worse. Cold Canadian air swept in from the northeast, and as the last of the fleet passed Ten Pound Island they found the inner harbor frozen over up to the Gaslight Company's wharf. At Rocky Neck, gill-netters were forced to break channels through the ice, and dories and small gasoline boats had difficulty forcing their way across Smith Cove. The last of the returning fleet had no choice but to tie up at one of the wharves between Harbor Cove and the Gorton-Pew complex, or else to set anchors along the side of the harbor channel. But the men were home.

For those alone during the season, the Fishermen's Institute held its annual supper. One hundred and fifty fishermen feasted on oyster stew, crullers, coffee, apples, and confectionery, and received a muffler and comfort bag. For fishermen with families, Christmas was a magical time. Romaine Olson, then only three, remembers her father's return:

> I was lying in bed and Mama was standing in front of the bed holding Jack in her arms [he was less than a year old] and then I heard a Boo sound and I looked up and there was Papa, Aunt Svea and my sister Gerda. I thought the little tiny kids outside were trying to scare me, and I really jumped—I was scared silly. But once Papa was home, we were all so happy. He would always be singing and laughing and I can remember us lying on the bed and we were jumping all over him. He was just a happy person. He really loved us children. I remember sitting on his lap. It was so cozy and nice, and mother calling for us to come to have something to eat.

On Christmas Eve, the Olssons, like so many others in Gloucester, gathered with friends and loved ones. They joined their neighbors, Maud

and George Martin, and the Martins' boarder, the renowned artist Theo-
dore Victor Carl Valenkamph,* at an open house at their East Gloucester
home. Valenkamph was a good friend of Steve's, and Steve's daughter Gerda
remembered that he "gave each of the children a pound box of chocolates
and he gave Mama a big five pound box. I can still remember Mr. Valenkamph
singing with my father—Papa always enjoyed singing, and his voice could
often be heard in song. They were two happy Swedes."

Happy reunions were taking place in fishermen's homes all across the
community—in East Gloucester, Portagee Hill, the Fort, and Riverdale.
There were quiet times and family parties. Stockings hung with gifts for the
children. Fathers lit wax candles on the Christmas trees, and parlors were
bathed in a soft yellow glow. But then the all-too-brief visits came to an end;
the schooners pulled away from the wharves to see in the new year on the
grounds. It had been a fine ending to the season, as most of the fleet had
made money, exports had increased, and duty-free salt-cod fish from Newfoundland and Canada had kept the local processing plants humming. Above
all, America had avoided entanglement in the European war.

---

*Valenkamph was a native of Sweden. He studied painting at the Royal Academy in Stockholm, but also had an
extensive familiarity with the sea. He had been in both the Swedish navy and merchant marine and had sailed
around the world. He entered the United States in 1898, five years after Steve Olsson, and had earlier established studios in Lynn and Boston. He painted the sea in its many moods, and his pictures were much admired.

CHAPTER 22

1915—An Altered Landscape

As 1915 dawned, the war in Europe had yet to affect the day-to-day lives of the Gloucester fishermen. American newspapers shouted no warnings of impending doom, the President rattled no sabers, and the German submarines (U-boats) had sunk no American craft. Gloucester fishermen were optimistic about their future. Gorton-Pew Fisheries, the city's primary employer, had processed a record 47 million pounds of fish in the previous year. Upwards of 700 men and women worked at 1 of their 4 processing plants, while another 1,000 men went to sea on 1 of their 43 schooners.

With mild January weather, the gill-netters and dorymen brought in full fares and received good prices for their catch. But in early February, fierce winter gales were closing in and the mood changed. Floe ice covered much of the Gulf of St. Lawrence and the western Grand Banks, and on some trips most of the fish had to be brought in on handlines dropped over the rails of the schooners. Dorymen spent weeks at sea with little opportunity to set their trawl.

But the concerns on the northern grounds* soon took a back seat to the troubling news from Europe. In early February, Germany announced that a war zone existed around England and Ireland.† German U-boats were prepared to shoot at anything afloat, and the government in Berlin informed Washington that any protest by the Americans would be considered an un-neutral act.§

President Wilson was compelled to act, and he duly asked for written assurances that the 20 or more American vessels approaching the new war

---

*At this time, the crew of the *John Hays Hammond* set trawls on only the very last days of a 21-day trip to the Sable Island Bank, and with their supplies running low they came in with a light load. At the same time, the *Teazer* spent her entire trip fighting to escape from the ice floes on the Burgeo Bank. She kept her auxiliary engine going night and day to pull through the ice, and having used up most of her supplies, she headed back to port with few fish and her planking scraped bare by the ice.

†England countered by imposing a blockade on all German ports.

§The note also included an explicit complaint against the practice of allowing British ships to cross the Atlantic under an American flag, as had the British liner *Lusitania* on her most recent crossing. Some maintained that America's allowing such cross-registration was a major factor in the new German move.

zone would not be molested. No such written response was forthcoming, and in early May the unthinkable happened: A German U-boat sent a torpedo into the side of the giant passenger liner *Lusitania*. In 18 minutes, the 761-foot vessel disappeared below the surface. More than 1,000 men, women, and children, including 138 Americans, were lost.

Wilson was stunned. He sent a note of protest to Berlin and told the American public he was prepared to "act quickly and firmly, but deliberately." Before he could act, Berlin expressed regret for the loss of American lives, blaming Britain for the unfolding crisis. If Britain had not tried to starve the German civilian population through her coastal blockade, Berlin would not have had to resort to such retaliatory measures.

The fact that Germany had responded so quickly suggested that Berlin might be prepared to step back from the brink. Washington insisted that attacks on American craft and the taking of American lives had to stop. Wilson told Berlin that he would consider the next violation of American rights to be an "unfriendly act." Yet, with the staggering loss of life on the European continent, neither Wilson nor most of the American people were ready to enter the dreadful conflict.*

Finally, in late August, Germany sent the necessary assurances. The seas would be open to American shipping and American lives would be protected. In September, Germany even withdrew her limited fleet of U-boats from the English Channel.

The immediate possibility of U.S. entry into the war had passed, and the fishermen of Gloucester continued to have great faith in the future, as evidenced by contracts placed with boat builders.† At the same time, fishermen spoke of feeling isolated as they sat on the vast northern grounds. They carried no wireless and could go for weeks without hearing news of the war, or the actions of the Wilson administration, or the intent of the Germans. Many felt vulnerable as they fished.

Yet what counted was the continued success of the fisheries, and on a typical early fall day many different types of craft could be seen unloading in Gloucester. There might be a halibuter, a couple of Cape Northerners, a salt drifter, a mackereler, and even one or two British-registered craft landing duty-free salt cod at one of the processing plants. The firms, vessel owners, outfitters, and skippers were making money. Vessel owners no longer felt the

---

*By August 1915, losses were in the millions, with a similar number of wounded. By country, the war death count stood as follows: Germany, 1,630,000; Austria, 1,610,000; Russia, 1,250,000; France, 460,000; England, 181,000; Turkey, 111,000; and Belgium, 49,000.

†For example, Captain Archie McLeod let a contract with the Arthur D. Story boatyard for the largest schooner in the fleet, and called her the *Catherine*. She was a mammoth vessel, at 130 feet in length and 159.67 gross tons of displacement.

need to sell their craft, and orders for new vessels continued strong throughout the war.

At the same time, the men crewing the dories did begin to wonder about their future. Outfitting costs had begun to rise, and these costs were taken out of the "lay" of the trip before the balance of funds was divided among the owners, skipper, and dorymen. To many dorymen, it seemed as if there was less money to go around. A few openly admitted to flirting with the idea of joining the new American Federation of Labor (AFL)–sponsored Fishermen's Union in Boston. Men were listening carefully to the words of the union business agent as he spoke of changing the lay from the trip, improving working conditions, and ensuring job security. They were interesting ideas, but for now that was all they were, ideas: Collective action was a year and a half away.

If some of the dorymen were uneasy, the skippers were, by and large, content. Life was good, and between trips many men came together at the Master Mariners' or a favorite local saloon to play cards and talk of life on the grounds. On one such day, early in the fall of 1915, Captains Charles "Blossom" Colson and Jeff Thomas found themselves in the same Gloucester saloon, celebrating with their friends. But a Gloucester saloon could also be a tough place, and as the night unfolded, a fisherman who had had too much to drink began to pick on Blossom. He was teasing him and Blossom was getting upset. Blossom started to stammer a little, and still the antagonist persisted. Then, from the back of the room, came a loud booming voice. It was Jeff Thomas. He yelled out, "Leave that man alone. Stop teasing him." A hush came over the room, and Blossom was left alone for the rest of the night. Even though Jeff Thomas was mild mannered and didn't often get upset, he was upset that night and he meant business; Blossom was his friend.*

Skippers such as Colson and Thomas enjoyed one another's company, and in a time of new prosperity the bonds they forged were real and deep. But they were also highly competitive, and in 1915 Gloucester watched intently as they competed for highline honors in haddock and halibut. As Captain Colson left on his next trip, he noted with interest those who sat atop the halibut highline race. Colson was in the top five, but based on the most recent trips, the lead fell to the Gorton-Pew schooner *Teazer*, under Captain Peter Dunsky. She had made a fine stock of $24,500 since the beginning of the year. Second place went to the *Robert and Richard*, under Captain Robert Wharton of the John Chisholm fleet, and third place went to the *Oriole*, under Captain Daniel McDonald of the William H. Jordan & Company fleet.

---

*Personal communication from Jeff Thomas, the grandson of the skipper whose name he carries.

These positions would not stand for long, as each trip between then and December saw shifts in position.

As for Captain Colson, his next halibuting trip proved anything but stellar—only 7,000 pounds of halibut. A severe fall gale prevented the men from setting their dories during almost the entire trip. Running low on supplies, he returned to port and surrendered any long-shot hope of becoming highline of the halibuting fleet. The contest ultimately came down to two men: Robert Wharton, skipper of the *Robert and Richard*, and Gus Hall, skipper of the *Rex*. Captain Wharton made landings from the Cape Shore in October and November, bringing in fresh halibut on each of his trips. By November he assumed the mantle of highliner of the halibut fleet. But by the end of the year, Captain Hall had won, achieving a final stock of $30,500 for the year,[*] with each of his crew receiving a $740 share.[†] As for Captain Colson and the *Natalie Hammond*, he, too, achieved a fine stock, but it was made on halibuting in the spring and summer, and haddocking the rest of the year. His stock for the year came to $32,970, with each of his men receiving a share of $773. It was one of the best stocks and shares made by any vessel in any branch of the fisheries during the year.[§]

It had been a good year, and as the season drew to a close, the local paper took note of the success of the community's growing fleet of small Italian shore boats, still calling them the Ginny fleet. While some in the community might consider the operations of small boats of little importance, their accumulated annual landings, all of which came into Gloucester, were quite respectable. The Italian fleet had grown to more than 30 boats, each being fished with a three- to five-man crew. They went out for no more than a day or two, and sometimes only for a morning. Depending on the season, they brought in fresh herring, ground fish, and mackerel. Their total landings for 1915 ran into the millions of pounds, and during the year's mid-season mackerel run they had made big money. As a result of their success, one entire section of the Fort was now owned by Gloucester's Italian colony.

---

[*]If the fresh-halibut fleet had a reasonable year, the flitched fleet was all but nonexistent. Of the two craft that went out, the schooner *Senator* under Captain Alex Langer arrived back on September 15 with a small catch. She had 45,000 pounds of flitched halibut and 7,000 pounds of salt cod. The other vessel involved in this fishery, the *Atlanta* under Captain Richard Wadding, landed 110,000 pounds of flitched halibut and 20,000 pounds of salt cod.

[†]The highline of the fresh-haddock fishery was Captain Ernest Parsons in the schooner *Pontiac*. He stocked $31,000; she went single-dory fishing. The highline of the mackerel-seining fleet was Captain William Corkum in the steamer *Lois H. Corkum*. He stocked $33,200. Because these vessels went out with smaller crews, each of the men received a share of $1,100.

[§]Steve Olsson's wife, Hilda, banked much of the money toward the day when they would purchase their own home.

CHAPTER 23

1916—The Lull Before the Storm

With the start of the new year, Gloucester craft began to fit out for winter fishing on the Banks. Thousands of barrels of bait herring had come south on British- and American-registered schooners, including the *Flirtatious** under Captain Charles Daley, while the first cargo of Mediterranean salt reached Gloucester—1,700 tons from Spain on the five-masted schooner *Margaret*. Wherever one looked, dorymen could be seen stowing supplies, checking gear, and preparing for the first trip of the year.

But to the dismay of the fishermen, in early January, strong winds and a mixture of snow and heavy rain lashed the coast. Shore craft and gill-netters found it difficult to leave the harbor, and only the large offshore schooners made it out to Georges and the Gulf of Maine. Once there, men spent as much time chipping ice off the deck and rigging as they did setting trawl and bringing in a catch. But the fish were there, and among the first skippers to return with sizable fares were Captain Archie McLeod and the young Norwegian, Captain Carl Olsen. Each brought more than 40,000 pounds of fresh, iced fish into Boston, and with cod going for $8 a hundredweight and haddock for $5 to $7, the two skippers received top dollar for their fish.

Carl Olsen was a relatively new skipper, rising from a penniless immigrant in the late 1890s to command one of Gloucester's finest highline schooners. He had arrived in North America on a European square-rigger, jumping ship in Nova Scotia and walking south into the United States. On crossing the Maine border, he asked a local farmer to look after his heavy

---

*She was now under registry to Newfoundland's Bay of Islands Fisheries Company. In April she ran aground on Big Ledge at Bay of Islands, but was soon off, and Captain Daley took her into the salt-bank fishery. Robert C. Parsons in *Times, Tides, and Tales* (Vol. 1, Issue 6, 2001) tells of Captain Daley and the *Flirt* in 1915, when Charles and Joseph Daley went to the rescue of two men in an overturned dory. One man was on the bottom of the boat, the other in the water. The one in the water, Austin Breen, shouted to Charles, "Save Jim! I'm all right," but Breen was to die of hypothermia. Charles was taken by Breen, saying, "Fancy, a man saying he was all right, holding a dory's gunnel, which was under water. Breen could have easily gotten on the bottom of the dory, but he knew by doing so, he would probably drown his weakened dory mate and maybe himself too."

For the 1914–1915 winter season, Captain Daley brought the *Flirt* north from Gloucester to gather herring from the frozen Salmonier Arm of St. Mary's Bay. Holes were cut in the ice and nets were set to catch the herring. The fish were spread on the deck to freeze and then stored in barrels. Once the schooner was filled, the people of the bay used axes and saws to cut the vessel out of the ice, and the *Flirt* sailed to Gloucester with a load of herring.

bags, promising to return at a later date. He traveled south to Gloucester, signed on as a doryman, and began to learn his trade. He never retrieved his bags, nor saw the farmer again.[40]

But even for the best skippers, one of which Carl Olsen was to become, the early season storms made it difficult to set trawl and bring in a fare. For weeks on end, towering seas and blinding snow swept over the grounds, and craft returned with torn sails, damaged bulwarks, wrecked gurry pens, and snapped bowsprits. In late February, the *Corona* was dashed to pieces on Green Island, Nova Scotia, her crew lucky to escape with their lives. Three crewmen from the schooners *Benjamin A. Smith* and *Reading* were lost at sea.

At last, as February gave way to March, tranquility settled over the grounds and schooners returned with fine fares. Boston prices for haddock, cod, and fresh halibut remained high,* and it soon became clear that 1916 would be a much better year than the past one. But then, in late spring, the talk around the waterfront changed: President Wilson asked the country to focus its attention on the war in Europe. He sent a message to Congress criticizing Germany for the recent torpedoing of the French cross-Channel passenger steamer *Sussex*. Wilson equated her destruction to that of the *Lusitania*, and insisted that Germany constrain its U-boats. War hung in the balance, but Berlin responded positively to the President's demands: Neutral ships would now be warned before being sunk, and humanitarian precautions would be taken to safeguard the lives of noncombatants.

Wilson's fall reelection campaign was to pick up on this theme, his advisors building it on a simple six-word slogan: "He kept us out of war." Yet there were Americans in Gloucester and elsewhere who did not share the President's view of the world. Republicans saw the Democratic President as a vacillator, someone who stretched the notion of neutrality to its breaking point. By negotiating with Berlin, the President had failed to understand the patriotic spirit of the American people, and the Republicans promised a fierce campaign in the fall.

As prospects of entering the war faded, the economic changes brought on by the conflict returned to center stage. Profits, wages, and inflation were on the rise. The economy was growing as never before, and while many prospered, inflation was touching the lives of the average American family. If left unchecked, great harm would result.

---

*Only one vessel, the *Atlanta* under Captain Richard Wadding, made a flitched-halibut landing in 1916; she had also been one of only two flitchers in 1915.

In Gloucester, the Net & Twine Company was the first to grant a 10 percent wage increase to its employees, making the adjustment in late spring and promising a second increase in October—a promise that was kept. The fish-processing companies followed suit, saying they understood the need to help sustain the workers on the wharves, lofts, and offices in this most unusual time. Inflation was real, and the processing firms also increased the salaries of their workers in the spring and again in the fall.

Now it was the turn of the fishermen, and they looked to the vessel owners to match the actions taken by those who owned the skinning lofts and box plants—but they were to wait in vain. The vessel owners contended that there was no need for adjustment and that the fishermen had no need to worry. As the lay of the trips increased, they argued, everything would balance out, including the men's wages. The fishermen did not agree. They said that rising outfitting costs were already taking a bigger bite out of the stock. This was leaving less, not more, money to go into the men's shares. But the owners held fast to their position, insisting everything would come out all right in the end.

The men did not hold tight. By late spring, growing numbers of dory-men could be found turning to the Boston Fishermen's Union for help. They signed union cards, attended meetings, and spread the union message from one vessel to the next. Their demands were simple: an adjusted lay, all-union crews, and improved working conditions at sea. Organized labor had come to the fishing fleet, and Gloucester's fisheries would never be the same. Symbolic of the changed environment, in late May the eight union dorymen on the halibuter *Mildred Robinson* refused to go to sea without an all-union crew. They stood firm, and with the skipper anxious to get away, he agreed to their demand.

It was a symbolic victory, but important nonetheless, for to many in the union movement it now seemed possible that they might actually carry the day. If they stood together, who knew what could be accomplished? Yet 1916 would see only the first skirmishes in the contest between union and owners. The fishermen called no strike, there was no adjustment in the lay, and few vessels went out with all-union crews. True collective action was still a year away.

Sensing that something disturbing was in the air, many old-line skippers stepped forward to condemn union proposals. Skippers, they said, selected their own crews and had full control once at sea. Sure, dorymen grumbled, everyone knew that the language in the forecastle could be hot and pointed, but once on deck, skippers ran the show. Men's lives and livelihoods were in their hands, and Gloucester skippers took the responsibility seriously.

The environment was charged, and the resolve of the skippers was hardened by reports of an open revolt on the schooner *Esperanto*. The story began innocently enough, when the *Esperanto* was damaged on the Cape Shore and the skipper sailed her to North Sydney for repairs. The marine railways were already in use. Other vessels were lined up to come on the ways, and Captain Stewart decided to go to Port Hawkesbury, 77 miles to the south, to repair his vessel. For the trip to be made, the crew would have to agree to man the pumps. They demanded and received a $10 bonus for so doing. Then, once the repairs were complete, the men issued a second demand, a $25 advance before they would take *Esperanto* back to sea. Again the owner ordered the skipper to make the payment. Once back on the Cape Shore, the men toiled for days in a vain attempt to bring in a catch. One frustrating day blended into the next until the men came together in open revolt, ordering the skipper to turn for home. It was an unthinkable state of affairs, and Captain Stewart rejected their demand out of hand. But that did not end the story. That night the men threw the remaining bait over the side and cut the rigging. The fate of the *Esperanto* was in their hands, and Captain Stewart had no option but to turn for home.

The events that took place on the *Esperanto* polarized the community. Owners and skippers were now more determined than ever to maintain the status quo, and dorymen continued to worry about their future. Collective action seemed their only hope, and somewhat later in the year, Gloucester's Portuguese dorymen became the next to organize, forming an association to pressure owners to adjust the lay and permit all-union crews. The men went from ship to ship, talking about their concerns, and almost all skippers permitted men to come aboard and gather together in the forecastle, but not the skipper of the schooner *Jorgina*. Captain Alvaro Quadros stood his ground. He refused entry, chaos ensued, the Gloucester police were called in, and arrests were made.

In this heightened environment, about 200 Portuguese fishermen soon came together at the association's first fleet-wide meeting. Their leaders said the men would now go to sea on their terms, on all-union vessels and with an increased share of the lay. They were prepared to act as one. The owners were not impressed. They tied up the fleet, including two vessels that had already taken on bait and ice and were ready to go to the grounds. No further decisions would be made until all 16 Portuguese vessels were in port. Until then, the Portuguese fisheries were closed down, and while the association released statements affirming the willingness of the men to go to sea, the owners refused to talk. In their view, Gloucester's Portuguese dorymen were among the best paid, and outside elements had caused all of the trouble. The owners held out, and by month's end the association's embargo was broken. The status quo would be maintained.

Gloucester skippers, whether Portuguese, Nova Scotian, or Scandinavian, were unwilling to let others decide who went out on their vessels, what happened once at sea, or how the lay was divided. Masters and owners made these decisions, not fishermen. Nowhere was this proven more true than on the vessels of Captains Billy Thomas and Clayton Morrissey.[41] On his next trip, when Captain Thomas brought the *Thomas S. Gorton* into St. Pierre Island, he permitted each man to buy up to $10 worth of liquor, but he warned that if he caught "any man showing sign of liquor while . . . out on the banks the whole business goes overboard." The captain was known to be a man of his word, and the liquor remained stowed. At about the same time, on the schooner *Arethusa*, Captain Clayton Morrissey applied the same rule when his mostly French, Nova Scotian crew came into St. Pierre. But as they returned to Gloucester, two of his crew got "touched up" a bit, and Morrissey was good to his word: $500 worth of alcohol went over the side. Morrissey was of the old school, a rugged man, 6 feet 4 inches tall, and no crewman dared step forward to question his authority. They might grumble in the forecastle, but none thought it wise to take on the skipper.

But even as owners and skippers took pleasure in the accounts, Captain Billy Thomas's days of command were coming to an end. The newly emerging union environment troubled him, and he sold the *Thomas S. Gorton* to the Gorton-Pew Fisheries. He would spend the remainder of his days as manager of the Gloucester Cold Storage and Warehouse Company freezer.

Billy's was not the only craft sold in 1916. Vessels went for good prices to firms in Newfoundland and Nova Scotia, taking the place of Provincial craft that had moved into more profitable commercial activities closer to the European war. Newfoundland agents were in the market for 40 to 50 vessels, and it was an opportune time for Gloucester skippers and owners to sell off their wooden schooners. Fifteen vessels were sold in the first half of 1916, including the *Clintonia* and *Frances P. Mesquita*, and another seven were sold by year's end, including Lemuel Spinney's *John Hays Hammond* and Robertson Giffin's *Conqueror*.

But the fleet was far from being in decline; rather outfitters and skippers stepped forward to commission a dozen of the most modern of vessels. Big stocks had become an everyday happening, and demand was on the rise.*

---

*For example, the new schooner *Catherine*, under Captain Archie McLeod, stocked $42,348; the *Arethusa*, under Captain Clayton Morrissey, stocked $49,482 in haddocking and shacking; and the *Sylvania*, under Captain Jeff Thomas, stocked $49,850 in fresh haddocking and shacking. Similarly, the *A. Piatt Andrew*, under Captain Wallace Bruce, broke all records for a 12-month period, stocking $53,395 in the fresh-trawl fishery.

Two of Gloucester's largest firms ordered the city's first otter trawlers from the shipyards in Essex. These immense wooden vessels would begin to slowly change the face of the Gloucester fisheries. But mostly what stands out is not that otter trawlers were ordered, but rather how long it had taken Gloucester outfitters to initiate the change. The first American otter trawlers had come to Boston a decade earlier—large vessels, with costly steam engines to manage immense nets. But only now, with the changed economic conditions brought on by the war, did Gloucester firms commit to the huge investments required for these vessels.

Otter trawling represented a departure from everything known in the past. A new day was dawning, a day in which fish were caught in huge nets managed by a few men from the deck of a trawler. They used a mechanical winch to lower the net into the sea from the deck, and it was then dragged behind the vessel, bouncing along the bottom, gobbling up all in its way. After an appropriate period of time, the skipper called for the net to be closed, and power winches raised it to the surface and lowered it onto the deck. When opened, the bounty of the sea came spilling out, ready to be sorted, stored, and boxed. There was no longer a need for bait, hooks, or dorymen. Otter trawlers went out with smaller crews, fished in all kinds of weather, and brought in substantial catches.

Such trawling in America went back to 1905, when the Boston-based Bay State Fish Company commissioned the *Spray*—an all-steel, 150-ton craft, 127 feet at the waterline, and with an engine that was able to make 10 knots. She first went to sea in December 1905, fishing off Chatham and on Georges Bank, and an onlooker said that as the skipper prepared to fish, the trawl lay on the deck in a tangled mass. Forward and aft stood two iron frames, called "galluses," not unlike inverted V's, and through blocks attached to these ran the wire from the winch that ended in the "otter boards," one hanging from each gallus. The boards were 11½ feet long by 3½ feet high, heavy and steel bound, weighing 750 pounds, and were used to keep the trawl net open as it was guided along the sea bottom. On the skipper's command, the winch commenced to turn, the otter boards were lowered into the water, and the net and boards sank beside the motionless steamer. When sufficient wire had been let out to ensure the trawls would drag on the bottom, a long hook-tipped line called a "pbon" was run down one wire, crossed over the other, and then the two were brought together and made fast.

Two toots of the whistle and the *Spray* made full speed. Once the skipper decided the trawl had been out long enough he gave the word, and the winch reeled in the wire until the end of the trawl rose dripping from the water. Then, by means of a guide rope, the crew hauled the edges of the trawl over the rail and, laying hold, yanked the heavy trawl to the deck. Out poured a mass of fish on the deck—cod, haddock, large and fat flounder, as well as shark-like dogfish, huge ugly skates, and green-eyed monkfish whose mouths were lined with needle-pointed teeth. There were also squid, butterfish, and odds and ends of fauna and flora. Then the skipper called for the trawl to be again "shot" out, and as it went into the sea, the crew used pitchforks to sort the fish—cod, haddock, and flounder were kept, all else went overboard, many with "an unfriendly remark on the part of the fishermen."

The catch from the first trip was slight—8,000 pounds of haddock, 1,500 pounds of cod, and 2,000 pounds of mixed fish; but on the *Spray*'s fifth trip she brought in 97,000 pounds of fish after 12 days on Georges. On her next trip, in early March, she was among a group of vessels overtaken on Georges by the winter's heaviest gale, yet she rode out the storm without damage, and returned with haddock, cod, flounder, and quite a number of deep-sea scallops. She then went to Western Bank, returning with 130,000 pounds of fish. Unable to sell all her catch at T-Wharf in Boston, for the first time she brought part of her catch (90,000 pounds) to Gloucester, and hundreds of fishermen and ordinary people came out to give her a most thorough inspection.

As her luck improved, protests began to be heard from the fishing community. Captains circulated a petition claiming that otter trawling would tear up the seafloor and kill off the fish. They said that on a recent trip, the *Spray* brought in 23,000 fish so small that they sold for only 1 cent apiece. In July 1909, the issue escalated to violence when Captain Vincent Nelson of the schooner *Senator Gardner* fired on a French otter trawler—otter trawlers had been introduced in Europe long before they arrived in the Americas. Nelson had struck a "fine spot of fish" on the eastern shoal of Quero Bank when a French steam trawler appeared, carrying off the dorymen's trawl lines as it swept over the fishing ground. Captain Nelson spoke with the French skipper and thought the Frenchman would henceforth drag clear of the ground trawl, and for the remainder of the day the trawler did, indeed, keep about two miles from the *Gardner*'s dories.

But imagine Nelson's surprise the next day when the otter trawler was nearer than ever, dragging over the ground where the dorymen had just set their lines. The dorymen vented their anger in strong Anglo-Saxon expletives, but Captain Nelson went further: He brought up his rifle and fired five or six shots at the French trawler. No one was hurt, there was no damage, but the trawler quickly left the grounds. Captain Nelson reported that each otter trawler had two trawls, and while one was being hauled the other was lowered and put to work; there was no let-up. He believed the French masters did not know the fishing grounds and kept close to the schooners to find the prime spots.

As the years rolled on, there were other confrontations and many calls for reform—the pleas represented more than simple self-interest; the dory fishermen sensed that the fish stock was in danger, as they had seen the mackerel and halibut disappear from the grounds. Suggestions included limiting the times when otter trawlers could go to sea to prevent the disturbance of the spring haddock spawning season; limiting the number of otter trawlers permitted on the Banks at any one time; setting a size limit on otter trawlers to reduce the chances of any one vessel over-fishing an area; and a full prohibition of these vessels on the grounds. All proposals failed: The tide could not be turned. The large otter trawlers were a fact of life, and they would bring in a substantial catch during the war, both in Boston and in Gloucester.

For much of 1916, Gloucester's focus had been on financial matters, but the community could not fully ignore the ongoing war in Europe. In the fall, the people would be called on to speak at the polls, and the people of Gloucester watched from a distance as the two parties selected their presidential candidates and marched toward the fall presidential contest. Gloucester was a largely Republican city, and while there was broad support for President Wilson's focus on global neutrality, there were those who questioned his economic and other policies. In a speech before the Master Mariners' Association, Charles D. Smith, the city's Democratic postmaster, spoke of the unique international circumstances through which they were living. He was sure that no matter their party, the skippers wished the President "Godspeed and success in his efforts to keep our loved country out of war." While Europe was being torn apart, America was enjoying unbounded prosperity, and the postmaster wanted all to remember that they had the President to thank for that.

In response, the local paper, whose politics were Republican, editorialized that while Wilson had, of course, done good things, the President's "personality" was in question. The writer said that Wilson shifted positions several times before taking final action, indicating the President was unable to trust his own first judgment. Local Republican politicians went even further. They said the United States had become soft under Wilson's leadership, repeatedly seeking assurances from Berlin and then waiting patiently for the Germans to respond. More decisive action was required.

As the campaign rolled on, both the Germans and the British took actions that confounded their American supporters, calling into question not only Wilson's leadership but also the trust placed by the Republicans in the

British. For months, the British had been intercepting American merchant vessels at sea, preventing delivery of supplies to Germany, while the Germans sent a huge U-boat to America on a so-called "commercial" venture. It was a confusing time. Most in Gloucester looked on Germany as the enemy, but the fishermen were distressed with Britain's continued harassment of neutral shipping.

In August, the Mannes schooner *Lucinda I. Lowell*, commanded by Captain Fred Thompson,* was detained for inspection in the Shetland Islands. The British said the cargo was Canadian herring bound for Germany, not American herring bound for Norway as claimed on the manifest. The British were always leery of Norwegian cargos, saying such goods all too often went south to Germany. Gloucester craft from the Fort had, indeed, taken the herring within one and one-half miles of the Cape Ann shore; Gorton-Pew Fisheries had packed the fish, and it was stowed on the *Lowell* for delivery to Norway. The documentation was immaculate, and after a prolonged wait the British permitted the *Lowell* to go on her way. In a second August incident, a British patrol craft escorted the schooners *Lizzie Griffin* and *Maxine Elliot* to the Shetland Islands, where they went through a similar charade. Papers were checked, cargos inspected, and the vessels made to wait for a prolonged time before being released. Gloucester ships fishing for herring off Iceland faced a more daunting challenge. If their herring could not be shown as destined for the British Isles, there was little chance of it getting through. One Gloucester skipper reported that the wharves and fields of Iceland were filled with barrels of British-confiscated herring, rotting in the sun.

As the outrages escalated, more and more Gloucester people expressed their dismay with the British position: Maybe Wilson was right to keep America on the sidelines. Then the German merchant submarine *Deutschland* arrived in Baltimore, Maryland. She had gone undetected in her 4,000-mile trip across the Atlantic, having cruised 1,800 miles under the sea. She had eluded British and French naval vessels, and showed up without notice in an American port—a remarkable feat for any vessel. She brought a cargo of dyestuffs and picked up a cargo of crude rubber and nickel. The trip took 16 days, and Germany extolled the prowess of her merchant vessel, both in America and back at home. Everyone got the real message: U.S. shipping would be at risk if the country entered the war on the side of the British; German U-boats could reach American shores.

---

*In taking this command, Captain Thompson was combining business and pleasure, intending to visit his native home while in Norway.

In mid-October, as the presidential campaign drew to an end, the *Deutschland* again appeared off the U.S. coast. She bobbed up alongside a fishing sloop, the *Little Marguerite*, off Massachusetts. Joseph Brager, the lone man on watch, saw her periscope break the surface, followed by the U-boat herself. Her searchlight scanned the sea, and she disappeared below the water.

Then America went to the polls. In Massachusetts interest was keen, but no one doubted the state would go Republican. Gloucester's only Republican rally came right before the election, with 600 people filling three blocks of Main Street. They came to hear Gloucester's Republican Congressman, and Augustus Gardner did not disappoint. In his view, Wilson's claim of having kept the country out of war was "an appeal to our faint hearts not to our gratitude. In plain English, Vote for Wilson because he can't be kicked into war. . . . To the sorrowful bewilderment of our friends and to the malicious glee of our enemies in the cap of the Goddess of Liberty paltering fingers have stuck a white feather."

While it was evident where Gardner and the majority of his Gloucester constituents stood, it would take days to learn where the country stood. This was to be one of the closest elections in the country's history. Wilson finally carried the day in the Electoral College with a vote of 276 to 255. He had won his second term.

It was generally assumed that America would remain at peace, and in Gloucester most believed the city would continue to build on its prosperity. Here the vote had gone Republican, but by less than the expected margin: 1,918 for Hughes, 1,510 for Wilson. The President might have kept the country out of war, but many felt humiliated by the cowardly abandonment of American rights, and who knew what the next year might bring? Eleven months earlier, inflation had been in check, the President had had the respect of his people, and the country had been at peace. By year's end, inflation had risen to an unimaginable 12 percent, the President had been reelected by the slimmest of margins, and a German submarine had crossed the Atlantic without being detected.* Now the question facing Gloucester and the country was whether it would all change. Wilson said no, but the key decisions of 1917 would be largely out of his hands.

---

*One other significant change occurred in this year. By January 1916, 18 states had voted for a prohibition on alcohol, including Maine, many of the southern states, and Colorado, Arizona, and Oregon.

CHAPTER 24

Winds of War in Early 1917

In the opening days of 1917 nothing seemed to have changed. America was not at war, Germany was not systematically attacking neutral shipping, and Gloucester's fishermen were receiving fine stocks for their first trips of the new year. The *Natalie Hammond* had received a stock of $2,600 and the *Sylvania* $4,600, while for three quick trips, Captain Joe Mesquita had stocked $12,000 on the *Joseph P. Mesquita*. Gloucester was also excited to see that Gorton-Pew had bought yet another of its competitors, the Cunningham & Thompson Company—the second largest fisheries firm in the city, and the major presence in the Fort section of the waterfront. To finance the purchase, Gorton-Pew sold $750,000 in preferred stocks, and gained 12 schooners, docks, and plant facilities, including a concrete building said to be the last word in fish-plant architecture. On balance, it would seem that the positive economic environment of 1916 was carrying over into the new year, but there were troubling reports about the war in Europe and rising inflation in the United States.

Nevertheless, with fine January weather, life went on as usual. The winter fleet was at sea and the stories filtering in had a familiar ring to them. There was an account of the first man who was lost at sea in the new year— Joseph Daley, the brother of Captain Charles Daley and the nephew of Joseph Bonia, disappeared when his dory overturned. On a more positive note, with good early season weather, skippers were bringing in fine first and second trips. The catch was high, prices were holding, and the fishermen had money in their pockets. In mid-month, when skippers tied up to attend the annual Master Mariners' Association banquet, the dorymen enjoyed a few extra days ashore with their families. Romaine Olsson remembers being asleep when she "heard a loud 'boo' in the hallway, just outside her bedroom. There was Papa trying to scare us children." Steve Olsson was home, and Romaine sheepishly told him that she wanted a penny "in the worse way" to buy candy at Savages' store. Steve grinned, and not only gave her a few pennies but joined her the next day in selecting just the right pieces. It was a simple act, but it typified what was happening all across the community as the men spent a few wondrously average days with their families.

As Romaine and Steve went out to choose candy, and Captain Colson and the other skippers came together at the annual Master Mariners' Association banquet, Germany announced that she would rescind her promise of the previous April to safeguard neutral craft and lives. From February 1, German U-boats would no longer forewarn neutral ships before firing upon them, nor would U-boat crews take special steps to safeguard neutral lives. The period of American inaction was at an end. Wilson's valiant balancing act passed into history, and the sweet stories of Gloucester fishermen and their children soon seemed but a dream of a world long lost.

On February 3, German Ambassador Von Bernstorff was ordered to leave the country, diplomatic relations were severed, and Secretary of the Navy Daniels announced a new policy of restricted access to U.S. Navy yards. The sounds of war began to sweep across the country. In Gloucester, the Master Mariners' Association telegraphed its support to the President, and on the waterfront, the first sign of the new economic reality was witnessed. The price of halibut was bid up to 27 cents per pound for white and 16 cents for gray, an overnight increase of almost 50 percent.

Three days later the central committee of the American Red Cross telegraphed all chapters placing them on a war footing, and the Gloucester chapter was quick to respond. Even so, its actions were overshadowed, for one day at least, by the concern for the fate of one of the schooners of the fleet. The local paper reported that the schooner *Joseph P. Mesquita* had been hoved down on Roseway Bank off Shelburne, Nova Scotia. The onrushing water swept the decks clean of dories, chain, and the substitute skipper, Captain Peter Richards. The schooner was lucky to have survived.

On the following day, the paper's lead story shifted back to the imminent threat of war. An official from the Wilson administration said the government was unimpressed with the German offer to discuss differences; there would be no discussions as long as Germany refused to reverse its new policy on neutral shipping.

The Wilson administration and Congress moved swiftly to establish new regulations and pass new laws to help bring the country through this time of peril. Among them was a presidential order that forbade the sale, charter, or lease of American craft to noncitizens without the approval of the U.S. Shipping Board. It applied to all craft, including fishing vessels, and some in Gloucester wondered whether the vessels of the local fleet might now be safe from sale during the course of the war. About 200 vessels remained in the district: Gorton-Pew owned 65, while the remaining vessels were owned by 106 different individuals and firms. On the very next day, the U.S. naval authorities announced plans to confiscate Boston's entire fleet of 14 steam-fishing trawlers. They were to be converted to minesweepers and assigned to patrol duty along the eastern seaboard. The action came without warning, and Boston's fish dealers protested mightily. Gloucester outfitters, however, remained silent. The removal of the Boston trawlers provided a competitive advantage to Gloucester and its sailing fleet.

In late February, the government announced plans to mobilize reserve units, and the men of the Massachusetts Naval Militia were among the first to be examined to determine their readiness for sea duty. The Naval Reserve also called for men to enlist, particularly those who had knowledge of the many power craft that were volunteered for use during the crisis. Just as these events were being digested, word reached Gloucester of the loss, off Europe, of three craft with direct ties to the port.

Two went down in the war zone surrounding Great Britain, the third in the Mediterranean. But it had been the weather, not German U-boats, that had been responsible. The largest of the lost vessels was the British steamer *Ronadalen*, a salt transporter that had made regular trips into Gloucester; the second and third were Newfoundland-based fishing schooners, the *Rose Dorothea* and *Flirtatious.**

Then the government agents came to look over the Gloucester fleet. They had already stopped in Essex to inspect the two large wooden Gorton-Pew beam trawlers then nearing completion. Their task was to see what might be available should America go to war, but no decisions were announced as they headed back to Boston. Meanwhile, the fishing continued, and the same weekend that the agents came, the *Sylvania*, *Natalie Hammond*, and *A. Piatt Andrew* were among a large number of Gloucester's wooden sailing schooners to land a total of 1.25 million pounds of fresh fish at Boston. The Lenten season had just begun and fresh fish was in high demand. Prices were rising, but the dorymen still believed that their wages were lagging behind

---

*The *Rose Dorothea* had been sold out of Gloucester within the year, and she and the *Ronadalen* were lost while transporting supplies to Great Britain. The schooner *Flirtatious*, under the command of Captain Charles Daley, had made regular trips from Newfoundland to Gloucester through the previous fall. But over the winter she sailed to Genoa, Italy, with a cargo of Labrador salt fish, and was returning with a cargo of salt when she went down. She ran into a squall off Cape Finisterre, Spain, and as the winds picked up, her cargo shifted and she rolled over, her masts went under the water, her decks were swept free of boats and rigging, and the load of salt was dumped into the sea. Afloat, but in the turtle position, she could not right herself. Three men were lost, including one from Gloucester and two from Newfoundland, while the rest of the crew either escaped into small dories or joined Captain Daley on the hull of the overturned vessel. For several days they held on and were finally rescued by the schooner *Annie M. Parker*, out of Gloucester. The news of her loss reached Gloucester on Ash Wednesday. Friends of the crew thanked God for sparing the lives of Captain Daley and his mates, and many wondered who might be next. All knew there would be real dangers once America entered the war, for Gloucester had just received a first glimpse of the risk to her fleet and men.

inflation. For the union it was the perfect time to strike. Thus, even as the shore workers unloaded the fish and the dorymen of ships such as the *Natalie Hammond* stowed fresh food and bait for their next quick trip to Georges, the union called out its members.*

The next scheduled trip of the *Natalie Hammond* and the other Gloucester craft then in Boston did not take place—the fishermen heeded their union and came out on strike. As Romaine Olsson remembered, "Papa brought home oranges and other good things, he and his mates having divided up the food that they had just brought on board the *Hammond*." Captain Colson let them take it home, and the men all sat back to see what might happen. The dorymen had never before gone out on strike, but these were trying, uncertain times. There was no way to tell how prices and shares might change once the country entered the war on the side of the British, and faced with such uncertainty the men demanded a change in the formula used to divide the stock among the dorymen, the skipper, and the owners—a formula they claimed to be unfairly weighted in favor of the owners.

The union focused its demands on the price charged to the men for upkeep on gear used by the dorymen, which seemed to be rising uncontrollably. Nor was this all: Skippers provided the oil for running the engines, and these charges came out of the lay; they owned the trawl gear and even the foghorn used on the vessels, and the crew had to pay a percentage use fee on each trip to cover wear and tear. While dorymen said that money was being unfairly taken out of their pockets, the skippers retorted that they were the ones running the risk; without this arrangement, skippers said the vessels would continue to be sold north to Nova Scotia and Newfoundland. In their view, it was the poor fishing practices of the men that resulted in gear being damaged or lost, and without the extra share from the stock, owners would be ruined.

It was a collision of wills. The strike was set for March 1, but in reality it began on February 18. Three hundred and fifty haddock fishermen were in port, and they left their craft at the docks, refusing to go back to sea. As soon as a vessel docked, the crew quit. By the end of the day, 20 Gloucester schooners lay idle in Boston. With the arrival of the next group of 11 vessels, the union believed the strike would affect the entire fleet. Late on March 28, 6 of the craft were towed from Boston to Gloucester: the *Natalie Hammond*,

---

*The Fishermen's Union of the Atlantic was chartered in Boston in 1915. It started small, but within two years claimed a membership of 4,200 fishermen in Gloucester and Boston, with a branch being started in New York. Their goal was to safeguard the welfare of the fishermen, increasing pay, lightening labor, and reducing hazards. They also said they wanted to help the owners by producing more fish and bringing them to market in the best possible condition.

A. Piatt Andrew, *Robert Sylvania*, *Richard Sylvania*, *Elsie*, and *Thomas S. Gorton*. Their dorymen had refused to sail back to Gloucester, and some, including Steve Olsson and John Lund, boarded a train at the North Station for the trip home. Steve was soon back with his wife and children, and his daughter Romaine remembered this as a happy time. "Papa was always working, making money for the family, and now he was home. What a treat! Even though they were on strike, Papa was smiling; he was always in a good mood. He just seemed happy to be home—even if the whole fleet was on strike." Gerda, Steve's oldest daughter, concurred:

> He was so glad to be home. He would hug us. . . . Papa did a lot of walking, then as always. He walked all over town. He would say, "It is good to be on solid ground." He went downtown to "fisherman's corner" (in the heart of the downtown). There was a bank and the striking fishermen met out front. The fishermen got together and talked. Right down the side street (Duncan Street) they had the Fishermen's Institute—the dorymen could get books or other things they needed at the Institute. He went in there to meet his friends during the strike. His friends were the other fishermen.

Few of the men were willing to discuss the situation with the press. They limited their comments to simple assurances that they would "hang it out." It was clear, however, that the men from the more successful vessels hoped for a speedy resolution. They had begun to make big money in 1917, and most saw that they had nothing to gain by a protracted tie-up of the fleet. One doryman on the *A. Piatt Andrew* said he "was prepared to go out (to sea), strike or no strike. That (Captain) Wallace Bruce was good enough for him."

Two days into the strike, 17 vessels were laid up in Boston, 6 in Gloucester. Some 500 fishermen were idle, and many of the Nova Scotian fishermen were boarding trains to go home, their bags piled high in the depot. During those early days there was talk of arbitration, but nothing came of it. By March 6, 30 vessels were on strike.

As Gloucester dorymen were picketing the wharves, America was driving toward war. On March 10, President Wilson gave merchant vessels permission to shoot at trailing U-boats, and called for a special session of Congress on April 16. War with Germany was inevitable. In Gloucester more than 40 vessels were tied up, nearly the entire offshore fresh-fish fleet.

Within a matter of days, the union's resolve was tested as the schooners *Atlantic* and *Louisa R. Sylvia* prepared to sail on salt-banking trips. On the *Sylvia*, Captain Newman Wharton, one of several skippers who turned in their union cards, planned to sail with a nonunion crew. On March 15, two salt-bankers were away: the *Catherine Burke* with a nonunion crew and the

James W. Parker with a mixed crew. The men could not chance letting the salt-banking season go by. To make the season pay, the salt-bankers would be gone for months, and it was now or never. In any case, the men felt that the strike would be over long before they returned.

Although two schooners left, none followed. The union established a regular, orderly schedule of picketing the waterfront. Frequent patrols passed along Main Street. All wharves were covered night and day and all vessels monitored to ensure they remained tied up. The union wanted no other vessels leaving for the Banks—salt or fresh. Talk of war continued, but Gloucester's focus remained on its fleet. Two police officers stood watch at the Sylvannus Smith wharf when a large delegation of union men gathered to protest the outfitting of the schooner *Eugenia* for a Georges handline trip. They prevailed upon the crew to quit, but several remained adamant in their intent to sail. While most of the union men were holding fast, a few were feeling it in their pockets and were ready to go back to sea.

At the Cunningham & Thompson Company wharf (now owned by Gorton-Pew), Captain Porper's *Cavalier* was being fit out for a halibuting trip. This time the men were persuaded to quit the vessel by a delegation of union pickets. Farther up the harbor, Captain Joseph Mesquita became the next skipper to attempt to evade the strike, and he succeeded. The *Joseph P. Mesquita* did not come under the provisions of the union, since her Portuguese crew had a separate association and its own labor agreement. They were not out on strike and there could be no objection if the *Joseph P. Mesquita* went to sea. But the strikers knew that this was unusual for Captain Joe. It was true that if someone from the old country couldn't find a place on a schooner, Joe would take them, but it was also true that he normally sailed with a mixed crew—Newfoundlanders, Nova Scotians, and Portuguese.[42]

One month into the strike, the two sides finally agreed to air their positions in a public forum. Mayor Stoddart arranged to use the Municipal Council rooms as a setting to hear both sides. Representing the owners and skippers, Frederick H. Tarr was the first to speak. He insisted that his group would not arbitrate on certain matters, specifically those related to gear, engines, and the recognition of the union—that is, almost everything in dispute. He had no suggestions to make, and several of the Aldermen were surprised and angered. They asked Mr. Tarr to take a message to those he represented to see if they would agree to come together, for under the present circumstances nothing could be accomplished. But the masters and owners were not about to give in. They said the period of the last two years was the first in a long while that the industry had been prosperous, and the owners were justifiably concerned about rising costs. In fact, in recent months sail prices had doubled, rope had gone from 8 cents to 27 cents per pound, and so forth. Captains Morrissey, Thomas, Stream, and Thompson insisted the fishing fleet had declined because owners had not been making a fair return on their capital investment. They would not have sold their schooners and abandoned the business if they had been making money. These skippers feared the proposed changes would drive more owners out of the business. If the costs of outfitting for a trip were not recovered, owners would see little profit. The owners were making money, but they knew costs were rising and feared that to tip the balance of the lay more toward the needs of the men would harm their industry.

For the fishermen, the union's business agent said that terms had been sent to the owners significantly before the March 1 strike date, but they received no reply. In fact, he said, the owners had taken away the craft of two Gloucester skippers who signed on to the union demands. The union had wanted a working agreement, but there was no communication.

The union's resolve was again tested on April 2, when a group of men began fitting out the *Veda M. McKown* at the Davis Brothers' wharf. Word spread quickly and union men congregated outside the gate. Police refused entrance, but the men on the *McKown* were in plain view. Strikers knew who they were. The resulting confrontation was generally peaceful, although some union men called those aboard the vessel "scabs." Some shouted, "We'll get you," which proved too much for the men on the *McKown*, who quit the vessel. The union men dispersed.

A mass meeting was then convened at City Hall, the largest gathering of organized labor ever held on Cape Ann. In his remarks, Secretary Brown stated, "It's a case of a showdown between the fishermen and the skippers and owners. We are on the level and all we ask for is justice." In his opinion it was not the stated issues that most bothered the owners, but rather it was recognition of the union. Notwithstanding Brown's hyperbole, there were major differences of opinion on the substantive issues dividing the two sides. Brown's underlying message, however, was clear: The men had to hang together. They were now five weeks into the strike. The dorymen's last wages were running out, and the union had little money with which to help the men. Union men insisted that they had never wanted to call a strike because they had known how devastating it could be on the wives and families of their members, but they had been forced to act; what else could they have done?

At last came a glimmer of hope. On April 3, the owners and skippers invited a Fishermen's Union committee of 15 to sit with them and the State Board of Conciliation to review their positions. The union jumped at the offer and the two sides met at the Master Mariners' Association building.

After an hour's discussion the owners and skippers agreed to provide a written response to the union's February resolutions. The document was swift in coming: The union demands were officially rejected. The owners would make no concessions, but at least the two sides were talking and were now taking steps to work toward a resolution. Two days later the schooner *Suanto* tried to break the strike. She sailed on April 5 with a nonunion crew under the watchful eye of a patrol boat from the Massachusetts District Police. The union was awaiting the report of the State Board of Conciliation, and they let the ship leave port.

On April 9, rather than continuing to wait for the Board's decision, Gorton-Pew tried to import strikebreakers into Gloucester. Forty-one fishermen arrived in West Gloucester by train at 9:45 in the evening. They were shuffled into a bus and two cars, and headed toward town. After a mile they were met by a large crowd at the Western Avenue Cut Bridge, the only land route into Gloucester proper. Strikers hurled rocks and other missiles at the vehicles as they passed, and as word spread, other union men headed for the waterfront. When they reached the Pew property, the fishermen seized the strikebreakers and paraded them to union headquarters. The union had gained the upper hand, and the city called out its entire police force to rescue the strikebreakers, removing them to the safety of the police station. No vessel left Gloucester Harbor under these men.

Two days later, the tone of the strike changed again. The Board of Aldermen placed a compromise proposition on the table, and the owners and skippers appeared more receptive. But the agreement soon broke down, with both sides offering counterproposals. Several days later, another settlement offer was placed on the table by state officials, and it was accepted. Under the plan, fishermen agreed to continue to pay for a variety of the things they had paid for in the past. These included one-half of the cost for engine oils; tow bills incurred when on trips (but not when in the local harbor); a fair percentage of the cost for electronic hoisting engines (if used) and existing diesel engines; and a 10 percent charge for gear until it was paid for, after which the expense for lost and condemned gear would be taken out of the gross stock.* However, the men would not be required to pay for foghorns or for new engines unless an arrangement had been agreed upon by the own-

*Marine engine technology had improved to a point where it provided benefits to schooners engaged in dory fishing, not only to seiners and swordfishermen. Older schooners were being retrofitted with engines in increasing numbers, and virtually all new schooners went into service as auxiliaries. At the same time, even though engines were an important fact of life, these craft were still primarily sail-driven and would remain so for another decade. As for the hoisting engines, they were 7-horsepower single-cylinder engines that ran on just about anything (kerosene, gas, diesel) and were employed for working ground tackle, hoisting sails, and emptying fish holds.

ers and the crew. In addition, union membership would not be required for a fisherman to go out on a vessel—union men would go out with nonunion men. The last clause of the agreement was the most telling. It stipulated that the settlement was operational only while America was at war with Germany, for in the interim the country had gone to war. The union leader William H. Brown said that as patriotic citizens the men wanted to do their duty to the nation, and were therefore settling for a less than favorable resolution to their demands. Thus, the men said, they waived what they believed to be their just rights, not wanting to be or act any less patriotic than other citizens of Massachusetts.

DURING THE LAST DAYS of the strike, President Wilson had asked Congress to declare that a state of war existed between the United States and Germany. He urged Congress to accept "the gage of battle with all the resources of the nation." On April 6 he signed the war act, and the United States formally entered the conflict on the side of the Allied forces of Britain and France.

Within days, the Commandant of the First Naval District urged captains and masters to maintain a lookout for German submarines. In Gloucester, the city awaited word on the mustering of reservists, the naval militia, and reserve ships. In Boston, work had begun on converting the three trawlers *Foam*, *Spray*, and *Ripple* to minesweepers, the workers adding three-inch bow guns, wireless, and searchlights. But for now, Gloucester's wooden schooners were unaffected, and with the strike over, more than 40 craft were soon being fitted for sea. In Yarmouth, Nova Scotia, 125 once-striking fishermen boarded a steamer for Boston. The *Cavalier* and *Natalie Hammond* were among the first to leave for the Banks, followed by the *Sylvania*, *Arethusa*, and *Ingomar*. The *Hammond* baited at Edgartown, Martha's Vineyard, where the men sent penny postcards home telling of their safe arrival. Steve Olsson told Hilda and the children that the trip across the Bay had been uneventful and he was fine. It would be a short trip of less than a week, the *Hammond* returning with 80,000 pounds of fresh cod and 17,000 pounds of halibut in Boston. Once again men had money in their pockets.

Within a week, all but 15 of the striking craft had been fit out and crewed, and were off to the Banks, even as local registrars began enumerating all military-age men subject to the wartime draft. The Adjutant General identified 2,027 eligible Gloucester men between the ages of 21 and 31 who, unless otherwise disqualified, would enter the draft rolls. Many of these men were disqualified: More than 700 were aliens who had come to Gloucester to work in the fisheries. One man who was not disqualified was Gloucester's Congressman, Augustus P. Gardner, who resigned his seat, was assigned the

X *A view of Harbor Cove looking toward East Gloucester as a schooner heads out under foresail, staysail, and jib. Atlantic Maritime Wharf (left); the steam ferry wharf (foreground).*

rank of colonel in the Army, and headed south to train with his unit. He had championed the U.S. entry into the war in the election last fall, and now was following through on his personal conviction.

In Essex, Gorton-Pew's two new steam-powered, wooden-hulled otter trawlers, *Walrus* and *Seal*, the first for the Gloucester fleet, were about to be launched. Captain Clayton Morrissey would skipper the *Walrus*, while Captain Lemuel Spinney took command of the *Seal*. Although not mentioned at the time, it was ironic indeed to see two of Gloucester's most respected dory skippers now responsible for Gorton-Pew's new otter trawlers. Two thousand people came to see the launching of the *Walrus* at the James & Sons yard on May 19. But in Boston, the Navy Department had taken control of two more of Boston's steel-hulled steam trawlers, and the Gorton-Pew Fisheries was fearful that their new trawlers might be next. At an Executive Committee meeting, the company voted to preempt such a move, arguing that the new vessels were important to food production, but the company was more than willing to spare other vessels.[43] In addition, as a sign of patriotism, the company subscribed $25,000 (with the possibility of another $25,000 if warranted) to the local Liberty Loan drive, which was then working its way across the country, with a national goal of raising $2 billion for the Allies to purchase American raw and manufactured goods. The government never did commandeer the new wooden trawlers; metal vessels were the ones in demand.

By early June, as a Gloucester company of the state's guard was mustered into service,* the Gloucester fleet was setting new stock and share records: The fishermen were doing quite well in wartime America. On a shacking trip, the *A. Piatt Andrew* stocked $8,735, with each doryman receiving a $208 share, both shacking records. And it set the tone for the second half of 1917, as stocks of $4,000 or more became commonplace. The shaky economic future predicted at the time of the strike failed to materialize, but it was still a troubling time, with rising apprehension lest the Germans appear on the coast and attack the fleet. In July, when a blazing meteor left a fiery path across the sky, with smoke and sparks filling the air, crewmen standing watch on the Gloucester schooner *Thalia* believed the Germans had arrived. Vigilance became the watchword of the day.

In early August, the draft of 21- to 31-year-olds was in full swing on Cape Ann and across the country.† Of 176 Cape Ann men examined over a three-day period, 77 were rejected, and of 99 accepted, all but 29 claimed an exemption, with one Gorton-Pew fisherman asking the company to support his exemption on the grounds that he was "imperatively necessary for the industry." Although the company did not support the request, within a week they had made just such a petition on behalf of one of their senior managers. Men who were accepted and had no legitimate exemption were scheduled to be called into the U.S. Army within the month.

Among the civilian community, many soon found ways to support the war effort. Women organized to knit clothes for the departing soldiers—mittens, sweaters, and muffs—while others donated money to a nationwide tobacco fund or the Library War Fund. One dollar covered the cost of the purchase, maintenance, and circulation of one book for the soldiers.

In the meantime, the Gorton-Pew Fisheries signed over a $12,000 check to Captain Lem Spinney to purchase gear for their new beam trawlers. The *Walrus* was ready for her trial trip, the *Seal* was taking on engines and machinery, and the government agreed to commandeer neither.

In the midst of these diverse preparations for war and concerns for the vessels of the fleet, the scope of the conflict was suddenly brought home to the community when word came that the schooner *John Hays Hammond*,

---

*The first few U.S. troop transports reached France in late June.

†When the United States entered the war it had only 107,641 men in its standing army, giving the country the seventeenth-largest army in the world.

one of the foremost Gloucester vessels of the age, had been sunk. Captain Lem Spinney had only recently sold the *Hammond* to parties in St. John's, Newfoundland. Reports said that the schooner had been sunk on a voyage from Newfoundland to Ireland while under the command of one of the most respected of Gloucester masters, Captain Joseph Bonia. There was little information, and for Bonia's loved ones the wait was almost unbearable. Then came word that he was safe. On August 13, he sent a telegram to his wife saying that he was in a New York hotel and would be in Gloucester in a few days. Once home, Bonia had much to say, and the newspaper used his words to recount the story of the *Hammond*'s loss. "When Captain Joe tells anything," they wrote, "he goes at it direct with no frills, no fuss and no 'feathers.'" It was a story of pure grit and seamanship. "'Well you know,' says Captain Joe, 'we left here January 20, and went to Halifax, N.S., where we loaded with salt codfish for Cork, Ireland.'" The trip was unexceptional, the schooner going from Cork to Fleetwood, England, to take out, and thence, loaded with salt, to Iceland. There she loaded salt cod and headed back for Cork. Captain Bonia remembered the trip down as being

> ordinary until we reached off Fastnet, where we plainly heard the report of guns for some time, but we saw no submarines. We knew, however, there was something doing for somebody, and sure enough when we got to Cork we learned that we were the lucky fellows for within 10 miles of us the subs had put down crafts that very day. Some luck for us, what? Just think of it, there we were becalmed nearly all that day, helpless, and they never saw us.
>
> Well they didn't give us a load at Cork, but sent us up to Fleetwood in ballast to take on salt. That was about June 28. Well, sir, we fetched there all right, but they were slower than cold molasses in giving us a load and it was July 22 before they finished up with us and sent us off—once more for Iceland. Now here's the part I suppose you fellows want to know—this is where we got ours. We was going along on Friday, July 27, just as slick as a bean and figured we were pretty near clear of trouble for this trip any-how—we were 360 miles from land then you know—when all of a sudden at 12:30 p.m., Zip—and then zip again, and again, and the shells, for that's what they were, began to drop around us thick and one went through the main rigging. That was too close for comfort you can bet. One of the gang just looked up and said 'submarines.' So I see, says I, and that's all there was said, only we got the boat launched and dusted into her and pulled away from the *Hammond* just as soon as the Lord would let us.
>
> And mind you, up to this time we hadn't seen the sub [U-44] at all. She was out of sight when she fired. Soon, however, she came into view and we got a glimpse of her conning tower. As she came closer she plugged

some more shells at the good old *Hammond* and pretty soon it was all over and down below for her. Then the German chap comes up and gives us the once over. The skipper, he didn't do any talking. The crew—18 I counted—were on deck, and one of the petty officers did the talking and relayed our answers up to the boss bully standing there stiff and silent at the conning tower. I don't remember it all, but our little confab ran something like this: "Where were you going?"—"Iceland."—"What do you have for cargo?"—"Salt."— "Where did you land?"—"Fleetwood."—"What nationality?"—"American." Then the officer chap looked up to the captain and with one of those queer smiles and a wink that told a whole book full, said, "He says he's an American." "See any steamers?" was the next question, and that was the end of our little confab, except that as he was leaving he sang out to us—he talked good English all right—"Steer east, there are plenty of trawlers on Rockall bank."

That cuss didn't intend we should ever reach land if he could help it and he was trying to set us astray. I didn't steer his old east course. There was grub and water in our boat. I had our reckoning, knew where we were, and steered southeast for the Irish coast, and we would have made it too, under oar and sail, but after we had made over 100 miles toward the coast in 28 hours an English patrol boat pokes her nose in sight and yours truly and six other thankful ones were soon safe aboard the *Transient*, that was her name. Say, they used us fine. We were five days aboard with them and they landed us safe at Lough-Swillie. I don't know as that's how they spell it, but that's how they pronounced it on the north coast of Ireland. The rest of it don't amount to much. It was the evening of the 28th when they picked us up. After we landed we were hustled to Londonderry and then to Liverpool via Dublin and the American Consul sent us home.

Oh, by the way, coming over on the liner—I can't tell you her name, we saw a submarine and one of the gun crews had a shot at her, but I guess he didn't hit her. We didn't see her again though. The craft we came home on carried six guns. There boy, I guess that's the story.

Through Bonia's recounting, the war had become personal for the Gloucester fishermen, but it was still far enough away for the men to pursue the fisheries as always. The *Natalie Hammond* (sister ship of the *John Hays Hammond*) and the other schooners took no special precautions as they went back to sea and continued to make prosperous trips to the Banks. By the end of the summer, the *Natalie Hammond* had achieved an impressive record in the halibut fishery. She stocked $25,000 for 6 trips, with each of the crew receiving a share of $644. On her very next trip, the local paper reported that the *Natalie Hammond* under Captain Colson had set "a record, which probably will not be broken for some time," stocking $7,601. Colson was highline

)(*The men of the* Natalie Hammond, *c. 1920. From left to right: two uniden-*
tified men, Albin Lund, Steve Olsson, Nils Lund, and an unidentified man.

of the halibuters, having stocked $32,680 in 4 months and 10 days. At the time of the strike no one could have anticipated such results.

But now the federal government began to get involved in the country's commerce. In September, representatives of the major fish processing firms conferred with the Food Administrator, future President of the United States Herbert Hoover, looking for ways to increase the production and consumption of saltwater fish. They discussed the lingering effect of the strike, the confiscation of the steam trawlers, and the differences in state-by-state restrictions on fishing within three miles of shore.* Executives from the Gorton-Pew Fisheries sought permission and security to send craft across the Atlantic with salt fish for Europe. Permission was denied on the grounds that the risk was too great and the promise of success too small.

---

*Fishing industry negotiations in Washington also bore fruit, with the Herbert Hoover-led Food Administration stating that "every effort should be made as a war measure to speed up the saltwater fisheries on both coasts and that in this campaign of speeding up it is highly essential and in fact vital, that insofar as possible the restrictions embodied in these state laws be removed for the duration of the war (including) the removal of all restrictions on the purse-seining operations within the three-mile limit on the shores of all the Atlantic coast states where restrictive laws are now in force and we are prepared to recommend that torching regulations wherever present be fully removed for the duration of the war." These changes were acted on quickly by the Massachusetts legislature and the Governor.

With this avenue blocked, and with large amounts of fish on hand, the Gorton-Pew Executive Committee instructed Mr. Benjamin Smith to wire its agents in the north at Gaspé, Caraquet, P.E.I., Magdalen Islands, Bay of Islands, and St. John's to cut back on operations—the company had secured all the fish it could handle and finance. Indeed, Gorton-Pew was overextended; credit was tight, and the company decided to keep the schooner *Arkona* at St. John's and to terminate the charter of the schooner *Cecil* at the same port. No Gorton-Pew vessel took cargo to the Mediterranean.

In France, the first two American soldiers were wounded by a German artillery shell. But at home, even though all now knew the war was real, life went on much as before. Around the Gloucester waterfront, the men of the fleet continued to bring home great shares. For a nine-day haddocking trip on the *Edith Silveria* under Captain Joaquin Silveria, each man received a $144 share. In one month, Captain John G. Stream's men received a share of $390—high numbers even in a time of unprecedented prosperity. And then the beam trawlers came on line. Hundreds of people crowded the Fresh Fish Company wharf to see *Walrus* for the first time. She was immense—479 tons gross, 246 tons net, and 173 feet in length—and she left on her first trip on September 15.

Less than a week later, thousands of people again came out, lining the streets as 53 young men left for the Army training facility at Camp Devens in Ayer. They carried their personal belongings in suitcases, assorted bags, and even old newspapers. Each man also had a small khaki comfort bag provided by the local branch of the Red Cross. Within days of their leaving, the *Walrus* came into Boston from her maiden trip with almost 300,000 pounds of fresh fish, and on the same day the *Natalie Hammond* unloaded an immense catch. Steve Olsson rushed to make the Gloucester train. He arrived in the evening and spent two full days with Hilda and the children. As his daughter Romaine remembered:

> Over a four-week period, Papa would be home four or five days. It was several weeks out and by the time he got home he was ready to go out again. We were so used to Papa going. We did not take a lot of interest in the details. We were much more interested in when he came home each time. Oh yeah, that was a happy moment. Mama would say—Guess who is in Boston. And then he would bring those Concord Grapes, those purple grapes in an oval shaped wooden container. I still remember how good they tasted. He would also have a halibut in his case and once home he would cut it into pieces to give the neighbors and dear Aunt Bertha in Rockport. One of our tasks was to deliver the halibut.

The delicacy of a box of grapes was now well within the means of Gloucester dorymen. It was no extravagance; while this was a time of great inflation, the shares of the fishermen were rising at an even faster clip.* Many of the more than 200 Gloucester and Boston fishermen who had gone to the West Coast at the time of the spring strike began to return. In all, about two-thirds of those who had left returned, even though many feared that they might become liable for military duty. West Coast draft boards had granted exemptions to the fishermen, but the Gloucester board had yet to act.

In mid-October, the *Seal* returned with her first fare, but she had run into engine troubles and landed only 125,000 pounds of fresh fish. Gloucester now had two otter trawlers taking huge fares of mixed fresh fish into Boston, but their presence had little effect on Gloucester's sailing fleet. With rising demand, and the absence of many of the Boston beam trawlers, prosperity reigned for all. On the *Elk*, the young Norwegian Carl Olsen was typical of the sailing masters. In mid-October, he brought in 25,000 pounds of fresh halibut, 25,000 pounds of salt fish, and 115,000 pounds of fresh fish. His fare broke all records for a halibut trip. Captain Olsen stocked $11,700 for the mixed trip, while each of his men received a share of $301. Meanwhile, the *Natalie Hammond* came in with a fine late October fare, reaching a year-to-date stock of $39,577 in the halibut fishery. Each doryman had already reached $1,030 in shares, and Colson now shifted over to winter haddocking. The haddockers, too, were having a good year. Although the *Mesquita* had been laid up seven weeks for repairs, the skipper's seasonal stock had already reached $44,389, each doryman having received $1,408 in shares.

As Christmas approached, most of the craft were in port, and Gloucester's Local Exemption Board gave the younger dorymen an early seasonal present. The Board assigned them to the Class 2 draft category, which meant they would not be called for military service until all Class 1 men had been called. Effectively they were considered mariners, as they worked the vessels on their passage to and from the grounds and as such, the Board said they were exempt from military service.

For fishermen with families, Christmas of 1917 was a time for being thankful. America was at war but the fishermen had not been, and would not be, called. No Gloucester craft had run afoul of a German U-boat and only

---

*But along the waterfront, the loft and other workers had been falling behind, and the plant owners and the Fish Workers' Union formally ratified a 5 to 20 percent increase in pay.

⚓ *Schooners rafted up four deep at a wharf in Gloucester's inner harbor, pre-1920. The schooner* **Smuggler** *is second from left.*

20 fishermen had been lost at sea. Although there was concern that Germany might yet declare New England waters a war zone, for now the families of Gloucester could relax. Dorymen and their families came together to celebrate a very personal Christmas. Of the *Natalie Hammond*'s crew, Steve Olsson had relocated his family to Rockport, close to Macker and Bertha Simmons. More precisely, Hilda had moved the family, for Steve was at sea at the time. They moved into a small, tidy apartment with an old organ belonging to the landlady. The children picked out tunes on the old instrument. The rent was $10 a month. These were good times for Steve Olsson and the men of the *Hammond*, for over the 10-month season, Captain Colson had stocked $59,216 and each doryman had received a share of $1,542. It was a splendid return, and Steve and Hilda banked much of the windfall, waiting for the day in the middle of the new year when they were at last able to purchase a house of their own.

For the Olssons, the move to Rockport brought Hilda and the family into more immediate contact with her sister Bertha, brother-in-law Macker, and their children.* There was a small Christmas party, and Romaine remembers her uncle Macker bringing a bag of little round peppermint candy, with red stripes in them. "Oh, we thought he was so wonderful." But in the midst of the Christmas festivities there was a disquieting report—another Gloucester captain had lost his vessel to a German U-boat. Stephen Black, captain of the three-masted schooner *Jennie E. Righter*, had been carrying lubricating oil between European ports, with a French port as her final destination, when off the coast of Spain a U-boat approached and put 25 shells into the vessel. The crew boarded small boats, but according to the captain the wind increased so much that they had to put out drags. As the gale increased, it seemed impossible the small boats could stay afloat, but they weathered the worst of it, and when the winds decreased, they made sail and proceeded to land.

Then, on December 31, 1917, as if to remind the community of the new realities brought on by the war, the *Gloucester Daily Times* published pictures of the first three American prisoners of war—men from the four American divisions to reach France in 1917.† The article said that "the names of the three whose heads are shown are not known. Perhaps some American father or mother will recognize her son, and know he is not dead though reported missing."

---

*Steve's two oldest children, Gerda and Romaine, were enrolled in the George A. Tarr school in Rockport. Romaine was in the first grade, Gerda the third, and each has a distinct memory of their father in that winter in Rockport. For Gerda, the memory is a simple image of an often absent father encouraging a loving child to do well in school, although all knew that the responsibility for the well-being of his small family sat squarely in the hands of Hilda. As Gerda remembered, "When I came home from school, Papa said to me: 'Who was the smartest one in the class?' looking at me with a big grin on his face. Not to disappoint him, I would say: 'I was.' I know that was what he wanted to hear and he always knew how to make you feel wonderful." As for Romaine, she remembered that there was a granite step outside of the front door of their Rockport home, and she broke a glass on that step, and then "being so afraid, that I thought I would really get the dickens for being so careless as to break the glass. So I hid in back. There was some kind of a room in the back of the house and I stayed there for hours. Mama was so careful with our money and Papa was home. But when I owned up to it, there was no punishment, Papa and Mama understood."

†On November 3, a German raid on a line near Vosges resulted in the first three American casualties of the war.

CHAPTER 25

1918—The War Comes to America

On New Year's Day, the temperature at the Essex car barn of the Bay State Railroad Company dropped to −21°F. The inner harbor froze over, something that almost never happened, effectively locking the fishermen and vessels of the fleet either in or out. Traveling in unheated streetcars was almost unbearable. Coal was in short supply. The Public Safety Committee, a wartime relief group, distributed six hundred 35-pound bags of "nut coal"—anthracite coal about 1⅝ inches in diameter—the coal dealers having put on extra teams and staying open on New Year's Day. Three days later, people from East Gloucester walked on the ice across the inner harbor to the downtown section of the city. Large ice floes extended into the outer harbor, ending at the Eastern Point breakwater. By January 5, the solid freeze reached into the outer harbor, although not yet to the breakwater. On that same day, the schooners *Hammond* and *Catherine* arrived at Boston, having spent the previous night anchored in an ice field off Graves Island. With Gloucester iced in, the two schooners offloaded in Boston and sailed directly to Western and Georges banks. It would be some time before any of the men again got back to Gloucester.[*]

News came of the death of ex-Congressman Gardner. He had been training with his unit at Camp Wheeler in Georgia when he contracted a severe cold and died of pneumonia.[†] He was but one of more than 2 million

---

[*]As the *Hammond* left Boston, Steve Olsson's brother-in-law Macker (Norgren) Simmons fell from the rigging of a Story Company schooner. The spray from the sea formed ice on the mast, booms, and rigging, and while freeing it from the rigging he fell to the deck. His hip and leg were damaged, bones broken, and he was unable to walk. The injuries were serious and Macker was to experience a protracted in-patient stay, followed by months of painful therapy in his small home on Pier Avenue, Rockport. He lived near the end of a quaint lane, off Rockport's main street, with a view of the sea. But for Macker, and the two women who were committed to his care, these were trying times, and at its end, the injuries were to cost him his life, after months of intense suffering. He was 47 years old, having come to America 17 years earlier. Flags on the fishing vessels and at different homes in the vicinity were set at half-mast in his honor. The first grader Romaine Olson remembers accompanying her family to the funeral. "The teacher was quite upset because I went. She said, I was too young. I remember being in the Hack, and there was a man and a woman opposite me. They were friends of my mother. They smiled at me, it was just a polite smile, and I thought how rude they were to smile at a funeral. I took it all wrong." Romaine also remembers that she and Svea (Macker's daughter) went to the cemetery a couple of years later. "Svea had on a ponchy dress and there were little blue robins' eggs lying around on the grass. We went there on the trolley car—from Rockport to Lanesville—to the Seaside Cemetery."

[†]In March 1920, Governor Calvin Coolidge led an effort to name the auditorium of the State House in Boston after Gardner.

American men now in arms, about a quarter of whom would be in France by month's end. Wilson's Secretary of War, Newton Baker, said that Germany was planning a powerful submarine thrust at the American lines of communication, and the people of Gloucester could only wonder and fear what this might mean for the fleet as it fished on the Georges, Scotian, and Grand banks. Germany had already shown that she had long-range U-boats capable of reaching targets over 3,000 miles distant. The only question now was, would Germany loose them on America?

On the home front, Captain Lem Spinney left the *Seal* to assume command of the much larger French-built Nova Scotian steam trawler *Baline Boy*, and marketed his catch in Halifax. Captain Carl Olsen moved over to the *Seal*. The men were following the money.

In mid-January, the gill-netters and shore fleet were still unable to leave port. On January 16, only one small trip of 800 pounds came into Boston. Gale-force winds swept the coast and Gloucester's harbor remained iced in. Every dock was frozen solid. The *City of Gloucester* steamer could not make her regularly scheduled early-morning cargo trips to Boston. Towboats and

In the worst of winter conditions, men off-load their catch at Boston's Commonwealth Pier, resupply, and head back to sea.

)(*In the severe winter of 1918, three schooners are icebound off Ten Pound Island in the outer harbor—it was an opportunity for curious townsfolk to visit and walk all around the vessels. The* Kineo *is to the left, the* Sylvania *is in the middle, and the* M. F. Curtis *is to the right.*

gill-netters were frozen fast at the docks. The processing firms were quiet. In Boston, only a trickle of fresh fish reached the market. On January 19, the city fathers asked for government help to break up the ice sheet. One week later, the *Natalie Hammond* sailed into Boston with 53,000 pounds of ground fish. Haddock prices had risen to $10 a hundredweight (as compared with $2 prewar and $5 a year earlier), and the U.S. Food Administration called for increased fish landings. But the weather remained bad, a winter unlike any in recent memory. Shore craft could not go to sea and offshore craft could not return to Gloucester between trips. By the first week of February, the Gloucester ice sheet extended into the middle of the outer harbor. The inner harbor had been frozen for a month, and the ice was now 20 inches thick.

On February 4, two schooners, the *Kineo* and *Sylvania*, returned from Boston, anchoring off Ten Pound Island at the edge of the ice. They hoped to make a quick change of sails before returning to Georges. By the next morning, the ice had reached them and they were frozen tight; all the crew could do was walk ashore. With the temperature at −18°F, and winds blowing at 45 miles per hour, the windchill was –87°F, with below-zero weather predicted for two more days. The *Kineo* and *Sylvania* would remain locked in the ice. On February 6, the *Sylvania* secured a "pung," a sleigh on runners, and six of the most robust of Captain Thomas's crew hauled it from Pavilion Beach to the ice-encased craft, took down the wind-torn sails, and hauled them back across the ice for repairs.

The citizens of Gloucester came out in great numbers. They visited the wharves, tramped around the two icebound schooners, and for the first time in anyone's memory walked from the inner harbor to the outer breakwater at Eastern Point. Hundreds of people made the journey, while many young people cut holes in the ice and fished for eels. Another oddity was the sight of a horse and sleigh carrying provisions from Pavilion Beach to the two trapped vessels.

On Friday, February 8, the wires between Gloucester and Congressman Lufkin's office at Washington were kept humming, as ever more people asked for help to open the harbor. Secretary Reed of the Board of Trade contacted the Navy yard and the Office of Food Administration. In turn, they contacted the Assistant Secretary of the Navy, Franklin Roosevelt, asking him to send relief to Gloucester. Finally, help was ordered, but Navy officials said the available minesweepers were not up to the task—their hulls had been damaged in an earlier attempt to open the harbor. On February 9, a steel-bowed vessel, the converted Boston beam trawler *Crest*, was brought in. By noon she had forced a channel from the outer harbor to Halibut Wharf in the inner harbor. At the same time, the tug *Blanche*, out of Beverly, was at work freeing the two schooners trapped near Ten Pound Island.

At last the inner harbor was partially opened, and two Newfoundland schooners landed herring trips from the Bay of Islands. The *Crest* remained for another day, opening channels. With a warmer turn on February 12, and with men chipping ice from the wharves and sawing ice into cakes to flow away on the outgoing tide, Gloucester at last had reason to hope. For the first time since February 1, a half-dozen gill-netters sailed out of the harbor to retrieve their nets. Most of the netted fish had spoiled.

On the very next day, however, the prevailing wind blew the floating ice back into the harbor, choking the channel that had just been opened. Navigation in and out of the port was difficult. Captain Colson once more headed back to Georges directly from Boston. The crew had accumulated a considerable sum of money, but there was no way to get it to their families. On this trip, the catch of the *Natalie Hammond* was again substantial: 62,000 pounds of fish; the stock came to $4,764 and the per-man share was $144. It was a very successful trip.

On the next trip, when Steve Olsson finally got back to Gloucester, he was pleased to at last share his earnings with Hilda and the children. Like others, he used a small part of his funds to buy presents. It was like a second Christmas. Steve bought a doll for his daughter Romaine to use in a school recital, while other fathers bought sleds and wagons for their small children.*

Along with the ice, war had been creeping into the lives of the fishermen. The fishermen found their regular routines now under the scrutiny of the military. Signs appeared forbidding "alien enemies" from going beyond specified areas along Rogers Street and portions of the waterfront. Employees of the wharves and freezers, railway workers, vessel fitters, and fishermen were required to show a permit on entering the wharves and piers. Crewmen could not return to their vessels without the signed permit. But the vessels of the fleet were soon back to sea, and returned with banner catches. On February 18, Boston had its biggest "fish day" in some time: 13 vessels landed nearly 1 million pounds of fish, including 220,000 pounds from the beam trawler *Tide*, 66,000 from the *Natalie Hammond*, and 82,000 from the *Sylvania*. Prices remained high; increasing the fish supply by 50 percent had become a priority goal of the U.S. Food Administration, and Herbert Hoover would let market forces drive the actions of the outfitters and the setting of prices. His Food Administration had earlier supported efforts to supersede restrictive state laws, and word now came that the market would be open to northern imports. Secretary Redfield of the Department of Commerce issued an emergency decree permitting Canadian and Newfoundland fishing craft to market their fresh catch in American ports, even if it meant they came directly from the fishing grounds. They no longer had to report to their home ports to shift the registry. Redfield said that Americans must eat more fish if the country was to keep its fighting men abroad sufficiently supplied with wheat and beef.

The directive was accompanied by little public comment, but the notes of the Executive Committee of Gorton-Pew Fisheries include the assertion

---

*Romaine would remember this doll for the rest of her life. "When I was in the first grade I was scheduled to say a little piece in front of the class using a doll in the talk. But I didn't have a doll and another little girl agreed to lend me her doll for the recital. So I rehearsed at home with my mother, and my father once he had returned after being away so long. Finally the day came when I was to go in front of the class and make my little speech. I held the other girl's doll, said my words, and sat down when I was finished. And about five minutes after I finished, there was Papa and Mama standing at the doorway of my classroom with a doll—they had bought it for me. They knew what I was up to, they knew I didn't have a doll, and they were hoping that they would get there in time. But I had already said my piece, but thanks to a very sensitive Rockport teacher it was not too late. The teacher had me do my piece all over again, only this time with the new doll from Mama and Papa, and with Mama and Papa in the room. Oh, how I can remember that doll, how I can remember the smile on Papa's face. I loved him so much and I loved the doll Papa and Mama gave me. It was a boy doll. It had a boyish hairdo, red and white striped shirt, and brown khaki pants. And every night before I put it to bed in the dining room closet, I would kiss it. I loved that doll, and I loved my parents for giving it to me."

that the Wilson administration had, in fact, responded to Gorton-Pew's suggestion for overcoming artificial barriers that restricted production by the fish companies. Gorton-Pew now had unfettered access to supplies of northern cod, no matter the registry of the vessel landing the catch. The company was also in an acquisition mode. With plants in Newfoundland, Nova Scotia, Labrador, Maine, Cape Cod, Gloucester, and elsewhere, they responded quickly when a fish-processing business at Southwest Harbor, Maine, came up for sale. The purchase assured the company of an ever-increasing supply of fish. Back in Gloucester, Gorton-Pew purchased additional shares of the Gloucester Cold Storage and Warehouse Company, becoming its principal owner.

In late February, the final 15 percent of Gloucester boys called up in the first draft left for Fort Devens. There were now 800 Gloucester men in service, and the newspaper began to reprint letters from locals already in France. A full page was set aside in each edition of the paper.

Reports of losses continued to be rare, and by March all that mattered locally was that Gloucester's outer harbor was free of ice; for the first time in a month craft hauled out at the local marine railways. Vessels were again sailing to the offshore grounds, and both schooners and steam trawlers brought large fares of fresh fish to Boston, Gloucester, and Portland.* In fact, Portland was experiencing a significant upturn in its fisheries, due largely to landings by the beam trawler fleet and the landings of Provincial craft that now came south under the ruling by the Secretary of Commerce.

As for the Gorton-Pew Fisheries, their two steam trawlers brought in banner trips. On March 5, with the most remunerative fresh-fishing trip on record, Captain Clayton Morrissey brought the *Walrus* into Boston with 180,000 pounds of haddock, 40,000 pounds of scrod, 40,000 pounds of cod, and 10,000 pounds of pollock. He stocked $18,000. Only one week later, the *Seal* came in with the largest fare ever landed in Boston. It took two days to off-load the catch, and after the trip her command passed from Captain Henry Atwood† to Captain Fred Thompson. Regardless of the vessel or the skipper, however, record stocks and record shares remained the order of the day.

On April 1, as nearly 2.8 million pounds of fresh fish were landed in Boston, prices finally slumped as demand slowed after the end of Lent.

---

*The U.S. Bureau of Fisheries report on fish quantities landed at Boston, Gloucester, and Portland by American fishing vessels in February indicated that fares were brought in by 158 steam and sail vessels: 167 trips at Boston, with 8,609,850 pounds valued at $618,346; 86 trips at Gloucester, with 2,078,649 pounds valued at $108,885; and 125 trips at Portland, with 549,328 pounds valued at $43,672.

†For many years, Captain Atwood had been the local manager of the Atlantic Maritime Company, resigning that position to take command of the *Seal*, on which craft he stocked over $60,000. He now became manager of the Gorton-Pew steam otter fleet and assistant manager of their entire fleet.

Many fares were brought to Gloucester to sell to the splitters. The biggest catch was on the *Walrus*, some 275,000 pounds. Three other beam trawlers brought in large catches: the *Billow* with 142,000 pounds, the *Wave* with 196,000 pounds, and the *Heroine* with 200,000 pounds. The sailing fleet, which had been fishing on Georges, also returned with large cod fares. No one was complaining.

On May 2, the *Spray*, now under the command of Captain Carl Olsen, landed 152,000 pounds in Boston, followed by a late May fare of 400,000 pounds. In this same period, Captain Morrissey on the *Walrus* brought in two fares of 450,000 pounds each. From early April to late May, each of the dorymen on the *Natalie Hammond* cleared $795 in the halibut fishery. This was on top of earlier season earnings in the haddock fisheries. In four trips, over the 51-day period, the *Hammond* had stocked $26,952. The local paper said it was "a clever performance, accomplished only by the liveliest kind of hustling, but the stock achieved on the trips in that length of time is most unusual and it is believed never excelled in the fresh halibut fishery."

With the unprecedented windfall, Steve Olsson became one of a number of fishermen in 1918 to buy a home, but it was his wife who made the decision and carried out the task.* Hilda had found Rockport confining, and she and Steve located a house for sale in East Gloucester, a couple of doors from Captain Colson at 143 Mt. Pleasant Avenue. It was a quiet, residential street, just off the trolley line and near Smith Cove, where Captain Colson anchored the *Hammond*. The purchase price was $2,400, which Hilda and Steve paid in cash. The move itself was planned and executed by Hilda. As always, Steve was at sea. Hilda purchased used furniture from a Rockport guest house that was going out of business. She scheduled and supervised men to paint and wallpaper,† and the move was a great success. "Mama and Papa loved their new Mt. Pleasant house—it was quite a place to come to. They thought it was a real big, great house."

---

*This was also a sad time for the Olsson family, as Macker Simmons passed on, his death coming two days prior to the *Hammond* making port, and once again Hilda had to confront a major life event without Steve by her side. But he was soon in port, and joined Hilda in comforting his close friend and sister-in-law, Bertha Simmons.

†As the day of the move approached, Gerda was upset—she had yet to see her new home. "Mama would go to Gloucester to see how things were proceeding and she left us children at Aunt Bertha's in Rockport." Gerda, who had just finished third grade, was determined to get to Gloucester, and on one of those Gloucester days, as her mother went to the trolley, Gerda chased after her. But Hilda had other ideas. In an irritated voice she cried, "You go back to Pier Avenue with Aunt Bertha. I don't want you to bother me given all I have to do." But Gerda was adamant, she would not go to Aunt Bertha's, so she laid down right in front of the trolley, across the tracks. "I would not move. I could hear people inside saying, 'If she were my child.'" But it was too late, Gerda was Hilda's child, and Hilda had no choice but to take her to Gloucester. As Gerda went through the house for this first time she was struck by how clean and crisp everything looked, and she remembers in particular the new wallpaper in the little room upstairs—"it had children marching with paper hats on their heads."

But even as the Olssons rested comfortably in their new home, the community as a whole looked east to a new danger: U-boats had finally arrived. The terror so long feared was about to become a reality. The news headlines over the past two months had described Germany's latest drive toward Paris. Tens of thousands of young men had lost their lives, and with American troops being released for use by the Allies, American deaths had begun to climb. Included among those lost to enemy fire were two Gloucester boys: Charles B. Knutson from East Gloucester and Roland E. Cole from West Gloucester. Reports of individual deaths on the battlefield were soon to be followed by reports of Gloucester schooners lost on the North Atlantic.

U-151 sailed from Kiel, Germany, on April 14, and in late May began laying mines in the Delaware Bay.* From June 2 to 8, she sank 10 craft between Cape Cod and the Chesapeake Bay, and the Navy Department soon closed the ports of Boston and New York. One of the first vessels lost, the 1,971-ton schooner *Edward H. Cole*, sailed out of Boston, and was 75 miles off the Atlantic Highlands when she went down. Her skipper was given seven minutes to leave the craft, after which the Germans blew it up. But no one was lost, as the crew was picked up by a passing steamer. Of the other vessels sunk by U-151 at this time, the 5,093-ton passenger steamer *Carolina* experienced the greatest loss of life: 16 persons drowned at sea.†

The situation had become desperate off the Atlantic coast. Merchantmen and fishermen sat unprotected, easy targets for the U-boats. Fishing schooners had no wireless and many did not know of the U-boat attacks. Gloucester's Board of Trade telegraphed Secretary of Commerce Redfield, asking the Navy to warn the fishing fleet of the danger. The Navy agreed.

On June 7, as the *Natalie Hammond* entered Boston Harbor, the Navy announced it would close the gate and net crossing the entrance of the har-

---

*U-151 was one of six large *Deutschland*-class, mercantile submarines in the German fleet. She had been converted for use as a long-range attack submarine, and could remain at sea for up to four months. Her armament consisted of two large 5.9-inch deck guns, two smaller 3.4-inch deck guns, 19 to 30 torpedoes, and many mines. On reaching the American coast in late May she laid mines in the Maryland and Delaware bays.

†The skipper received a wireless message warning him to be on the lookout for submarines, and telling him of the loss of the *Cole*. At 6 o'clock Sunday evening the U-151 surfaced, raised the German flag, and ordered the *Carolina* evacuated. The sea was smooth, order was maintained, and within 20 minutes 10 boats had been lowered. When the last boat cleared, the Germans poured seven shells into the steamer, and the *Carolina* sank slowly as the sun was setting. According to an unnamed survivor, "The sea was still smooth and there was no danger." Boats kept close together, everyone taking a turn at the oars. Then there were faint flashes of light on the horizon, and before long a storm came up, kicking up a nasty sea. Boats became separated, and the one holding the man telling the story capsized, throwing all 35 of its occupants into the sea. They righted the boat but it capsized again, the passengers and crew clinging to its sides. When the storm finally passed it was pitch-dark, and as the survivors righted the launch once more, they clung to her for dear life as they slowly bailed out. "Some could not stand the strain, became exhausted, let go their hold and sank. It was terrible." The survivors scooped out the water, permitting one after another to get into the launch. They kept this up all night until all those who still clung to the side of the launch had climbed back in. There were 19 survivors; 16 had drowned.

bor from 8 P.M. to 4:30 A.M. each day.* All fishing vessels and coastal traffic would be permitted to go on as usual, but overseas shipping had to await instructions from Washington.

Over the next few weeks, returning fishing craft brought no word of unusual adventures. Some reported having received word of the earlier raids and had put out extra watches as they continued to set trawl. Many said they hugged the coast on their journey. What they did not know was that U-151 was on her way back to Germany. The skipper said he had sunk 23 ships and 61,000 tons of shipping.[44]

German U-boat activity was next reported on July 13, and it was in the middle of July that Gloucester vessels finally came under fire. The first word of the U-boat activity came when the survivors of a sunken Norwegian barque reached an Atlantic port. The bark had been captured on July 6 and the crew of 19 ordered to their boats. About the same time, the Gorton-Pew Executive Committee decided to procure war risk insurance on all their vessels and cargos. In mid-July, the *Gloucester Daily Times* stopped posting notices on the arrival and departure of the vessels in the fleet. The war had hit home, and the change in *Times* policy responded both to a directive of the Secretary of the Navy and to the paper's own sense of the danger to Gloucester vessels on the Banks.

In only a matter of days, U-156 sank the tug *Perth Amboy* and four barges off the coast of Orleans, Cape Cod. At the time of the loss, the sea was calm and a fog bank sat four miles off the shore. In the midst of the bank, the 11-year-old son of the skipper was the first to see the U-boat as she emerged from the fog. He ran to the cabin, grabbed a small American flag, and waved it at the Germans. The Germans closed on the vessel and ordered the crew to leave. Hundreds witnessed the attack, some from shore, others from sea. Captain Marsi Schull of the small seiner *Rosie* was five miles off Orleans when "suddenly we heard the report of a big gun. It sounded like one of Uncle Sam's navy ships at target practice. We didn't give submarines a thought. We looked toward the tug and her tow and was startled to see a submarine. . . . She looked like a big whale, with the water sparkling in the sunlight as it rolled off her sides. Then we saw the flash of a gun on the U-boat and saw the shell strike the pilot house of the tug."

The U-boat next shelled and sank two of the barges, and then turned its attention to the *Rosie*, and in a flash a shell was skipping along the water. Captain Schull yelled for full speed ahead, and in his words, "the *Rosie* jumped ahead through the brine (making) us feel a bit more comfortable. The Germans must have fired as many as five shots at us, nearest coming

---

*At about the same time, on June 5, the second round of selective service registration began around the country.

within 10 feet of our stern. But we were traveling pretty fast, and when the submarine crew saw their shots were falling short, they gave up."

In an article in the Nantucket newspaper *Inquirer and Mirror*, Edward G. Thomas described how the sound of the German guns first broke the stillness of the morning air. "The guns of the submarine were banging, then the men folk of the Cape, some of them hatless and coatless . . . grabbed their guns and hastened to the shore to give the Germans a warm reception should they dare to set foot on land. . . . The commander of the German U-boat had picked out the only spot along Cape Cod where there is deep water and no shoals. Off Nauset deep water extends for miles—and the German commander must have known it."

During the attack, the crew of a Portuguese fishing schooner, believing their vessel was being fired on, abandoned ship and headed for shore in their dories, leaving the empty schooner in the vicinity of the doomed tug and its tow. In fact, it was said, "the crew of the fisherman jumped overboard at the first shot and swam ashore. They thought that their own vessel was being attacked. The fishing vessel was within a few hundred yards of the sunken barge. . . . The abandoned ship floated aimlessly in the calm sea. The submarine, however, paid no attention to it. Finally, after the submarine had withdrawn, the fishing crew went aboard their craft and brought it into Orleans bay."

Two more U-boats, U-157 and U-117, were now roaming along the New England coast, and within three days U-156 attacked another ship, the Gloucester schooner *Robert and Richard* under Captain Robert Wharton, highliner of the 1918 halibut fleet. Bound for home from the Western Bank, the schooner was about 100 miles off Thatcher's Island, east of Portsmouth, when the skipper and crew reported hearing a bang, followed by a shot directly across the bow of the vessel. It fell into the water on the leeward side, and the men on deck raised the alarm, shouting down the hatchway. They first thought it was a coastal defense craft, but they were not long in doubt. U-156 was about two miles away and steaming on the surface, making good speed. Three or four men were visible on her deck, and the forward gun was trained on the schooner. Captain Wharton ordered his craft brought up into the wind, and he and his men threw up their hands. There was no possible escape. Wharton gave the command to abandon ship, and the men got six dories into the water in "record-breaking time and without waiting to gather up any of their personal belongings or even getting water or provisions." Wharton said, "If they were going to shell us we didn't propose to get killed." U-156 approached from the north, and a man in the conning tower beckoned Wharton to come alongside. Speaking good English, he asked the schooner's name, where she had come from, and her destination. He ordered Wharton to pull alongside with a dory for his use, saying there were other dories on the schooner that Wharton and his men could use. Three Germans got into the dory, but only one was in uniform and Wharton gave him the schooner's papers and flag. The German told the fishermen "to make ready to leave," and as one of the dorymen put it, "We were frightened some when the Germans came aboard, as we did not know what action they would take. When we got the orders to take to the dories we lost no time."

When the three men from the U-boat boarded the schooner, they acted as though it was an old game to them, one they thoroughly understood. Two took a bomb with a time fuse attached to it. They cut some of the trawl lines, bent them to the fuse, and "swung it over the stern of the schooner after they had lighted the fuse and then whip sawed it aft under the keel until they got it about amidships. There they made fast the two ends of the whip, hastily boarded the dory . . . and pulled back towards the U-boat." Now all waited for the bomb to explode, including the Americans in dories one mile away. When it exploded the "vessel settled deck to the water, then listed, dipped her bow under and stern in the air and sank in two minutes with all her sails set." The men in the dories rowed toward land. Three of the boats stayed close together, while a fourth, with two sails, was quicker in making for shore. The men were adrift from late morning until 6 P.M. the next day, some 30 hours later, when Snug Harbor came in sight.

Captain Wharton said that "while the Germans were inclined to say but little, they were civil and did not allow any bullying tactics." They spoke good English. One of the men said that he owned a house in the States, in Maine, and had lived in the United States for a long time prior to the war. They did not say how many ships they had sunk, where they had come from, or where they were going.

Now, the unthinkable had come to pass; one of Gloucester's premier fishing vessels had been lost and others soon followed. In only a matter of days, the schooners *Rob Roy* and *Muriel* of Gloucester, the *Annie M. Perry* of Boston, and the *Sydney B. Atwood* of Nova Scotia were sunk off Sable Island on the Cape Shore. In each instance, the U-boat commander first replenished provisions, taking eggs, butter, and fresh vegetables from the North American vessels. He ended with enough supplies to keep him at sea for a number of weeks.* The commander then had his men plant bombs

---

*In the same week, off Burin, N.F., the British-registered schooner *Gladys M. Hollett* was sunk by a German U-boat, and the German commander stripped her of everything movable, including the skipper's clothing, watch, and nautical instruments. The U-boat captain told the Newfoundland fishermen that he once sailed out of Gloucester and was familiar with the whole fishing situation. His orders were to sink the entire Lunenburg fleet, to destroy vessels, but not to drown the fishermen.

amidships below the keel, the fuses were set, and the fishermen made to row the boarding parties back to the U-boat. It was then up to the fishermen to put water between themselves and the schooners, for as the bombs exploded debris was thrown into the air and the schooners quickly disappeared below the waves.

Three days later, another Nova Scotian schooner, the *Nelson A.*, was sunk, bringing to five the number of vessels lost to U-156 in a four-day period. However, this was but a warm-up to the events of August 10, when U-117* surfaced on Georges Bank in close proximity to 25 to 30 small swordfishing vessels. The U-boat commander, in all likelihood not believing his good fortune, ordered the skippers of the fishing craft to stand by and await their destruction.[†] In all, nine vessels were lost, some only after their provisions had been removed—a side of beef, a sack of potatoes, a barrel of flour, fresh vegetables, and clothing of all descriptions. One crewman on the *Lena May* looked on in amazement as a German sailor paraded around in the $3 straw hat the fisherman had just purchased. Andrew St. Croix, mate on the schooner *Kate Palmer*, saw the sinking of seven vessels while standing aloft on the crosstrees of his vessel. Even before the submarine came into view, he counted 11 explosions at 15-minute intervals. As each swordfisherman carried two or more dories, at least 18 boatloads of men were now adrift on Georges. Some would row for up to three days, with little food and water, before making land or being picked up by other vessels. They were 125 miles offshore, and with little wind, they were unable to rig sails with swordfish gaffs and blankets.

Having completed her task, U-117 disappeared under the sea. She did not attack the men in the dories, and in the days to follow all of the fishermen made their way to shore. Over the next week U-117 was not to be seen.[§] Only one skipper, Captain Clayton Morrissey of the *Walrus*, reported a close encounter with a "sea viper." The *Walrus* was off Highland Light, outbound to Georges, when a U-boat appeared on the surface, no more than 150 yards dead ahead. Captain Morrissey had little time to think. The sub had seen the fisherman and was readying to fire as Morrissey turned his bow toward the submarine, setting his craft on a full-steam collision course with the "viper."

---

*U-117 had left Germany one month earlier. She was a large mine-laying submarine.

[†]The vessels included *Anita May*, Captain Frank Lance, gas screw, 31 ton gross; *Cruiser*, gas screw, 25 ton gross; *Earl and Nellie*, gas screw, 24 ton gross; *Katie L. Palmer*, Captain Robert Jackson of Edgartown, gas screw, 31 ton gross; *Mary E. Sennett* from Rockport, built at East Boothbay in 1904, gas screw, 24 ton gross, 11 ton net; *Old Time*, gas screw, 18 ton gross; *Progress*, gas screw, 34 ton gross; *Reliance* from Gloucester, carrying a seven-man crew, gas screw, 19 ton gross, 10 ton net, built in Gloucester in 1903, and uninsured; *William H. Starbuck* from Provincetown, gas screw, 53 ton gross.

[§]Unknown to the Americans, a serious malfunction caused U-117 to head back to Germany.

The sub got off a shot, but it went wide, and as the *Walrus* closed the distance between herself and the stationary U-boat, its commander had little choice but to submerge or risk a collision with the approaching otter trawler, which would have cost him his boat. Once the submarine disappeared under the waves, Captain Morrissey reversed his course and took his craft off into the mist.

It was but a brief respite; U-156 had steamed north in search of the immense fleet working on the Scotian and Grand banks. More than 250 craft from all nations were on these grounds, and U-156 inflicted heavy losses on the Gloucester fleet, taking down the schooners *A. Piatt Andrew*, *Francis J. O'Hara Jr.*, *Sylvania*, and *J. J. Flaherty*. These were large, splendid vessels, commanded by superb skippers: Captain Wallace Bruce on the *A. Piatt Andrew*, Captain Joseph P. Mesquita on the *Francis J. O'Hara Jr.*, Captain Jeff Thomas on the *Sylvania*, and Captain Charles T. Gregory on the *J. J. Flaherty*. Their loss put a big dent in Gloucester's remaining schooner fleet.

U-156 had begun her campaign by first seizing the Halifax steam otter trawler *Triumph*, putting a German crew aboard, and then setting out in a two-vessel convoy in search of Gloucester and Nova Scotian schooners.* The German captain armed *Triumph* with two light guns and transferred a number of bombs to sink the schooners. Using her, the U-boat commander could get close without raising the alarm.

The attack on the *O'Hara* took place on Middle Ground off Sable Island as the men were preparing to make their first set. Crewmen spotted a schooner and beam trawler steaming toward their craft, but nothing seemed out of place. Captain Joe Mesquita decided to make a run down to the two craft with a view toward getting a line on where to set the next day. As he drew near, a voice from the trawler sang out, "Heave to, we're going to sink your vessel." Mesquita thought it was a joke, as he and his men had been joking about submarines on the way out. Laughing, he told his men he knew the steam trawler captain perfectly well and thought he was having fun with them. But it was no joke, and two shots soon came whizzing over the schooner's bow. Captain Mesquita did not comprehend the situation, but he gave orders to lower the jib and jumbo and let her jog.

A voice then boomed out, "Come aboard and bring your papers and quick too." The U-boat commander was speaking. A dory was put over the side and Mesquita rowed to the trawler. On reaching the U-boat, he found an angry captain demanding to know why Mesquita had not stopped when ordered. Mesquita quickly explained that he thought the vessel was a

*The *Lucille Schnarees*, *Pasadena*, and *Una P. Saunders* of Lunenburg, N.S.

Provincial trawler under the command of a friend. He often took mail ashore for the *Triumph*, and Mesquita gave his friend water, bread, or whatever he needed when running low on supplies. The U-boat commander asked, "Didn't you see the flag of Germany afloat?" Mesquita responded that he had not seen the flag, even though it was raised as the German craft approached, and he was sorry. As Mesquita said later, the U-boat commander then "did not seem so mad and smiled." The men of the *O'Hara* were given five minutes to go back to their vessel, take off bread and water, and row off to a safe distance. The Germans then took their bomb out of its black bag, threw it under the keel, and blew up the craft. There was barely a sound; the craft lifted out of the water and then toppled from one side to the other before going to the bottom. Captain Joe was no more than 150 feet away. He turned from the sinking vessel and began a four-day row until he and his men were picked up by another vessel.

Before they approached the *O'Hara*, the Germans had stopped and sunk the *A. Piatt Andrew* and the Mahone Bay fisherman *Pasadena*, followed by a French schooner from St. Malo.

Captain Jeff Thomas of the *Sylvania* was the next to be approached, and he reported first sighting the *Triumph* an hour and a half before his vessel was destroyed. He, too, had found nothing suspicious about the trawler.[45] His crew was on deck, getting ready to go out to haul, when they saw a steam trawler in the distance. They paid little attention, even as the Canadian trawler began to approach. Jeff believed she was coming over to find out about the fishing. It was not until the craft raised the German navy flag that he spied a machine gun on the pilothouse. As the *Triumph* came alongside the *Sylvania*, the officer on the bridge used a megaphone to announce that he was the captain. He ordered Captain Thomas to come aboard and bring his papers and flag, and to be "quick about it." A dory was launched and Captain Thomas, with two of his men, rowed to the U-boat and went aboard. They knew the Germans were sinking craft along the coast, and were aware of what was about to happen. After examining the papers, the sub's commander said to the skipper, "Sorry Captain, I'm going to sink you." He then gave Captain Thomas 10 minutes to abandon the vessel. In Thomas's absence, the crew of the *Sylvania* had not waited; they had busied themselves launching dories, taking clothing, food, and other effects that they needed. One doryman even passed down the schooner's two sextants. Then a bomb was placed under the stern of the *Sylvania*. When it went off, the masts were blown right out of her, and the *Sylvania* gradually settled into the sea. The submarine steamed eastward.

There were now fishermen in dories all along the coast. Some were picked up by other vessels. Some of the crewmen of the *Sylvania* were brought aboard the schooner *Catherine Burke* of Gloucester. Even so, most were still at sea in their small dories, and when word reached the Nova Scotian shore communities, help was soon on its way. The Nova Scotians were out to save the men of the fleet, Americans or Canadians, it made no difference. The lives of more than 100 men were at stake, and Canadians in motorboats went out from every Nova Scotian port to find the survivors.

The crew of the Nova Scotian schooner *Lucille Schnarees* was picked up and brought into Canso. They had been sunk by bombs and gunfire from the *Triumph*.* Seven of the crew of the *Sylvania* were brought to a Canadian port by a revenue cutter that had picked them up not far from where their ship went down. Eighty-seven men, the crews of four of the schooners, arrived at Halifax on Thursday evening. According to the paper, "The men were made as comfortable as possible at the immigration headquarters and the Sailor's Home. They lost practically everything they had with them on the vessels excepting the clothing they wore when attacked. Some were fortunate enough to take the best of their wearing apparel while others wore the clothes in which they fished as they were away from the vessel at the time she was sunk. As they marched through the streets in a body, passersby soon guessed what they represented and eager crowds besieged the victims of the Hun pirate."

As for the men of the *Sylvania*, one account reported that on arriving they went to barbershops to have their long beards trimmed.[46] Captain Jeff Thomas had the good fortune to land in Arichat, the village in which he was born. Marshall Barineau "recalls a knock on the door and in came Captain Jeff Thomas in his oil skins with five other crewmen. Captain Thomas and his men landed on the beach about two hundred yards from the house in which he [Thomas] was born." After his schooner was sunk, Jeff and his four crewmen made the trip in a dory, rowing part of the way and using the Evinrude motor that they had rigged on the small craft. One man on shore remembers hearing the little engine as they came up the middle of the harbor: "Putt putt putt putt, with their engine going."[47]

Back in Gloucester, the families had yet to hear of the loss. Gordon Thomas, the 12-year-old son of the skipper, had watched as the men of the *Sylvania* fit out the craft for sailing. He took special note of a pile of watermelons sitting on the

---

*On August 24, *Triumph*'s coal having run out, the Germans removed the radio, guns, and other valuables, and sent her to the bottom.

wharf, waiting to be stowed.[48] Everything had seemed normal as the *Sylvania* sailed, although everyone in Gloucester was aware the Germans were near. But now the *Sylvania* had been lost, and a friend of the family, a man named Eddie Nugent, came knocking on the door, saying, "They got Jeff." That was it, "They got Jeff," no more was known. Everyone was shaken, and it was days before the men were accounted for. Then came the news—Jeff was safe in Arichat, others were in Canso. Nova Scotian streets were packed with fishermen from sunken American and Canadian schooners.

Captain Jeff had owned half of the *Sylvania*, and she had been loaded with fish; now it was all gone. The crew stayed three or four days at Arichat, transferred to Port Hawkesbury with the assistance of the American Consul, and headed for home.

The last Gloucester craft lost to U-156 was the *J. J. Flaherty*, the largest sailing vessel in the fleet at 162 tons gross. On the August day she went down, U-156 also sank three Lunenburg craft and the new Boston schooner *Rush* (sailing with a crew predominantly from Gloucester). The *Flaherty* had left on a salt-bank trip on May 21, and had more than 200,000 pounds of salt cod in her hold when sunk. When confronted by U-156, 18 of the fishermen were out setting trawl and never returned to their craft, but all of the men were saved, landing at St. Pierre and Miquelon. Their ship had been sunk on St. Pierre's Bank, on the western end of the Grand Banks, 30 miles from St. Pierre. This was to be the farthest east the submarine operated in its attacks on the fishing fleet.*

In mid-August, the Navy Department ordered one vessel to patrol Georges and the nearby banks to safeguard the fishing fleet, while also asking the fleet to operate within a more well-defined area.† In addition, the Navy Department deployed the USS *Arabia*, a 113-foot auxiliary schooner, for use as a decoy ship to draw a prowling U-boat within range of a U.S. submarine. The bait was never taken. Though three additional U-boats had been dispatched, no more fishing vessels were attacked.

The U-boat attacks had been designed to coincide with Germany's continued thrusts along the French western front, where the fighting had been fierce. By early July, 1 million American soldiers were in France, and 300,000 were on the lines between the Germans and Paris. One in six of these men

---

*U-156 was lost on her passage back to Germany when she hit a mine off Bergen, Norway. She went down with all hands.

†In late September, one other fisherman was lost, the beam trawler *Kingfisher*. Captain and crew reached port safely. She was sunk 85 miles off the Atlantic coast, brought down by the usual combination of gunfire and bomb. The *Kingfisher* was worth $175,000 but was heavily insured against submarine dangers, the insurance being for about $250,000.

died or were wounded in the second Battle of the Marne. Yet, by late summer the pendulum had swung to the side of the Allies, and the expanding American presence was key to the success. American and French troops worked together to push back the Germans, and as they did so reports came in of the loss of several more young men from Cape Ann. By September, the Allies had driven the Germans back to where they had been in March when they began their massive spring campaign.

In Gloucester, skippers, including Colson and Mesquita (the latter in a new vessel) made successful trips to Georges and Brown's banks. They were having the best year of their careers; Captain Colson had already stocked $70,000, with each of his dorymen receiving more than $2,000. But concerns with stocks, shares, and German submarines were about to be overshadowed by news of a more somber nature—the Spanish flu was about to hit the United States. Its effects were first felt in Boston,* and by the time it ended, American losses to the flu far exceeded those in the war. No Gloucester fisherman was lost in the war, but many were taken by the flu.

THE DEADLY WAVE of Spanish flu swept into Boston in late August and early September, the first place in the United States to be affected by the disease.† On September 6, the Gloucester paper reported that an influenza epidemic was running through the sailors stationed at Commonwealth Pier, Boston, with more than 350 cases having been reported in a week. The Massachusetts State Department of Public Health held out hope that it would be in the nature of an old-fashioned grippe, no deaths having yet been reported. Surgeon General Blue of the U.S. Public Health Service described the ailment as having a "sudden onset," people being stricken on the streets; at work in factories, shipyards, and offices; or elsewhere. Symptoms included a chill, then fever with temperatures from 101° to 103°F, headache, backache, reddening and running of the eyes, pains, and prostration. The recommended treatment consisted principally of bed rest, fresh air, abundant food, and Dover's powder (a drug containing ipecac and opium) for the relief of pain. Convalescence required management to avoid serious complications, such as bronchial pneumonia.

---

*Earlier, the first report of flu among the fishermen of North America came from Belleoram, Newfoundland. The time was mid-June, and the report stated that *la grippe* had seriously hampered the fishing fleet of that Fortune Bay community. Two vessels had lain at anchor for a number of weeks, owing to crews being sick, and when these craft went back to sea they could man only seven dories each. Many people were laid up with *la grippe* in the communities that lined Fortune Bay, and several deaths were reported. But this first wave of the disease was mild and few people died.

†The more deadly flu did not hit St. John's, Newfoundland, until early October, when 14 infected seamen were transferred to a hospital from a ship in the harbor.

In the second week of September, a Gloucester letter carrier named Samuel E. Curtis and his wife became the first people on Cape Ann to die of the flu. They left a daughter barely 10 days old. At the time of Samuel's death, his brother Chester, a teller at the Gloucester National Bank, was seriously ill with the same disease, while his sister, who taught in the Revere schools, was sick with pneumonia. It was just the beginning. By mid-month whole families were stricken. So many postal workers were sick the postmaster ordered the main office and annex thoroughly fumigated. With 636 pupils and 19 teachers out due to illness, Gloucester closed all schools and public amusements. On September 18, the YMCA closed its gym and pool because two of its workers had taken ill. There were now hundreds of cases across the city. The old armory was taken over as an emergency hospital.

The death count rose steadily—2 on September 19, 4 on the twentieth, 5 on the twenty-first, and 11 on the twenty-third. Ambulances worked throughout the day bringing patients to the hospital. Then came word that men on five fishing vessels had the flu. After getting only as far as Provincetown, the schooner *Laverna*, under Captain Robert Wharton, had 11 sick men, so he turned home for Gloucester. At the same time, the schooner *Edith Silveria* came in with two sick men, who were taken to the Red Cross hospital by police ambulance.

The school committee kept the schools closed, while the Board of Health and the Selectmen extended the closure of motion picture theaters, dances, and all other forms of amusement, pledging to do what they could to prevent contagion among children. Sawyer Free Library closed. Eight operators were out at the telephone company.

The "neighborhood" sections of the local paper listed sick and recovering people all over Cape Ann. Churches closed their Sunday schools. On September 23 there were so many new cases that exhausted ambulance drivers had to be relieved by doctors and interns. Nursing shortages and overcrowded hospital facilities were widely reported. By September 25, there were 11 more deaths.

A call went out to the Adjutant General's office asking for field tents and other field hospital equipment and associated medical personnel. Equipment and personnel began arriving on September 26. Addison Gilbert Hospital was soon surrounded by a sea of 100 tents, with some 240 patients receiving care in the makeshift facilities.

A Portuguese schooner, the *Adeline*, was tied up because the skipper could not find sufficient crewmen. The flu was spreading through Portagee Hill. Gloucester's death count on September 26 stood at 49, 8 having been recorded on that day alone. Young men from Gloucester were also dying at Army bases at Fort Devens, Camp Upton, and Gettysburg, Pennsylvania.

Local regulations to prevent the spread of the disease now came forth faster than ever. One order prohibited the display of fruit and vegetables intended for sale or consumption on the sidewalk or in stands outside stores. All hacks were required to be fumigated and cleaned after each funeral. Milk dealers were forbidden to deliver milk in bottles to households where there was sickness. The National Guard was called out. Companies L and K, both Gloucester units, were ordered to help at the military hospital on the grounds of Addison Gilbert Hospital.

The schooner *Pollyana*, under Captain John G. Stream, returned with the skipper and crew sick, having laid over for four days in Provincetown. The seiner *Arthur James*, Captain John Shavey, came in with several crewmen ill with the grippe, while at Mrs. McCarthy's fishermen's boardinghouse on Main Street the situation was desperate, with more and more sick fishermen having returned to port.[49] On the second floor, 20 men were housed in two bedrooms, and 10 were sick. The owner's daughter, Nina, ministered to the needs of the men, bringing medicine, blankets, and food. Up and down all day and night, she worked to pull them through, but they continued to decline. Ambulances soon pulled up to take the men to the tents set up on the grounds of Addison Gilbert Hospital. Nina was particularly close to an old Frenchman, Joe LeClare, calling him a "darling." When he was well, he did anything he could for her. If she asked for a few cherries, Joe brought her a whole carton, about 20 pounds. This "dearest old man" died at the hospital.

At the height of the Spanish flu epidemic of 1918, Addison Gilbert Hospital was overwhelmed, and many patients had to be cared for in makeshift tents to the south of the hospital building.

On September 28, the community received new reports of schooners returning to port with more crewmen suffering from influenza. The *Hjela J. Silva* and *Henry L. Marshall* had gone into an Atlantic port, not Gloucester, because of illness. This was followed by word of sickness aboard the *Natalie Hammond*. Six of the crew had taken ill, and two had died. Three weeks earlier, when the *Hammond* was only one day out of port, Captain Colson had landed three sick men at Provincetown. Wallace Doucette died there. A week later he landed two men at Yarmouth, Nova Scotia. It was there that Soren Bjerm died. Both were married. Doucette was about 38 years of age and lived in Lynn. He had filled out the necessary seaman's registration forms only the day before going to sea, jumping on board just as the craft left port. Bjerm was 23 years old and left a wife and young daughter in Gloucester. Among those who had a mild case of the disease was Steve Olsson.

Meanwhile, nurses were arriving in Gloucester to help—10 from Canada and 5 from the State Department of Public Health; 10 more were requested. The Citizen Aid Committee to the State Hospital Unit called for "the volunteer assistance of every woman in Gloucester who can give even a few hours each day to the nursing of the sick at the State Hospital Unit on Washington Street." Experience in nursing was not necessary. Several hundred women were required.

By the beginning of October, nurses and physicians were scouring the city to check on the spread of the epidemic and the needs of the people. People were asked to secure the dogs of those who were sick, so they did not accost and threaten the physicians. To cope with the paperwork burden, Boy Scouts rode with the physicians to take dictation on the conditions in the house just visited.

Efforts to check the spread of the disease invariably left the specter of death over the community. For many, the idea of being taken to a treatment camp became equated to being taken away to die. In the Fort section of the city, among Gloucester's flourishing Italian community, Jennie Auditore recounted the actions of her family to "save" her sick brother.[50] She remembered the men going from house to house looking for those who were sick, and forcefully taking them off to the camps. "Whoever was taken away never came back." She called these men the "national guard" and said they entered homes without warrants. Although her young brother was very sick, the family was prepared to do anything to prevent his being taken away. Thus did Jennie's father arrange with a neighbor, whose small flat was on the same floor as the Auditores', to move her brother when the guard came knocking on the door. If need be, the neighbor was prepared to use a shotgun to protect the boy. But it didn't come to that: No one beyond the family and neighbor ever learned that the boy had double pneumonia. No one came to the house, and he remained in the flat and survived. Even so, many others in the Fort were dying. On October 2, a community survey identified 306 people in the Fort with the disease and 9 who had died. In the Portagee Hill area, out of a population of 2,082 people, 229 were sick, 465 had recovered, and 14 had died.

After the return of the *Hammond*, Steve Olsson was to spend one of the longest periods at home during his married life. Romaine remembers that her father came home with a mild case of the flu. It ran its course, but then the rest of the family came down with it, and "Papa took care of us." Gerda remembered:

> As Papa got better, the whole family got it—Mama and the four children. For the children it was not too bad, but my mother had it pretty bad—and remember that she was then pregnant with Olive. My father had to do all the chores in the house—he washed our clothes and ironed our dresses. I remember our neighbor Mrs. Nelson came in and thought that was so wonderful. He said, "Well, I should be doing it with my wife so sick. I have to take care of the kids." A lot of people were dying all over the place from the epidemic, but we made it through. It was a terrible time. We later received word from Sweden that as they went through our ordeal two of my cousins had died of the flu—the children of Sam Andersson who had come to America with Steve so many years earlier.*

On October 2 came word that only seven deaths had been reported in the previous 24 hours, and the number of new cases had decreased throughout the city. Yet, there were still large numbers of people sick, and in some areas it was still so desperate that men, women, and children were suffering from lack of attention. A sad case was reported where a father, mother, and

---

*March 4, 1919, from Gullholmen, Hilda's mother: "Many thanks for the letter that I got from you now after Christmas, it spread such joy when we got it. To see how you have it in the western country. . . . Enos and Sam [who had come to Gloucester with Steve in 1893 and then went back] much depressed over their [two] small children [who died in the Spanish flu epidemic]. But I know that you live well, so I thank you very much for the Christmas present that you sent me [20 Krons], for which I send my sincere gratitude, for times are now very pressing. It is a lot for those who have nothing. You promise me dear Hilda, a cozy room and a good house with you if I were close to you, so that is nice to hear. I believe you dear daughter, for you were always for me a nice and good daughter. I miss many times your kindness. I have nothing to complain about, I am very well. . . . I am so glad that you all kept your life [in the Spanish flu epidemic], so we should only get strength enough to work. . . . Here we are more or less well now, even if many have been sick with the Spanish flu. Nils Olof was sick but he went well, but as soon as he got better he got a cold again and then he got diabetes, which is a difficult disease, now he is at the hospital to see if there is a cure for him. Me and Elma [Hilda's sister] worked in the home . . . other wise we are more or less well. I miss my little Jacob [one of her two grandchildren, and Hilda's nephew, who died from the flu] he was so sweet. Now we console ourselves that the war is over, but still we don't see any comfort. On the contrary, poverty is worse. Hilda greet Lars so much [a nephew of Hilda's who was visiting in the states], tell him that uncle Olar is very sick. I don't think he will ever get well again. Ask him that he also write a letter to him, that would console him very much. I send the dearest greetings to Steve and the children. Thank you dear children for the letters grandma got. . . . From us all . . . Hilda how old is the little one, is it two or three years this coming May [Harriet was born in May 1916]."

five children were sick with no one there to even give them a drink of water. The matter was referred to the Red Cross.

There were still new reports of illness among the crews of the Gloucester fleet. The schooner *Athlete* under Captain Thomas Benham docked at a Canadian port with all crew sick. Some of the men had been too ill to stand their watches or assist in sailing the vessel back from the Banks. Even so, in Gloucester, the number of new cases continued to decline. On October 4, drugstores along Main Street closed at the normal hour for the first time in two weeks.

The number of new cases dropped more by October 7; only 12 deaths occurring over the weekend as compared to 24 deaths over the prior weekend. On October 16, for the first time since the start of the epidemic, Gloucester had no influenza-related deaths. There were only 176 total cases of all levels of severity, with 12 new mild cases reported. Only 35 patients remained at the State Emergency Hospital post.

Public schools opened on October 23, and the ban on public gatherings was lifted. As the epidemic ended, the *Natalie Hammond* and other vessels of the fleet prepared to go back to sea. Gloucester had weathered the crisis, but at a significant cost in lives and suffering.

In France, much had happened during the two months that Gloucester struggled with the flu. September saw the American forces coming together to launch their first unified offensive of the war at Aisne-Marne; 600,000 American soldiers moved up, and in a two-day period sustained 7,000 casualties.

By early October, the Germans raised the possibility of an armistice, but President Wilson would hear none of it as long as German troops were on foreign soil. The Allies were advancing in France, pushing the Germans back toward their border. But with no armistice in place, late October saw heavy fighting, and the American battlefield losses climbed. Germany again asked the Allies for an armistice. On November 1, Turkey was granted an armistice, followed by Austria three days later.

Finally, at 6 o'clock in the morning, November 11, 1918, Gloucester received word that hostilities had ceased in Europe. The war was over.* The German armistice had been signed. Bells and whistles sounded across the community, and the mayor of Gloucester scheduled a parade for later that afternoon. All draft calls were cancelled immediately, and 151 young men

---

*By war's end, 2.08 million American soldiers had gone to France, 1.39 million of whom saw active duty; 53,523 of these men were killed in action and 205,690 were wounded. Yet in a few months in late 1918, 675,000 Americans had died of influenza, including 43,000 soldiers.

from Gloucester never left for camp. The victory parade wound through the packed streets of the city. People cheered, shouted, and waved the flags of all the Allied nations; the festivities lasted until well after midnight. Auto horns and bells helped keep up the noise and rouse the spirit. On November 13, the *Gloucester Daily Times* resumed publishing the arrivals and departures of the vessels of the fleet; its Fish Bureau—which provided a wealth of information on the fleet and its movements—was back in business.

Within days, the submarine net guarding Boston Harbor was taken up and Captain Wallace Bruce, previously of the sunken schooner *A. Piatt Andrew*, assumed command of a Boston beam trawler, the *Roseway*. The Navy Department said it planned soon to release nine of the confiscated steam trawlers, returning them to their original owners. The Gloucester schooner fleet would now face new competition, and a lengthening list of Gloucester skippers announced that they planned to move from dory fishing to steam trawling. Among them were Captains Spinney and Morrissey. Steam otter trawling seemed to be the way of the future—long hours baiting hooks and pulling trawl had lost its luster.

Despite the impact of the epidemic, the 1918 fishing season ended on a positive note. The *Natalie Hammond* came in on December 16 with 17,000 pounds of haddock, 30,000 pounds of cod, 1,000 pounds of pollock, 2,000 pounds of hake, and 3,000 pounds of cusk. The headline in the local paper read: "Record Year's Work for Hustling Skipper." Colson's stock for the year was among the top four of the fleet—$85,328—with each of his dorymen receiving a share of $2,425. It was the second-best share of any craft in 1918, and the best share on record for a double-dory schooner.

Thus did prosperity reign as the men came home for Christmas. Men set fir trees into simple cross-timbered stands as their children placed apples and snap-on wax candles on the branches. Parents filled Christmas stockings with oranges, blocks of paper, pencils, crayons, games, small dolls, and toy soldiers. Christmas morning was for the children. Later in the day there were turkey and songs, with voices raised in French, Portuguese, Italian, Swedish, Norwegian, Finnish, and English. The war was over, and most of the people had somehow escaped the scourge of the flu.

Christmas that year was a special time for one father and his children, and Gerda Olsson remembers the day. She walked through the neighborhood with her father, Steve, and her sister, Romaine. As they did, one of Steve's friends stopped and said, "Steve, this has to be the happiest time of your life." Steve smiled, looked at his two girls, and responded, "Yes, I realize that, I know how fortunate I am." He then turned to Gerda and Romaine and said in Swedish, *"Dona en bra men hemma et bast,"* which translates, "To be away is good but to come home is best." The men were home, on firm land, and once the children were in bed, a husband and wife could sit talking quietly in their native tongue—peaceful talk, personal talk, meaningful talk.*

But they were also fishermen, and the Christmas interlude was over all too soon. Just two days after Christmas, the *Natalie Hammond* left to go haddocking along with two other vessels. The men were back to sea.

*In the Olsson household, the only exception to this desire for closeness was Hilda's requirement that Steve not smoke his pipe in the house. "When home, Papa would sit in the cellar to smoke his pipe. Mama didn't like the fragrance and that was the only place that Papa would smoke in the house—in the coal cellar. He might start by sitting in the kitchen talking with Mama, cutting his tobacco into square hunks, while also cutting apples from the barrel in the cellar into bite-size pieces. But finally the time would come, Papa had to smoke his pipe, and he would go to the cellar and we small children would join him. But never Mama." Papa was home.

CHAPTER 26

1919—An Uncertain Transition

With war's end, America no longer had a huge army to feed, and firms found themselves holding huge stockpiles of processed fish. During the war years, Gorton-Pew had seen dramatic increases in sales and profits. Net profits rose from $208,621 in 1916 to $563,039 in 1918, while sales volume went from $3,102,066 in 1916 to $9,376,821 in 1918. But now the firm was unsure of how best to move forward. The military contracts had ended, and for most of the year Gorton-Pew sought ways to sell its inventory of herring, whiting, and flaked fish to the civilian population. As Gorton-Pew went, so too went the economy of the Gloucester fisheries. Having gobbled up most of the competition, Gorton-Pew had become the one major fish processor left in the community.

The firm was both persistent and innovative in seeking to expand markets. Within the United States, Gorton-Pew launched a series of advertising campaigns. It asked a consulting firm to select a suitable label for their new line of 10-ounce canned codfish cakes. It was yellow and blue, with fish cakes sizzling in a frying pan.* The campaign pushed the image across the country. Ads were placed in the local newspapers of six large cities. The company gave out 100,000 copies of a recipe book featuring codfish cakes and other Gorton-Pew products, and they let a contract with the Household Products Distributing Company to conduct cooking demonstrations in 5,000 schools. In a more traditional vein, the firm spent $6,159 to paint and repair Gorton-Pew billboards, including 70 signs along the tracks of the Pennsylvania Railroad.

Outside the United States, Gorton-Pew moved on three fronts: Cuba, Norway, and southern Europe. In Cuba, it commissioned a study to identify a blend of fish and tomatoes that could be canned and sold on the island, hiring a Spanish cook from Palm Beach to create the recipe. In Norway, they scheduled a six-month advertising campaign to sell fish balls and broadened it to include other Scandinavian countries. Finally, they signed a distribution

---

*These colors have remained a Gorton brand mark to this day.

agreement with a firm in Greece to "dump" the company's vast stocks of salt cod onto the European market, but this last campaign failed; the salted fish went to Europe but remained largely unsold.

Despite the company's efforts, Gorton-Pew remained overstocked with old fish and was undercapitalized to address its most immediate needs. Its fate lay in the hands of its creditors, with management doing its best to ensure central control over the firm's many branches. Among other steps, they assigned their top men to run the Canadian and Newfoundland operations; individuals who were familiar with the Gloucester office and its ways of conducting business.* Only with accurate and constantly updated information, and Gloucester's control of the buying practices at these distant stations, could the firm hope to succeed. For most of 1919, given their focus on selling existing fish, Gorton-Pew bought little new fish. Managers reasoned that by reducing the capital tied up in inventory, the firm could improve its bottom line when they closed the 1919 books. To this end, they ordered their northern agents to stop buying fish and reduced the operations of their own fleet. Over the winter, vessels in the Gorton-Pew fleet had to sell their catch at Boston's fresh-fish market. The Gloucester plants took no new fish, whether from their own or other craft. From then on, market demand became the driving force behind all business decisions, rather than the previous approach of buying as much fish as possible.

Even with the new orientation, Gorton-Pew's fortunes continued to decline. The company was becoming desperate, and its predicament had a marked effect on Gloucester's fisheries and the men of the fleet. Wherever one looked—in Boston or in Gloucester—owners looked to sell schooners, and once sold, the vessels were not replaced.

In their desperation, the dory fishermen's only hope seemed to be another strike, but many feared that even so dramatic an action would have little effect on falling demand and prices. Until the oversupply disappeared, the firms would have little need for new fish and prices would continue to fall.

With an expanded fleet of otter trawlers† and schooners bringing in banner trips, haddock prices soon fell to prewar levels, selling at $2.73 per hundredweight. But vessels and men continued to go to sea, and over a two-day period in May the *Seal*, under Captain Carl Olsen, landed 200,000 pounds

---

*At St. John's, the man to carry this responsibility was Phillip Sellow, an ex–Gorton-Pew executive who had just returned from 18 months' service in the Army. The company also employed another returning veteran, Clinton N. Haynes, as the "superintendent of northern branches," providing a direct link between Gloucester and the stations in the north where fish were produced.

†Including the new otter trawler *Gloucester*, an enormous vessel at 175 feet and 250 net tons.

⟩⟨ *Fish are off-loaded into a cart, c. 1920.*

of fresh fish; the *Marechal Foch*, under Jeff Thomas, landed 128,000 pounds; the *Gloucester*, under Lemuel Spinney, landed 375,000 pounds; and the *Sea Gull*, under Fred Thompson, landed 127,000 pounds. Such enormous fares glutted the market, and the skippers were forced to sell much of their catch to the Gloucester splitters at less than 2 cents per pound. Seeking a solution, a council of five fishermen's unions from Boston and Gloucester came forward with a unique wage proposal. Believing that buyers were taking advantage of the fishermen, they proposed to impose price floors on all fresh fish brought to market. The union proposal set minimum wholesale price floors based on the inflated wartime experience in Boston and Gloucester. Failing an agreement by the producers to buy at these levels, they were prepared to go out on strike. In Gloucester, 30 men, representing almost the entire industry, reviewed the proposal.* Speculation was rampant that some firms might welcome a strike. With no fish coming in, they might at last move the large stock on hand.

---

*According to Gorton-Pew, Captain Morrissey of the *Walrus* made $7,079 in 1918, while Captain Olsen of the *Seal* made $5,664. The net profit of the *Walrus* was $19,662 for the year; the *Seal* ended the year $7,327 in debt because of accidents. If the proposed new rates had been in effect during that time, Gorton-Pew asserted that there would have been a difference of $30,374 for the *Walrus* (a net loss) and $23,443 for the *Seal*. Under the new proposed rates, Captain Morrissey would have made $10,446.

In any case, the firms refused to negotiate on minimum price levels, insisting that supply and demand should regulate the price of fish, rather than artificial guarantees based on inflated wartime prices. In mid-May, the War Labor Board discussed the union's petition. Present were the Fishermen's Union, the Bay State Fishing Company, and the East Coast Fisheries Company (a New York concern). Absent were the Gloucester firms. The Board's provisional decision suggested that they had no power to fix minimum fish prices. Nothing remained on the table for the 80 percent of union men who sailed on schooners. Fish prices had dropped dramatically. Haddock was now going for $1.75 per hundredweight, and hake for $2. Fishermen found themselves working for almost nothing.

Secretary William Brown of the Fishermen's Union argued that the men should strike. They now made less than before the war, while the cost of living remained high, and on some trips the stock did not even cover the cost of the outfitting expenses. The men had little to lose by going out, and in Gloucester the strike vote passed by a slim margin—240 to 214.

Gloucester's close vote reflected the nearly unanimous "No" vote by the Canadian fishermen who had come south for the summer. They had no cash reserves, no permanent homes in Gloucester, and as the strike began, many Canadian fishermen had no option other than to head north to Nova Scotia and Newfoundland. In less than a week, nearly 300 men had returned to the Provinces. Meanwhile, 100 vessels sat idle in Gloucester and Boston, representing almost the entire fleet except for the few salt-bankers that were away until the end of the summer. The masts and spars of the striking fleet encircled the piers in Gloucester and Boston—a sight reminiscent of days long past.

On July 10, a hearing before the State Board of Arbitration and Conciliation sought an avenue for negotiations. For half a day the two sides argued over a single point—the key union demand of including a minimum price floor for fish in the list of grievances. There was no agreement, and the strike rolled on. In Boston, the Italian Fishermen's Protective Association joined the strike, tying up their 200 small craft, each with its three- to four-man crew. The supply of fresh fish in Boston was effectively cut off.

By late July, Gorton-Pew had sold the schooner *Senator* to parties in Newfoundland, and with the strike in its fourth week, other Gloucester owners declared that they, too, were looking to sell their vessels. One firm advertised its entire fleet for sale, while Gorton-Pew was said to be considering placing many of its boats under foreign registry. In this climate, no one seemed willing to take the first step, but by July the union voted to let engineers return to work on the schooners' engines. Several days later the small Italian boats from the Fort took on ice and supplies in preparation for fishing.

However, heavy winds kept all but two in port. In Boston, the weekly receipts of fresh fish climbed to 1,200,000 pounds, but only because of the arrival of a large steamer from Nova Scotia.

Commissioner Henry J. Skeffington called all parties together in Boston. He was hopeful the strike could be settled, blaming continuance squarely on stubborn Gloucester firms. He accused them of "acting like a dog in a manger." According to counsel for Gorton-Pew, the owners based their decision "on the ground that the fishermen's demands involved the proposal of illegal price fixing." Skeffington's response was that "it did not appear to be illegal for the fishermen to fix the price of their labor, based on the kind and amount of fish they caught." By mid-August, while nothing had really changed, the parties agreed to a two-month stopgap solution. Under this time-limited schedule, fishermen were to get an increased percentage on the wholesale price regardless of whether the wholesaler made up the difference by raising the retail price. On the day of the agreement, Gorton-Pew sold another of its craft, the *Grace Otis*, to parties in Cape DeVerde, as the Gloucester sailing fleet remained at the docks. Two days later Gorton-Pew sold three more craft. Gloucester was bleeding vessels, and the new labor agreement only deepened the owners' overwhelming sense of helplessness.

Little seemed to have changed. Some fishermen remained ashore, and others went to sea for a chance to make money. For the first time in weeks, fresh fish was plentiful and business was about normal. But as soon as fresh fish came into Boston, prices fell: Haddock dropped from 12 cents to 7 cents per pound. On August 20 the fishermen met with Gloucester owners. The owners continued to insist that the minimum price proposal was illegal, but they did not object to the time-limited price schedule that had been agreed upon. A willingness to restore the business became the spirit of the day.

With the fishermen's strike settled, the schooner fleet resumed fishing. Along the waterfront, skippers were driving to be the first away. Sails were bent on, ice and supplies were loaded, and offshore craft prepared to head east to secure bait.

A late August storm kept the shore fleet tied up for several days, even as offshore craft got away. Sixteen craft left on August 26, and the immigration department at the Custom House was crowded with fishermen and skippers signing shipping articles. On August 27, another 16 vessels sailed, followed by 14 on August 28. On the following day, the price paid for fresh fish hovered at the agreed-upon minimum price. The haddock quote fell from $4.50 to $4.00 per hundredweight, the minimum.

Over the next three days, another 19 craft left Gloucester; while in Boston, the returning fleet flooded the market with more than 1.5 million pounds of fresh fish. Only a few craft sold at the union minimum of $4.50, most sold

※ *Crew of the* Natalie Hammond, *c. 1920—Nils Lund is kneeling in the middle of the group in a white shirt. Steve Olsson stands behind with his arm resting on the main boom, wearing oilskin jacket and promotional cap from Tarr & Wonson marine paint company.*

for less, and some half-million pounds eventually went to the Gloucester splitters. Within days, fresh haddock fell to $3.00 per hundredweight, but there were few sales even at the lowered price level, and late in September the price of haddock dropped to $2.50.

On October 15, the temporary minimum price agreement ended. Minimum price floors had not worked, owners had lost money, and the experiment would not be repeated. It had seemed like a good idea to the fishermen, but as the owners had forecasted, the laws of supply and demand governed the market. Both sides spoke of alternatives, but no immediate action was taken.

Gorton-Pew continued to sell its vessels. With warehouses full of unsold fish, they needed cash. The older schooners *Ralph Russell*, *W. H. Rider*, and *Pauline* went to Cuba for $20,000, less 5 percent commission. Next was the *Elsie*, sold to W. C. Smith in Lunenburg, Nova Scotia, for $12,500—Gorton-Pew having earlier taken majority ownership of the vessel from Captain Alden Geele. One year after the war had ended, vessels were being sold out of the district, no one was making money, and the future looked as gloomy as it had in 1913.

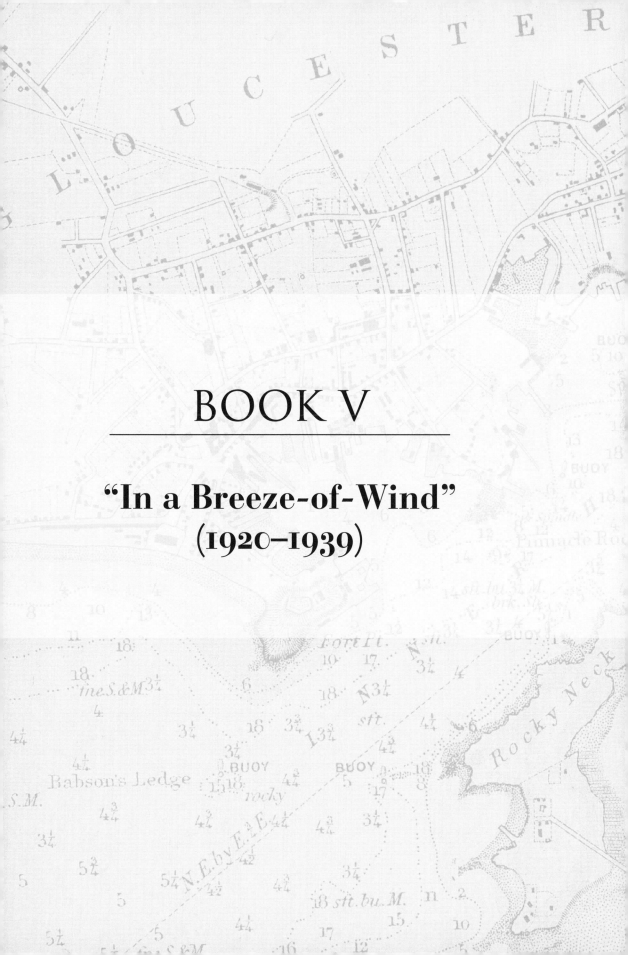

BOOK V

"In a Breeze-of-Wind"
(1920–1939)

CHAPTER 27

Economic Decline and the First of the International Fishermen's Races: *Esperanto* versus *Delawana*

It was 1920, and a sense of uncertainty hung over the community. Domestic markets were falling, foreign markets were demoralized, and local warehouses were bursting with last year's catch. Outfitters showed little interest in sending schooners on extended trips to the Scotian and Grand banks, and after Lent fresh fish prices dropped to prewar levels. The otter trawler fleet was tied up for months on end, and with the sale of 17 schooners in 1919, hundreds of jobs had been lost.

Captain Lovitt Hines became the first skipper in the new year to sell his craft, going north to try his luck on a salt-banker out of Lunenburg. Within days, Captain Alden Geele had followed, and Captain Clayton Morrissey announced that he had signed on as master of the new 700-horsepower Nova Scotian steam trawler *Bernard M*. He, too, was going salt-banking out of Lunenburg; maybe Canadian profit margins would be better than in Gloucester.

These were tough times, and the fishermen of Gloucester were once more forced to make life-defining decisions. Many of the Nova Scotian and Newfoundland fishermen who had left Gloucester during the union upheaval did not come back. Others, like Captain Charles Daley, who left during the war, saw little reason to return. Daley held dual citizenship, and for now fished out of Newfoundland. He settled in St. Joseph's and skippered schooners out of St. John's and small two-man fishing skiffs out of St. Mary's Bay. He had a fine house,* had married Lucy (Doody)† in the last year of the war, and by May 1920 had three children: Edna, John, and Alban. For now, at least, Charles would remain a St. Joseph's fisherman.

---

*Neighbors considered his home to be a mansion. It had a 40-foot by 40-foot footprint and was adorned with elaborate trim that Charles brought back from trips to New England and Europe. It had fire-hardened, multi-colored glass tiles around the fireplace, hardwoods on the walls and ceilings, and a 6-inch-thick front door that opened onto the bay. The house also had a uniquely carved railing that rose from the first-floor entry to the second-floor landing.

†Like Charles, she had spent time in the United States, working in a Springfield laundry.

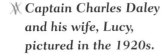

Captain Charles Daley and his wife, Lucy, pictured in the 1920s.

Back in Gloucester, other fishermen were looking to their future. Captain Jeff Thomas and doryman Steve Olsson had much to think about. Following the wartime loss of the *Sylvania*, Captain Thomas became a Gorton-Pew skipper, taking out their finest craft on fresh and shacking trips. But by 1920, the company's future was uncertain, its fleet had shrunk from 65 craft in 1918 to 38 in 1920, and more sales were imminent. Yet as long as the firm had a fleet there would be a job for Captain Thomas, and for Jeff's part he was content to be a Gorton-Pew skipper. The firm treated him with respect, his position was secure, and he saw no need to change.

For doryman Steve Olsson, the time had come to seriously consider returning to Sweden. As the *Natalie Hammond* prepared for her first trip of the new year, Steve and Hilda spoke of the unthinkable. He had been in Gloucester for more than a quarter of a century, Hilda for 16 years. They had five children. Their youngest child, Olive (Olivia), was less than a year old, and they talked of what would be best for their family. Steve's brother-in-law, Sam Andersson, had long ago returned to Gullholmen, and wrote of the prosperity of the local Swedish fishermen; he had his own boat and so could Steve. Both Steve's and Hilda's mothers were still alive, and Steve wrote to his mother after every trip. A move was financially possible, as Hilda had saved several thousand dollars and they could get a good price for their East Gloucester house.

)(*The* Natalie Hammond *docking at Gloucester Cold Storage Wharf on Rogers Street, c. 1920.*

Steve also spoke with Nils Lund, one of the Gullholmen dorymen serving on the *Natalie Hammond*. Nils had gone back to Sweden three times; his wife, Cecilia, had made the passage twice.* Nils and Cecilia were now in Gloucester, but Hilda's relatives wrote saying that if the Lunds could try to come home, why not the Olssons? But Cecilia and Hilda were of different character. When Nils went to sea, Cecilia found it difficult to bear. When Steve was at sea, Hilda watched over the family, tended the home, and paid little attention to reports of storms on the Banks, overdue vessels, and the loss of other women's husbands. She never feared for Steve when he was at sea. Although two of her brothers and her father had drowned at sea, she had the stoic makeup of a woman from Gullholmen. All she knew was the sea, and she could never imagine anything happening to Steve.

Even so, their siblings and mothers sent letters pleading for the Olssons to return to Gullholmen. The pleas were subtle but direct. Hilda's sister

*Lillian Lund Files tells of her grandfather helping to ferry vessels across the Atlantic in order to get a passage home to Sweden. "They sometimes went back to the old country by crewing a vessel that went across. . . . Gloucester fishermen helped to deliver this schooner to Holland. They had a free ticket back home because they were doing this. . . . Then it was just a hop, skip, and a jump to go back to Sweden."

Maria wrote: "It would be nice if we once again in life could talk to each other." Sister Alida was also blunt: "Hilda I dreamt this Christmas that you were coming home and had all your kids with you. . . . I dreamt about you so clearly. . . . I thought that you came in through the door here and what joy I felt, I cannot tell you." Steve's brother Albert was subtle: "I see by your letter that you and Nils Lund are on the same schooner. Greet him. Strange enough that he didn't stay in Sweden when he comes back here so often." Hilda's mother was the least subtle: "I long and wait after you so very much, but I know that it is a vain thing, so I will have to be satisfied with just having you in my thoughts. . . . I thought that when your girls become a little older you would have come this way sometime. Then I would have been able to speak to some of you before I die, but I have to accept this."

Steve had a deep love for his life in Gloucester, and he would not leave America.* He was comfortable on the *Hammond*, he was still making money, and he was close to Captain Colson. And, maybe of greatest import, Hilda and Steve would never leave Hilda's sister Bertha, who had only recently lost her husband, Macker.

Thus, on January 14, 1920, as three independently owned halibut schooners left on their first trip of the winter season, Steve Olsson joined Captain Colson and the Lund brothers, sailing out of Gloucester Harbor on the *Natalie Hammond*. During that winter, 16 schooners engaged in this most profitable of Gloucester fisheries. Most landed their catch in Boston, some sold their fish in Portland, but only one came to Gloucester—the *Natalie Hammond*. With a low supply and reasonable demand, prices remained high for halibut, while almost all other Gloucester fisheries faced poor demand and low margins.† Halibut skippers were even able to dicker for the price of the catch. In late spring, Captain Gus Hall of the schooner *Imperator*, while barely touching the end of Central Wharf in Portland, jumped ashore to see what the local buyers were willing to bid for his 55,000-pound fare. If the price was not right, he would continue to Boston. But the price was right, and Hall tied up and unloaded.

As Captain Hall unloaded his fare in Portland, Captain Morrissey brought the steamer *Bernard M* into Lunenburg with only 100,000 pounds of salt fish: The cost of the trip exceeded the stock of the catch. Neverthe-

---

*Steve and Hilda gave up thoughts of returning to Gullholmen. Toward the end of the decade, when Gerda graduated from high school, Steve toyed with the idea of taking Gerda to Sweden on a visit. But even that did not happen. Rather, he used the trip money to buy Gerda a car. Contacts with his Swedish family were through letters and photos, and the visits by nephews to America. Steve took one of these young men to sea on the *Natalie Hammond*, and they were to sit quietly in a dory, talking of life and the opportunities in America.

†Despite low prices, Gloucester's mackerel seining fleet had a good year, as they found themselves on the fish throughout the season. After Easter, 21 seiners fit out for the trip south and others followed.

)(*With the beam trawlers on the sidelines, the schooner fleet continued to go to sea. Here they sit five deep at the dock; men check the rigging and stow supplies prior to leaving for Georges and the Scotian Banks.*

less, the owners did not give up. After taking on coal, Morrissey steamed out on his second salt-bank trip of the season, and on his return he sold a more substantial 300,000-pound catch to Gorton-Pew Fisheries, which shipped the fish to Gloucester. But demand in both Europe and the Caribbean had largely dried up, and Captain Morrissey received a low price for his catch. The *Bernard M* had now lost money on her first and second salt-bank trips and was converted to a cargo ship; the Lunenburg fisheries were failing, and Captain Morrissey returned to Gloucester in search of another vessel.

In Gloucester, the beam trawler *Seal* had also been hauled out. Although her skipper, Captain Carl Olsen, had brought in good trips, his stocks had also failed to cover the costs. Soon, Gloucester's other two large beam trawlers, the *Walrus* and *Gloucester*, were hauled out, as were five of the Boston beam trawlers. Others followed, and any of them could be purchased for a song. A few days later, the first of Gloucester's small fleet of salt-bankers returned from its spring trip. The schooner *Hazel P. Hines* sold her 325,000-pound catch of salt fish for $18,591, and then converted to the fresh fishery. There was little call for second salt-bank trips.

In Lunenburg, many of the Canadian schooners also returned from their first salt-bank trip. Captain Alden Geele landed 300,000 pounds and

)(*The schooner* S. F. Maker *moves across Harbor Cove by tug.*

turned around for a second trip. But the trend was the same: International demand was on the wane, and many Nova Scotians saw little justification in continuing to send craft to sea. It was in this climate that the Halifax business community looked for a strategy that would put Canadian fish back on the international map. The local newspaper suggested sponsoring a schooner race to decide the best of the Canadian fleet. Vessel against vessel, these old-time majestic schooners, crewed by seasoned fishermen, would vie for top honors.

Within weeks, the Halifax community had subscribed $2,240 in prize money, and the race was assured. There was even talk of a 1921 match race between the best of the Nova Scotian fleet and the best of the New England fleet. But the men from Gloucester could not wait until 1921. Their fisheries were also in shambles, and Gloucester accepted the challenge in 1920. They wished to compete in the fall, and H. R. Silver of the *Halifax Herald*, head of the Nova Scotian race committee, jumped at the offer. What better way to increase visibility than to include the Gloucestermen? The business and fish-ing communities of Lunenburg, Halifax, and Gloucester were soon caught up in the excitement. In Gloucester, a local racing committee was formed to respond to the challenge.

Of the more than 120 Lunenburg schooners, 9 came forward to sail for the Canadian National trophy and prize money. The winner would have the honor of representing Nova Scotia in the match race with the New England challenger. The Canadian elimination race was held off Halifax in a moder-ate breeze of wind. The Nova Scotian salt-bankers sailed with the wind over the first half of the 40-mile course, averaging 10¼ knots. They sailed against the wind on the second half of the course, with the eventual winner, the schooner *Delawana*, averaging 7½ knots over the full 40-mile course. She was a beautiful vessel, a worthy Canadian champion, and Nova Scotians were certain she could beat any craft put up by the New Englanders.

In Gloucester, the local racing committee looked for a vessel to carry the American banner, but time was short, and in these hard times some began to wonder whether any of the local firms would step forward with a boat. And even if they did, how could the local community afford to prepare her for the race? There was no time to raise the funds needed to spruce up and fit out such a craft.

From Halifax, H. R. Silver soon wired Gloucester seeking confirmation that the town meant business. If so, representatives of the Halifax committee were ready to address all the details. Silver called for the race to be held off Halifax, and promised that future races would alternate between Halifax and Cape Ann. This was not to be a onetime race, but the beginning of a series of international races.

Fortunately for Gloucester, Benjamin A. Smith, one of the principals of the Gorton-Pew Fisheries, had caught the racing bug. On October 6, when the Gorton-Pew schooner *Esperanto* returned from a salt-bank trip, Smith announced that she was to carry the colors of Gloucester and America to Halifax. Gorton-Pew took full responsibility for fitting her out—the firm knew that it would be widely praised should she win. There was to be no local competition. Only one local vessel was available, and Gloucestermen were proud to send her north for the competition. The *Esperanto* was a real fisherman. Her pedigree was unquestioned. Launched from Essex in 1906, she was 107.4 feet long at the waterline, 25.4 feet in beam, 11.4 feet depth, and 140 tons gross. Although not the fastest of the Gloucester fleet, knowl-edgeable fishermen felt she could take *Delawana*. She had been built for speed and had shown her mettle in all conditions.

The *Delawana* was built along similar lines but was newer by seven years, built in Lunenburg in 1913. She was 106.6 feet long at the waterline, 26.3 feet in beam, 10.3 feet in depth, and 124 tons gross. She had been built for codfishing on the Grand Banks, and had always followed that fishery.

After unloading her 150,000-pound salt-cod cargo, the *Esperanto* lay at Gorton-Pew's D. B. Smith wharf showing all the signs of a hard season at sea. When Captain Martin (Marty) Welch of the schooner *Thelma* came over to inspect her, he found a craft in need of repair, and his opinion mattered, for he had been selected to skipper her in the race. Marty was a seasoned master mariner, and there had been general praise for his selection.

Interest in the race grew daily. Everyone had an opinion on its outcome. Some doubted the ability of the *Esperanto* to cope with the Nova Scotian defender. Others felt she would make it a runaway if there was a breeze— assuming, of course, that Gorton-Pew could get her in shape for the race. The International Trophy race was scheduled for November 1 off Halifax. A purse of $4,000 was to go to the winner and $1,000 to the loser.

People crowded the waterfront to look over the local contender. The watchwords of the day became: "Where is the *Esperanto*?" and "Let's go have a look at the *Esperanto*." The Smith wharf was besieged by an army of curious citizens. The local paper claimed that no incident since the raids of the German submarines had aroused so much interest in the fishing vessels and all they represented. People were taking notice—in Nova Scotia, in New England, and across America. Two true fishing schooners, with seasoned fishermen, were about to compete.

Gorton-Pew threw its men into the task of fitting out the contender. They put in long hours cleaning up *Esperanto* and stowing 85 tons of pig iron in her bowels for ballast. Taken to the Parkhurst railways, she was hauled and cleaned, and her sides were scraped and painted. She was then returned to the D. B. Smith wharf where her main-topmast was put in place, while the men in the spar yard shaped a new "rough stick" for the fore-topmast. The wheel and steering machinery were overhauled; riggers hoisted the new fore-topmast into place and bent on the sails. Finally ready, she barely resembled the vessel that had sailed into the harbor two weeks earlier. She looked to be a worthy contender.

On the day *Esperanto* was to sail, the normal haunts of the dorymen and skippers were deserted. The men joined the throngs from all over Cape Ann who lined the piers. Skippers, fishermen, families, and just plain well-wishers came to say "Godspeed" to *Esperanto*. Gordon Thomas, then a boy of 15, remembered that "there was quite a time when she left. I went down to Gorton Pew's wharf. She was there. You should see the crowds of people

)(*The victorious racing schooner* Esperanto *under full sail in 1921.*

down there. . . . They let the schools out and that was quite a thing when she sailed out the harbor. . . . Whistles blew. Gloucester right then caught on with the racing fever. It was a huge wharf down there and it was just like flies of people when she sailed out of the harbor."[51]

Motorboats flitted about and there was a mighty roar as the craft started on her way. The oldest residents of the city could remember nothing that rivaled the lively scene and enthusiasm. Foghorns bellowed a good-bye until *Esperanto* passed far down the harbor. At Eastern Point, a crowd lined the breakwater and the shoreline behind the lighthouse. A cannon boomed a final salute as *Esperanto* dropped below the horizon.

On reaching Halifax, Captain Welch told the reporters, including those from Gloucester who filed the account, of the grand send-off and of his uneventful trip up the coast:

> I'll never forget that sendoff at Gloucester, neither will any man on board. Why the whole town was out I guess. It seemed so, anyway. We put her through her paces out in the harbor . . . While we could still hear the cheers from the shore, we swung her off around the breakwater and were off on the big adventure. . . . Well, it was pretty moderate all along the afternoon as we glided down by Thatcher's and across Ipswich Bay and

out into old Fundy. What there was of the wind was coming steadily from the southwest. We seemed to be doing pretty well. We had her all dressed up and some place to go.

After signing on a pilot, Captain Welch sailed *Esperanto* on a trial run over the Halifax course. Captain Billy Thomas was appointed the official Gloucester observer on the *Delawana*, while Howard Lawrence was the Canadian monitor for the *Esperanto*. Harry Eustis, a teenager at the time of the race, but someone who spent a lifetime as a Gloucester doryman and cook, said Captain Welch found the *Esperanto* a hard vessel to put in trim.[52] Her usual skipper, Captain Charlie Hardy, told Eustis that the *Esperanto* was a "funny vessel," sailing by the "head a little bit." She was known as an "ordinary vessel," although in the wind she was good.

As the competition unfolded, real-time updates came in at Gloucester's Western Union office. Hundreds of people surrounded the office as manager Tim Harigan posted bulletins on a blackboard in the window. In race one, the final note said it all: "*Esperanto* wins first great race in fishermen's series— Leads Nova Scotia's hope by 20 minutes over the finish line." It was an enormous victory. *Esperanto* had started 27 seconds behind *Delawana*, but led by more than 3 minutes at the first mark and doubled her lead by Shut-in Island buoy. As the wind blew from the west, *Esperanto* made it a runaway.

A crowd gathered once more at the Western Union office to learn the fate of *Esperanto* in race two. The race opened with a slight wind that picked up as the race wore on. *Delawana* crossed the starting line a few lengths ahead of *Esperanto* and maintained a half-mile lead at Shut-in Island buoy. As word of the race came over the wires, Gloucester people first stood in silence— *Esperanto* was losing. At 2 o'clock, *Delawana* had opened a 12-length lead. At 2:10 both schooners were headed far to windward of the inner buoy and relative positions were unchanged. At 2:30 *Esperanto* had caught up to *Delawana* and was passing through her lee. At 2:35 *Delawana* was again in the lead, but only by three lengths. *Esperanto* had hauled in her sheets and was trying for *Delawana*'s "weather" (blocking her wind). At 2:40 the two skippers were said to be in a game of one-upmanship. One message stating "*Esperanto* now in the lead" was followed by another saying "*Delawana* now in the lead." At 2:42, *Esperanto* had again moved to a half-length lead. At 2:50 Captain Himmelman had his bowsprit abeam of Captain Welch's foremast. Then at 3:13, word came that *Esperanto* had rounded the inner buoy 30 seconds behind. With the wind holding, both craft were making over 7 knots; it would be a long tack, followed by a short tack up the harbor. But it was the *Esperanto* that made the better tacks. She crossed the finish line a few lengths ahead of *Delawana* just as the skies let loose a downpour of rain.* It was close; just the type of contest the fishermen had hoped for.

In Gloucester, Ben Smith received the message confirming *Esperanto*'s win and phoned the Gloucester Cold Storage with three words: "Blow your whistle." The whistle was sounded, and its message was picked up and reinforced by the boats in the harbor and the firms on the shore. The crowd at the Western Union office poured into the streets, spreading news of the glorious win. Children, who had just been dismissed from school, joined in the noisy festivities, falling in with the fishermen and the people of the city, who were dancing in the streets. The demonstration went on for more than an hour, and mischievous children kept it up well into the evening.

Congratulatory telegrams poured in from across the country. Senator Warren G. Harding wired from Marion, Ohio, even as he awaited the impending national Presidential plebiscite. His message was short and to the point: "Have just learned of the victory of your schooner *Esperanto* and wish to extend my warm congratulations. Such competitions can only have the effect of stimulating interest in the great work of reestablishing and maintaining an American merchant marine and redirecting the interest of our people toward the sea. That is a task worthy [of] the best efforts of all of us."

Before leaving Halifax, the men of the *Esperanto* cast absentee ballots in the Presidential race. Thirty-two (of 33) voted for the Republican ticket of Harding and Coolidge. Only one of the crew voted for Cox and Roosevelt.[†] The election itself was no contest. In fact, the Republican ticket carried Gloucester, the state of Massachusetts, and much of the nation.

Esperanto arrived back in Gloucester five days later, at 2 o'clock in the morning, the winds having held her up on the trip south. The streets were quiet; no more than a half-dozen people were out as she came up the harbor and tied up at the Parkhurst wharf. Captain Welch hollered to the watchman to let him know *Esperanto* was back in port, while his crew and the newspapermen who had accompanied him on the return trip went ashore. The only establishments open were the Savoy Hotel and Cameron's lunch cart. A small number of patrons rushed onto Gloucester's Main Street to greet

---

*Gordon Thomas ascribes the victory [material from Jeff and Gordon Thomas Oral History Collection] to the "smart sailing" by Marty Welch, and the mistake of the skipper of the *Delawana* in taking out all the ballast the night before the race—setting up his craft for a light-wind race. But it breezed up and that was where *Esperanto* won the race, the skipper getting permission from the Gorton-Pew representative to pass through shallow waters to pass *Delawana*.

†This was the first national election in which women had the right to vote. The Nineteenth Amendment to the Constitution passed in August of 1920, and in Gloucester the voter rolls included 5,121 men and 2,182 women.

the men of the *Esperanto*. Within an hour, the skipper was whisked home by automobile for a few hours' rest.

Later that morning, when Captain Welch returned to his craft, there was handshaking all around as plans were made for *Esperanto* to be received by the city. The tug *Eveleth* towed her out of the harbor, permitting a heroic return before all the people of Cape Ann. The Stars and Stripes and a broom were displayed on the masthead—the broom symbolizing her victory over *Delawana*—while multicolored signal flags flew from her rigging, fore and aft, from the topmast, to the end of the booms, and to the tip of the bowsprit.

Crewmen were joined by key principals from Gorton-Pew—Benjamin A. Smith, Thomas J. Carroll, R. Russell Smith—as well as many family and friends. At 12.25 P.M., in the middle of a beautiful day, the fire alarm signal belched forth a long blast, followed by whistles and bells around the waterfront. In an instant, the greatest crowd that had ever gathered on a Gloucester wharf swarmed toward Gorton-Pew's and the harbor front. Schools were let out and the youth of the city streamed toward the harbor. Thousands gathered on the wharf as the tug pushed the *Esperanto* into her berth, and people soon swarmed over the vessel. From stem to stern, a babble of voices filled the air, as well-wishers greeted the crew and owners. That evening a banquet for 700 people at the State Armory was followed by a gathering at the movie house, where pictures of the race were shown. Several hundred people attended, including the crew of the *Esperanto*.

The *Esperanto* had won a true fishermen's race. This might be a changing world, powered vessels might soon come into their own, but Gloucester and America could once more appreciate the worth of the men of a dory salt-banker. In years to come, as the fishermen of Gloucester looked back on this contest and the ones that followed, there was no question that this was the best. It represented the high-water mark of the international contest, but most important, it was a true fishermen's race. No one argued about gear or tactics on the racecourse; nor were the vessels newly designed for the competition: They were two salt-bankers that had been quickly scrubbed down and painted up, but they still carried their banks' sails. Great honor fell to these two craft and their crews. Lunenburg and Gloucester had done themselves proud, and the business interests of both communities basked in the momentary exposure.

CHAPTER 28

The Coming of the Racing Fishermen: *Mayflower* and *Bluenose*

As the thrill of victory passed, the fishermen came to realize that nothing had really changed. The grand spectacle had not insulated the fisheries from the economic woes engulfing North America. Gloucester's total catch tumbled from 120.4 million pounds in 1919 to 75.8 million pounds in 1920. Prices and wages fell, international markets remained unpredictable, and fish stocks in warehouses remained high. With little demand from abroad, recovery would depend on finding markets in the United States.

The situation was no different in the Provinces. In Newfoundland, the government stepped in to buy large quantities of unsold salt fish to lessen the blow to its fishermen. Even at the lowest of prices, much of Newfoundland's cured fish remained unsold. In this environment, many Newfoundland and Nova Scotian fishermen began to make the trek south in search of a berth on a Gloucester or Boston schooner. Among them was Charles Daley. He came to Boston, paid dues to the Fishermen's Union, and went out as a hand on a local vessel. Charles's union card described him as 5 feet 8 inches tall, with blue eyes and graying hair. For the next two years, he paid his union dues and sailed on Boston and Gloucester schooners. He was 1,000 miles away from his family, living frugally in a New England boardinghouse, and sending every spare penny home for Lucy and the children. It was the fate of men such as Charles, and they were glad for the opportunity to work, even as their children grew up without them.

Even so, there were some in Gloucester who thought, or really hoped, that the worst was behind them. Toward the end of 1920, a few skippers and small syndicates ordered new craft: the schooner *Laura Goulart* for Captain John Goulart, the schooner *Gov. Marshall* for Captain Matthew Sears, and the schooner *Oretha F. Spinney* for Captain Lemuel Spinney. In Lunenburg, Nova Scotia, a new schooner was commissioned, the *Bluenose*, designed near the limits of the size and other criteria specified in the deed of gift. The Nova Scotians were confident that she would claim the international cup for Canada. They had been astonished by *Delawana*'s loss to *Esperanto*, and in late

)(*Two men prepare to off-load a catch of ground fish onto their schooner,*
c. 1920.

November men from the Smith and Rhuland shipyard had already selected
the best timber to frame and plank the new schooner. Mid-December saw
her keel laid, and the people of Lunenburg talked incessantly of *Bluenose*.
If Gloucester had the best of the old-line schooners, Lunenburg would have
the best of the new: the first of the racing fishing schooners.

In Gloucester, although there was talk of a possible new racing fisherman in the fall of 1921, it was a distant hope. This was not a good time to
raise the needed capital for such a venture. None of the local firms or local
skippers stepped forward to take up the challenge. Gloucester's older banks'
schooners would contest one another in the highly anticipated trial race off
Cape Ann in the fall, and then sail north to vie with Nova Scotia's new racing
fisherman.

But in Boston, there were those of a different mind, and on December 10,
1920, a group calling itself the Schooner Mayflower Associates announced
plans to create an even faster racing fisherman. They had issued 200 shares,
rather than the more typical 16, and the shares had been quickly subscribed
by fishing, commercial, and yachting enthusiasts from around Massachusetts
Bay. The time had passed when purely fishing interests stepped forward to
underwrite a Gloucester salt-bank schooner. *Mayflower*, like *Bluenose*, would
go out as a salt-banker, and her designer, W. Starling Burgess, an established
naval architect who had designed his first fishing schooner in 1905, planned

for a vessel that would in every way be a match for *Bluenose*. The Nova Scotian vessel would face a staunch challenger, and from the earliest accounts, Canadians had little confidence that *Bluenose* could defeat *Mayflower*.

Mayflower's owners expected her to take on all comers, and it seemed likely that a Boston vessel, not one from Gloucester, would defend New England's honor. This troubled the people of Gloucester, and to quote the Board of Trade: "A great opportunity will be missed if these International series are not capitalized in the fullest degree, to the benefit of our leading industry." The local fish processors had no national marketing strategy, and this one event seemed the best hope for bringing the name of Gloucester into the homes of America. But there seemed little that Gloucester could do; none of her existing vessels would be a match for *Mayflower*.

These were troubling times, and despite the investment of some in new vessels, and their hope for the economy, 1921 would go down as one of the darkest, if not the darkest, years in the history of the fisheries in Gloucester, Boston, and Lunenburg. In early 1921, a number of Gloucester's fish-processing and vessel firms either filed for bankruptcy or were teetering on the edge of the abyss. When the fresh-fish fleet brought in an excess catch in late spring and summer, the Gloucester splitters took only the choicest of iced fish for use in the salt- and frozen-fish market: To sell on the American market one had to have a superior product. The fishermen had an unsettled summer.

Beginning as early as January, the Maine-based East Coast Fisheries Company had gone into liquidation, and its fleet of 17 steam trawlers lay idle at the wharves. In Gloucester, the winter gill-netters found prices so low that many hauled out, hoping that conditions might improve in the fall. In the hopes of strengthening its cash position, Gorton-Pew Fisheries once more began to sell off its vessels. The victorious *Esperanto* could be bought for $25,000, while in Lunenburg, the *Delawana* was also up for sale.* They were indeed troubling times.

Putting the best light on the situation, Gorton-Pew said they planned to sell some of their schooners in January, while there was still demand, and expected to buy vessels later in the year when the prices were lower. But they never did buy back in. The economy of the country was in free fall, and it did not bode well for selling Gloucester fish stocks. From July 1920 to January 1921, the cost of living in the United States fell by 9.1 percent, and all across the country business interests called for tariffs to protect America's industries.

---

*In May 1924, the *Delawana* was lost off Nova Scotia. Her spring stay snapped, causing the mainmast to crash down, smashing in the stern of the boat. Captain Cook and his crew escaped in dories and made it safely to shore.

The tariff rollback of the Wilson years was seen as a failure, and in New England, vessel owners demanded an immediate reinstitution of fish tariffs. At a mass meeting at the Young Hotel, representatives of the New England fishing industry voted to send a delegation to Washington to support new fish tariffs at Congressional hearings scheduled for later in the month. At a minimum, they hoped Congress would equalize the difference in the cost of production between American and Nova Scotian fishing vessels. Thomas J. Carroll, vice president of Gorton-Pew Fisheries, stated that because of lower wages and costs, Canadian competition was creating an impossible situation in the trade. Fresh Canadian fish sold for as much as 4 cents per pound less than the catch of native fishermen, and with the American market open and worldwide demand falling, the Canadians were dumping their product into America. Carroll feared that Canada might gain control of the industry, driving the Americans out of the market and thereby transferring the fishing metropolis to Lunenburg. Gardner Poole, representing the Atlantic Coast Fishermen's Union, supported these concerns, and asked for a duty of 1 cent per pound on all imported saltwater fish. He agreed with those who thought it was time to repeal the wartime measures that allowed Canadian vessels free entry into American ports. Finally, an editorial in the national *Fishing Gazette* supported the tariff.

On January 22, a Gloucester committee addressed the House Ways and Means Committee, advocating duties modeled along the lines of the former pre-Wilson tariff. Voices from Gloucester joined those from industries all across the country, manufacturers as well as farmers and fishermen, all seeking help from their elected representatives. But the outgoing Democratic administration was not in favor of tariff relief. It would be up to the Republican Harding administration to act: There would be no tariff relief in 1921.*

Without relief, the demand for locally caught fish remained low. Even with the large beam trawlers on the sidelines, Gloucester's schooner-based fresh-fish fleet overfished. The catch exceeded demand, resulting in "summer fish prices" in February. Haddock sold for $1.50 or less per hundredweight in Boston, and the Gloucester splitters paid as little as 50 cents per hundredweight for excellent quality iced fish. But, even though the prices were the lowest in years, skippers were often hard pressed to sell their catch.

---

*Canada was quite concerned with the possibility of losing its duty-free status with the United States. In April 1921, the Canadian House of Commons took up a motion to approve the old reciprocity agreement with the United States, which the country had rejected a decade earlier. The motion was defeated by a vote of 100 to 79. Sir Henry Dayton, Minister of Finance, believed that cordial relations with the United States had to be encouraged, and to try to put through the old reciprocity pact under present conditions would be counterproductive.

Who would have thought this could have happened, with New England's large fleet of beam trawlers remaining largely on the sidelines? By late spring, most beam trawlers had been hauled out, as owners and workers bargained over a new wage agreement. Owners proposed a 65 percent drop in pay—from $130 a month to $45 a month—to levels that had prevailed before the adjustment of only one year earlier. The union rejected the proposal, saying the new "lay" was too little to support their families. The owners, however, were equally clear, insisting they could not, and in fact would not, operate their steamers under the old lay. Craft did not go back to sea. Finally, in June, the union caved; an official characterized the action as the most drastic loss workers in any line had been forced to accept.

As the deliberations had dragged on, with the beam trawlers on the sidelines, Gloucester schooner fishermen, with support from a few Canadian schooners and beam trawlers, more than filled the void. During Lent, when demand was high, good prices had prevailed—$9 to $10.50 per hundred-weight being received for iced haddock and cod. But as soon as Lent passed, prices fell to $3 and continued to decline. Fish gluts became common, and with decreased foreign trade, only the best of the iced fresh fish was bought by the Gloucester splitters. Shackers were few. Most schooners made quick trips, landing an iced catch. Only six craft made salt-bank trips—three trawling and three dory handlining. It was the halibuters, as an incidental by-product of the fresh-halibut fishery, that brought in most of the salt catch. In fact, the halibut dorymen represented the only ray of hope in an otherwise dismal fishery. They prospered throughout the year, as they would through much of the decade. As had been true for over 40 years, the relative scarcity of the fish, coupled with steady demand and a limited fleet, supported good prices. In 1921, the halibut fleet, building on their success the previous two years, grew to 27 schooners, up from 16 in 1920. But even with the increase, the catch remained relatively small, never really exceeding demand, and there was no possibility of rescuing the industry as a whole.

Boston remained the primary port at which halibut was sold—and the editorialists in the Gloucester paper lamented the loss—but who could blame the skippers for seeking the highest price for their catch? They continued their efforts to stimulate an active price competition between buyers in Portland and Boston. As to the fishing grounds, although halibuters continued to fish on Georges and other nearby banks, the fleet had its greater success to the north, especially on the steep slopes of St. Pierre's and the Grand Bank. By season's end, nearly 6 million pounds of fresh halibut had been landed by the Gloucester fleet. The highliner was Captain Carl Olsen on the *Elk*,

)(*A giant halibut is off-loaded in 1921. It is*
extraordinary to think of such a fish being
caught by one or two men in a dory at sea.

with 418,000 pounds, followed by Captain Lemuel Spinney on the *Oretha
F. Spinney*, with 370,000 pounds.* The largest single fare of the season,
some 114,140 pounds, was brought in by Captain Archie McLeod on the
Catherine.

As this lone fishery prospered, Gloucester threw itself into preparations
for the 1921 International Schooner Race. What better way to remind the
public of this vital American industry? The desire was real, for there was no

*Other successful craft in 1921 were the *Hesperus* (3rd), 357,000; *Catherine* (4th), 331,000; *Pollyana* (5th),
322,000; *Waldo R. Stream*, Captain Robert Porper (7th), 282,000; *Natalie Hammond* (12th), 242,000; *L. A.
Dunton* (19th), 169,000; and last among the 27-schooner fleet was *Hortense* with 60,000 pounds.

national marketing strategy for Gloucester fish. The American defender was to be chosen in an elimination race off Cape Ann, after which she would sail north to take on the Nova Scotian champion off Halifax. The Halifax organizers would not schedule an International Race off Cape Ann until 1922.

The stage appeared to be set; the challenger from Nova Scotia would surely be the *Bluenose*, while the defender from America would be the *Mayflower*. A great race seemed in the cards, although there seemed little doubt that *Mayflower* was most likely to carry the day. But *Bluenose* and *Mayflower* would never race.

The Halifax organizing committee was not prepared to see *Bluenose* lose, and the Gloucester committee was not disposed to having a Boston defender. For Nova Scotia, it was imperative that a Lunenburg craft win back the cup. It was a matter of pride and national honor, and as the year unfolded, the committee went to extraordinary lengths to find reasons to disqualify *Mayflower*. They twisted logic and bore "false witness" to prevent a match race between the two great vessels of the day.

In terms of design and purpose, both *Bluenose* and *Mayflower* were configured for speed. They were "special" fishing schooners, not typical commercial craft, and some old salts in Nova Scotia and Gloucester were critical of them, claiming they were made to race, with only a secondary concern for commercial viability. Others disagreed, judging *Bluenose* and *Mayflower* to be the finest commercial fishermen ever launched. Both vessels met the conditions specified by the Nova Scotians in the "deed of gift": They were innovative, practical commercial fishing craft. They were large, true banks' schooners. Their principal innovation was a refinement in design to increase their speed in a breeze-of-wind. Although there was little commercial advantage for salt-bankers to return to port a few days earlier, the wish for speed harkened back to the 1860s. The *Bluenose* and *Mayflower* truly represented the pinnacle of schooner design, albeit at a time when vessels of their size and speed made little commercial sense.

As early as February 1921, when *Mayflower* was still a dream on paper, the Halifax Trustees pored over the deed of gift, looking for any reason to disqualify her. In Rule 5 they thought they had found it. They wrote to the Gloucester committee, saying that in their view *Mayflower* could not possibly be at sea by April 30. She would not be ready to begin the standard Lunenburg salt-bank season, which stretched from May to September. Thus, she would not be eligible to compete in the fall race.

When the note reached Gloucester it was supported by many in the local fishing establishment. Many on the Gloucester committee would be happy if *Mayflower* was disqualified. Yet, if *Mayflower* had been a Gloucester craft, there can be no doubt that the community would have insisted that she be allowed to race, even if she missed the deadline by a few days. The Halifax Trustees would never have received the "message" that if they ruled against *Mayflower* no one in Gloucester would object. Thus, the Trustees believed their race was not in danger; there would be an unspoken alliance between the Trustees and the Gloucester committee. The Gloucester committee, holding the cup as champion of the previous year's race, would put it up against *Bluenose*, even if *Mayflower* was disqualified.

Nevertheless, on February 25, with *Mayflower*'s ribs barely in place, the employees at the James shipyard were working six days a week, from sunup to sundown. In a triumphant telegram to the Trustees, the Boston syndicate announced that *Mayflower* met all the technical terms of the deed of gift and would be launched in time. She had all the bearing of a true champion. She was to be 131.6 feet long, 112 feet at the waterline, with a 25.8-foot beam, 12-foot depth, and 58-foot-long shoe. Meanwhile, at Smith and Rhuland's shipyard in Lunenburg, *Bluenose* was a week or more ahead of *Mayflower*. Locals took to calling her the "Big Ark," and she gave away nothing in size to the Boston schooner. *Bluenose* was 130.2 feet long, 110 feet at the waterline, with a 27-foot beam and 11-foot depth.

On March 24, the James crew chief announced that he expected to launch the new Boston fisherman by April 12, and Captain Henry Larkin promised to have *Mayflower* off to the fishing grounds before the April 30 deadline. The bowsprit had recently been fitted, planking was being smoothed, and the finishing touches were being made to the cabin and forecastle. At the same time, *Bluenose* was nearing completion and was launched within a matter of days. Nova Scotians described *Bluenose* as a staunch "real fisherman," something they were soon to say the *Mayflower* was not. They took to calling the Boston vessel a "freak"; while in Gloucester the local paper called her a "camouflaged yacht." Even so, the detractors of *Mayflower* had to admit that *Bluenose* was of a model that was a "little out of the ordinary run of the regular Lunenburg schooners." There were also those in Nova Scotia who said *Bluenose* was too much a "sporting type" of vessel and not enough a traditional commercial salt-banker. But all agreed she was beautiful.

On April 13, *Mayflower* was launched before more than 10,000 cheering onlookers at the James yard in Essex. It was "the biggest event" ever to occur in Essex."[53] Among the spectators were three members of the Halifax Trustee Committee, including the designer of *Bluenose*. In his book *Caught on Irons*,[54] Michael Wayne Santos reports that Wilmot Reed, secretary of the American Race Committee, met with the members of the Halifax Trustees and the *Mayflower* group, and a "brouhaha" followed. Feelings of ill will were present from the beginning, and only grew worse as *Mayflower* was made ready for sea.

With 16 days to fit out, *Mayflower* was towed to Boston and berthed on the north side of T-Wharf, where her masts were stepped, rigging installed, and sails bent on. While she sat in Boston, there came word that at least one of the key members of the Halifax committee had unexpectedly recognized *Mayflower* as a legitimate contender for the International Fishermen's Trophy. The stamp of approval came from none other than H. R. Silver of Halifax, chairman of the International Fishermen's Race Trustee Committee. He insisted that selecting *Mayflower* as cup defender rested entirely with the American committee. Presumably, if selected, she would be acceptable to the Trustees. On April 25, *Mayflower* left on her first salt-bank trip, days before the deadline specified in the deed of gift. The *Mayflower* syndicate telegraphed the Trustees, reporting that their schooner was at sea on her first salt-bank trip. Sailing north for the Magdalen Islands, *Mayflower* soundly defeated the newly launched Gloucester schooner, *L. A. Dunton*, in a match race from Gloucester to Shelburne, Nova Scotia. The result was not unexpected, as *Dunton* was smaller than *Mayflower*, at 104.3 feet long, 25 feet in beam, and 11.6 feet in depth. The time difference between the larger and smaller vessels was a staggering seven hours.

Captain Larkin continued to race the *Mayflower* from one Nova Scotian port to the next. From Shelburne to Canso the *Mayflower* came in at 7 o'clock in the morning, the *Dunton* at 5 o'clock in the evening; Larkin had kept the crew up all night working the sails. The *Mayflower* next beat *Dunton* in a match race from Canso to the Magdalens, and for eight weeks Captain Larkin beat every vessel, American and Canadian, he crossed on the fishing grounds. Word spread quickly that none could beat *Mayflower*. If there had been any doubt in Halifax, there was none now. *Mayflower* had to be excluded from the fall's International Fishermen's Race. The racing interests in Halifax and in Gloucester were determined to prevent the trophy—and the national and international recognition for the fisheries—from going to Boston, and they reacted with a vengeance.[55]

On May 12, Captain Larkin cabled the vessel owners that he was pleased with his new vessel. He had baited at the Magdalens, but the snow was so heavy that he was preparing to leave the Bay of St. Lawrence for the Grand Banks. This met yet another term of the deed of gift: staying at sea, no matter the weather, proving that *Mayflower* was not a fair-weather fisherman. Captain Larkin said he sensed that Nova Scotia was "losing courage."

Then, on May 14, there came another communication, one that colored all future deliberations on *Mayflower*. The commentator, a knowledgeable seaman, R. C. S. Kaulach of Lunenburg, reaffirmed the prowess of *May-flower*. More specifically, he said, "The Lunenburg schooner *Bluenose* would not have a ghost of a show in a race with the Boston schooner *Mayflower* over the course in and off Halifax harbor unless a gale with heavy seas, which he does not consider likely to happen." Kaulach made these statements in a public dispatch to the *Halifax Chronicle*. He had been on *Delawana* in the previous International Race, and had inspected the *Mayflower* on the stocks at Essex. In his view, if the race was to survive another year, the *Mayflower* had to be excluded. On May 24, the builders of *Bluenose* supported Kaulach's opinion, reinforcing what had become the 1921 "party line" in Halifax and Lunenburg: "*Mayflower* was not a fisherman." She had been built for the race, and the syndicate planned to sell the schooner to yachtsmen once the race was history. She should be excluded.

As the Nova Scotians settled into this position, there came word of the loss of the previous year's International Cup winner, the schooner *Esperanto*. Gorton-Pew had failed to find a buyer, and in late spring she was fit out for a salt-bank trip under the command of Captain Thomas Benham, formerly of the schooner *Athlete*, which had been sold the previous fall. While fishing off Sable Island, the *Esperanto* struck a submerged wreck, took on water, and as men scrambled into dories, she settled into the sea. The crew was picked up by Captain Geele, skipper of the schooner *Elsie*, and landed safely at Halifax. The *Esperanto* was gone, although there would be failed attempts to raise her: A noble vessel had come to an end. With her passing, the attention of the racing community would focus almost exclusively on the exploits of *Bluenose* and *Mayflower*.

On June 25, Captain Angus Walters, the skipper of *Bluenose*, returned to Lunenburg from the craft's maiden banks trip. He landed 900 quintals—100,800 pounds of salt cod—a light first catch, and one that did not cover the cost of outfitting the trip. Captain Walters reported passing *Mayflower* on the Banks, but as the vessels were sailing in opposite directions, there was no opportunity to test them against one another. But the two skippers had spoken, and Captain Walters reported that *Mayflower* had 175,000 pounds of cod on board. It is also interesting to note what he did not say after the meeting: He did not question the fishing prowess of *Mayflower*, nor did he suggest that *Mayflower* be excluded from the International competition. Two fishing vessels had met on the Banks, *Mayflower* had more fish than *Bluenose*, and the two craft continued on their trips.

Mayflower returned from her first trip on July 15, bringing 195,000 pounds of salt cod into Gloucester. She sold the catch to Frank E. Davis & Company, and three days later returned to Boston. But things were starting to get ugly. Within a matter of weeks, F. L. Pigeon of East Boston, managing director of the Mayflower Association and a Boston spar maker of some renown, charged that certain Gloucester interests had been doing their utmost over the summer to discredit his craft. He claimed that they hoped to persuade the Halifax committee to disqualify *Mayflower.* He chalked it up to "petty jealousy" and "personal spite." He further charged that the culprits behind the false propaganda were to be found on the American racing committee, all of whom were Gloucester men. Pigeon also accused Gloucester commercial interests of spreading false claims. They were saying that *Mayflower* was owned by millionaires and had not been constructed according to specifications in the deed of gift. They went so far as to say that she was a yacht and not a commercial fishing craft. He insisted that although the charges were false, the simple repetition of such lies served to give them credence. The *Mayflower,* capable of holding almost 600,000 pounds of fish, was a large, true fisherman, fully prepared to bring in enormous salt-cod fares, the equal of the typical Lunenburg or Gloucester salt-banker. Failing to acknowledge this reality required a remarkable degree of self-interest and local arrogance.

In his classic book on American fishing schooners, Howard I. Chapelle[56] says that *Mayflower* departed only somewhat from the model of a traditional schooner, and that the changes were not enough to "cause comment"; rather it was the "feud between Gloucester and Boston fishermen" that was the "real reason for banning the *Mayflower.*" Dana A. Story agreed that the Gloucester "collective nose was out of joint at the prospect of a Boston boat in the competition," and the *Mayflower* could not get a fair hearing from Cape Ann. There was the matter of local pride, fierce interport competition, and hoped-for national and international recognition: It was a lot to fight about.

In fairness, there were a few voices on the American committee who spoke out for *Mayflower.* In a personal letter, Mayor Wheeler of Gloucester described the *Mayflower* as a "splendid example of fishing craft"; and Captain Charles H. Harty,* the man who had built and sailed *Esperanto,* declared the Boston syndicate's new schooner *Mayflower* to be "one of the best he ever saw for the fishing business." Even as these exchanges took place, the *Mayflower* sailed on her second salt-bank trip, her syndicate assuring the Halifax and Gloucester racing committees that she would be back in time for the race off Cape Ann to select the American champion. This note also made it clear that *Mayflower* was now meeting yet another condition of the deed of

---

*Captain Harty, a Gloucester native, died on May 21, 1931, at the age of 80. He was serving as a buyer of mackerel for New York concerns. He was a master mariner for almost 50 years, a mackerel highliner, and a member of the Gloucester International Race Committee.

gift by being on the fishing grounds throughout the April-to-September fishing season. On August 19, the *Mayflower* syndicate formally entered their vessel in the American trials.

Mayflower returned to Gloucester from her second salt-bank trip on September 13, landing 180,000 pounds of salt fish, and was ready to race. Two days later, the American racing committee declared *Mayflower* as the defender of the trophy. They scheduled no elimination race, for no Gloucester schooner was willing to contest the best fisherman on the North Atlantic. But it was just a charade, a pretense, a humbug, orchestrated in full knowledge of how the Halifax Trustees would respond—Gloucester was not about to lose the publicity of an elimination race. The day after the *Mayflower* was declared the defender, H. R. Silver, representing the Halifax International Trustee Committee, sent a telegram saying that *Mayflower* was not a practical fisherman and was not eligible for the International race.

The *Mayflower* syndicate was outraged. She had just completed two successful salt-bank trips, landing more than 300,000 pounds of fish, more fish than *Bluenose* had brought in. The Halifax Trustees further stated that she did not conform to the letter of the deed of gift in regard to the date of sailing for the fishing grounds. But, they added, they "did not consider this point so important as the former."

It was a remarkable exchange. First, the man sending the telegram was the same H. R. Silver who, in mid-April, had called *Mayflower* a true fisherman and supported the right of the American committee to choose a defender. Second, the Trustee committee had been notified at the time that the *Mayflower* sailed on her first trip in April. They knew she had sailed within the time window specified in the deed of gift. Nevertheless, the Nova Scotians twisted reality—if ever a vessel was set up to win it was the *Bluenose*, no matter who was hurt, no matter what lies were spoken.

In nominating *Mayflower*, the Gloucester-based American committee had provided the Nova Scotians with the opportunity to exclude her from the race. The Nova Scotians had responded and their position was final. Now everyone in Gloucester could get on with the main game of selecting a local defender for the trophy, one of the port's old, true fishing schooners. It was not a bad objective, but the American committee had lost its honor. They never said a word for *Mayflower*, and as holders of the cup they would have been in a strong position had they spoken up.

Mr. Pigeon of the *Mayflower* syndicate was dumbfounded. He repeated his accusation that it was a case of "petty jealousy," and it was up to the American committee to defend *Mayflower*. But despite his cry for help, with little debate, the American committee voted once more to accept the decision of the Halifax Trustees. On that same day, for all of Gloucester to see, a note in the *Times Fish Bureau Column* lent support to *Mayflower* as a practical fisherman. On her second trip she had stocked $7,300—a most respectable fare in 1921 America.

When Fred Pigeon asked the Halifax Trustee Committee for an explanation, their response was arrogant and to the point:

> The Trustees are not bound to give any information unless they desire to do so. . . . We think everybody will agree that in the event of the *Mayflower* racing and winning the cup this year, no practical fishermen would build a vessel to compete with her next year because to do so the type would have to be even more extreme in design and thus be still further away from a practical fishing proposition, which would mean the end of international racing for all time.

The American committee read the words and repeated their support for the position, stating that no other American vessel would race against the *Mayflower*, for she was a racing machine. They now called for an elimination race. The plot by Halifax and Gloucester had succeeded; the Boston fisherman *Mayflower* was out of the race. For many in Gloucester, including the author James B. Connolly, the fishermen's race was a way for Gloucester to gain valuable publicity. He recounted his experiences on a western trip after *Esperanto*'s 1920 victory, saying that Gloucester was now heard of in places she had never before been known; this was what it was all about. If *Mayflower* was to race, Boston, not Gloucester, would be the community to benefit.

Five vessels entered the elimination race: the schooners *Elsie*, *Ralph Brown*, *Elsie G. Silva*, *Philip P. Manta*, and *Arthur James*. The *Elsie*, whose registry returned to Gloucester from Nova Scotia when the Frank C. Pearce Company purchased controlling shares in the boat, with Captain Welch at the helm, easily won the trial and sailed to Halifax to defend the cup against *Bluenose*—victor over seven other Nova Scotian craft in the Halifax elimination race. On October 22, *Bluenose* defeated *Elsie* off Halifax in race one of the two-race series. Her time of victory was 10 minutes. Two days later, *Bluenose* again outsailed the Gloucester vessel. *Elsie* was a fine, some would say superb, traditional fishing schooner, but she never stood a chance against the new, larger breed of racing schooner designed for the international competition. She was 11 years older and not designed as a racer, but in point of fact there was no economic reason for the outfitters of Gloucester to then commission a vessel of a size greater than *Elsie*.

In racing, with any kind of wind the larger vessel always wins. As Captain Welch noted, *Elsie* was as good as *Bluenose*, only she was smaller.

Bluenose was 130.2 feet long, *Elsie* 106.5 feet; *Bluenose* had a waterline of 110 feet, *Elsie* 103 feet. *Bluenose* had a mainmast of 96 feet and main-top-mast of 51 feet; *Elsie's* mast was 89 feet, her main-topmast 45 feet. *Bluenose* had a sail spread of 10,000 square feet; *Elsie* had a spread of 8,500 square feet.

Nova Scotians had achieved their goal. The International Trophy was for the first time in Canadian hands. There was no real honor in having won the race, but in the course of history only the winner is remembered. Nevertheless, at the time, all honorable, fair judges recognized that the race was a sham, a pretense. *Mayflower* should have been the champion. But no matter, the race was coming to Gloucester in 1922. It was time to win back the cup, and two Gloucester groups announced that they would construct special vessels for the coming year—*Puritan* and *Henry Ford*. The first announcement was made even as *Elsie* prepared to return to Gloucester, and reflected the yearlong efforts by Gloucester skippers and commercial interests to find the necessary capital to construct a racing fisherman.

Leading the *Puritan* syndicate, and skippering the vessel, was none other than Captain Jeffrey Thomas. He and his brother Billy had been bitten by the racing bug, as had many others in Gloucester, most especially "Captain" Ben Pine, who put up much of the money for the new craft and was its managing partner. Pine came to Gloucester from Newfoundland at the age of 10, went to sea as a doryman, had recently opened a major supply and vessel outfitting firm, and he and his fishing partners planned an economically viable craft— one that would enter the profitable fresh-halibut fishery.

The plan called for *Puritan* to be a 149-ton schooner with an overall length of 123.9 feet and 105 feet on the waterline—a good match for *Blue-nose*, although somewhat smaller, as the designer wanted to ensure that the vessel did not run afoul of the rigid design specifications stipulated in the deed of gift. She was to be launched in March and would go to the Banks as a halibut fisherman. W. Starling Burgess, the designer of the *Mayflower*, was selected to design *Puritan*, and he came up with a design that on paper was even faster than *Mayflower*, while giving the vessel the bearing of a banks' schooner the Nova Scotians could not reject. She could not be called a "more extreme vessel." The syndicate said *Puritan* was being built for purely commercial purposes, by commercial fishermen, but with fine lines for speed. Like the *Mayflower*, she was to be built by J. F. James & Sons of Essex. But despite the statements of the syndicate, had there been no race, the *Puritan* would never have been built.

With the first of the two new Gloucester vessels having been announced, on October 31 the so-called nonfisherman *Mayflower*, which had brazenly sailed off Halifax during the races between *Bluenose* and *Elsie*, left for the Banks on yet another fishing trip. She had not been sold to yachtsmen. *Blue-*

✕ *Guests crowd the deck of racing schooner* Mayflower *on her trial run. Men work on the bowsprit to raise a third head sail; they will soon release her main topsail.*

nose, on the other hand, sailed from Halifax to Puerto Rico with 600 casks of dried fish. Like other large vessels in the Lunenburg fleet, she became a merchant trader over the winter. The contrast was obvious but went unnoticed. Who was the true fisherman? There could be no question. It was the *Mayflower*.

By the end of November, the Gloucester syndicate behind *Henry Ford* announced the specifics of their racing schooner, a vessel to be skippered by a Gloucester legend, Captain Clayton Morrissey. She was being built with financial support from fishing and other Cape Ann interests, with no input from Henry Ford other than the use of his name on the vessel. Designed by Thomas McManus as a salt-banker, she was built for speed. She followed the general lines of the McManus schooner *Oriole*, a practical commercial fisherman, a craft considered to be one of the fastest vessels afloat in her day. McManus believed he had improved on the *Oriole* design, although no one could say whether she was capable of defeating *Bluenose*. She was about the same size as *Puritan* at 122.1 feet long and 105 feet on the waterline, and she was destined to be a fine vessel.*

*She measured 138 feet overall, with a 25.6-foot beam, a 12.5-foot depth, and a 15-foot draft. Her mainmast measured 88 feet.

CHAPTER 29

1922—"The Worst Has Passed" and the Races Go On: *Puritan* and *Henry Ford*

As 1922 dawned, most of the fresh-haddock fleet sat idly at the docks, their owners pressing for union fishermen to contribute more toward the cost of lost gear and installing new engines. However, with few fish coming into Boston and haddock prices soaring to $9 to $15 per hundredweight, 13 of Gloucester's finest schooners soon sailed under the old lay. But once on the grounds, poor weather set in; they returned with small fares, lost gear, and damaged spars. On two vessels, men were washed from their dories, and by late month fresh haddock still brought a respectable $6.50 per hundredweight.*

A week later, with improving weather and a rising catch, the environment began to change. Haddock prices had dropped to $4 per hundredweight, and there came word of the passing of the greatest Gloucester fisherman of the age, Captain Solomon Jacobs. In poor health, he had been ashore for several years and died of a massive stroke while tending his coal furnace. A man of vision, he had opened the northwest halibut fishery, introduced auxiliary-powered craft into the Gloucester fleet, and maintained an unequaled record as a highliner in the halibut and mackerel fisheries.† Gloucester lost a man who had contributed much to its golden age, a man whose like would never be seen again.

But Gloucester was a working port. There was little time for sentiment, and with Lent about to start, the fleet rushed out to the grounds. Skippers expected high prices over the next six weeks, and each sailed in the hope

---

*From the *Corinthian*, the lost men were James Gardner, age 47, and Avery Goodwin, age 45. Both were from Pubnico, Nova Scotia, and belonged to the Boston branch of the Fishermen's Union. From the *Teazer*, the lost men were Myron Lennox and Fred Thorne, both residents of Gloucester. Lennox was 28 and left a wife and five children. Thorne was 37, resided in a Gloucester boardinghouse, and was from Holyrose, Newfoundland. The local paper started a fund drive to help support the family of Myron Lennox, and within two weeks more than $2,000 had been raised.

†Gordon Thomas shared this view (Tape OH-48, Sawyer Free Library, 1978), saying that "Sol Jacobs was the greatest (Gloucester skipper) overall. . . . This man was more than a fisherman (he was a pioneer)."

of being the first to return. But with fine weather, the catch soon exceeded demand and haddock prices fell to summer levels. The best iced fish brought only $1.50 to $2.00 per hundredweight. Yet, even with the drop in price, the schooner fishermen did not despair, for the otter trawlers could not compete. Thus, as the economy began to slowly improve, the shore and offshore fleets had the market to themselves, and they were to have many good days in the months to come.* Meanwhile, for the fourth year in a row, the schooners in the fresh-halibut fleet, some 27 vessels in all, continued to find a ready market for their catch. Stable, high prices remained the order of the day, with Captain Carl Olsen bringing in the highline catch of the year, receiving $8,120 for a 65,000-pound catch.

The signs of an improving market were overshadowed by reports that Gorton-Pew Fisheries was teetering on the edge of bankruptcy. The loss of the company was unthinkable; Gloucester's success or failure was now tightly entwined with that of Gorton-Pew. Yes, Gloucester had other companies, and the Portuguese and Italian communities sent fleets to the grounds, but Gorton-Pew was in a league of its own.

Cash flow and accumulated debt were at the heart of the problem. Promissory notes in the amount of $2 million were to come due in the spring and no extensions were possible. To help keep the company afloat, four schooners were sold in January. Next the managers turned to the stockholders with a plan to raise needed working capital and pay off the noteholders. But the plan was rejected. The only option was to continue to sell off the firm's assets, and in February and March, they sold the schooners *Philomena* and *Waldo L. Stream* and the unused Reed & Gamage property in East Gloucester. However, by April, with the notes coming due, the company filed a receivership petition with the Boston court. Gorton-Pew's fate would now depend on the actions of a Boston judge, whose central responsibility was to protect the assets to the benefit of the creditors. He asked for comments, and the stockholders registered their outrage at the state of the company. The noteholders had controlled Gorton-Pew in 1921 and they had made a mess of it. More than $2 million in debt had been needlessly incurred, and the stockholders asked the judge to save the company. There was no need to liquidate the buildings, vessels, and other assets; Gorton-Pew had a future. The firm was well known across the country; many of their products were in demand.

For his part, the judge said he was "far from satisfied with the facts in the case." He did not believe Gorton-Pew should be liquidated and charged

---

*By October, when most of the fresh-fish fleet had refit for haddock or pollock, 22 of the otter trawlers unexplainably came off the docks. But they miscalculated the market: Supply again exceeded demand, prices fell, and the owners lost money.

the appointed receivers* to cut overhead and put the company on a profit-making basis.† He also blasted the prior management of the creditor's committee, saying they had built up debt, invested unwisely, spent too much on questionable advertising, and failed to find a way to move the enormous stocks of fish now languishing in the company's warehouses.

In fairness, although not noted by the judge, the deepening national depression had complicated the company's attempt to move into new markets. But regardless of how the problem had come about, there was hope for the future. Its ready-made codfish products continued to sell, and the company was able to hobble along throughout the rest of the year.

For the Gloucester fisheries, a crucial turning point had been reached. The company had been saved, jobs preserved, and as the country began to come out of the short-lived depression, Gorton-Pew would prosper. Even so, in 1922, the firm was unable to assume a central role in the life of the community, and it left the fate of the upcoming International Fishermen's Race in the hands of those who had so mismanaged the 1921 affair. The Halifax Trustees were as determined as ever to exclude *Mayflower*, while the American committee failed to appreciate the publicity value of *Mayflower*, *Puritan*, and *Ford* coming to the line together in the fall elimination race. No one of stature stepped forward to help the two groups come to a more enlightened understanding of how such a message might benefit the fisheries. Narrow, parochial racing concerns continued to drive the parties. Interport bickering ruled the day, rather than an informed discussion of which of the majestic fishing schooners would take the cup or how such a race might help the people of Gloucester and the country feel good about themselves.

Even with the continued sour words from the Trustees and the duplicity of the members of the Gloucester committee, a match race between *Bluenose* and *Mayflower* seemed a possibility. The *Mayflower* syndicate announced that they would put up a substantial cash prize to race the winner of the fall's International contest off Cape Ann: $3,000 to the winner and $2,000 to the loser. It was a significant sum, and in early February the Canadians agreed to the proposal. At the same time, not giving up hope of racing for the International Cup, the owners of *Mayflower* notified the Halifax Trustees that their craft had fulfilled all the requirements of the deed of gift. *Mayflower* had been at sea all winter; she had "proven herself the ablest sea boat

---

*The job of a court-appointed receiver is to execute the judgment of the court and to protect the property in dispute. A court-appointed receiver is an officer of the court and not an agent of the company, stockholders, or creditors, and is personally liable on contracts entered into in the execution of his functions.

†The salaries of key managers seemed high to the judge: Stuart Webb, president, got $25,000; Henry Gould, treasurer, $10,000; Thomas Carroll, general manager, $8,500; and Benjamin Smith, $8,500.

ever constructed," and deserved her shot at being the American challenger. More than anything else, they wanted to race *Bluenose*, no matter how it was arranged. *Mayflower* was a superb vessel, designed to win the cup, and her owners were willing to go to any lengths to see her prowess recognized.

In March, the first of Gloucester's two new racing schooners, *Puritan*, was launched. Thousands crowded into Essex as she glided down the ways at the J. F. James & Sons yard. There was no pomp: The owners, all of whom had the racing bug, insisted on a plain fisherman's launching. Though *Puritan* was somewhat smaller than *Mayflower*, old salts saw little difference between the two craft. What was clear, however, was that *Puritan* was no yacht. She had a graceful stern, not the sawed-off after end of *Mayflower*. She looked like an old-fashioned fishing schooner, and the Trustees could never say that she was not a fisherman. Nor would Gloucester have accepted her exclusion. If there was to be an elimination race off Cape Ann, *Puritan* must be in it. Gloucester was prepared to stand firm with one of her own in ways that she would never do for a Boston vessel.

On April 7, *Puritan* pulled out at Gloucester's Parkhurst railways, and many in the community came down to peek at her sleek hull. From Duncan Street there was a fine view of the rising spars and fine rigging of the new schooner; her topmasts stood high in the sky, rising above the neighboring rooftops.

Four days later, a second crowd came to Essex to watch as the *Henry Ford* was cut loose from the ways,* sliding effortlessly into the Essex River. But as she was towed to Gloucester, she ran aground in the Essex River and would not get to sea by the May 1 deadline. Only one of Gloucester's two new vessels would be technically eligible for the fall race.

When *Ford* finally arrived in Gloucester, *Puritan* was leaving on her trial trip. The breeze was light, and *Puritan* easily pulled away from the converted sub chaser that had accompanied her to Eastern Point. Gordon Thomas remembered that on that day as "the breeze freshened up . . . she started to heel over . . . [and] did that thing start to go. She vibrated and you could feel her go like that, trembling. . . . She acted stiff . . . she wasn't a tender vessel . . . [but] she took it. She took it well." *Puritan* was a special vessel, and on April 18 she left on her maiden halibut trip. She headed to Edgartown to bait up, and from there was scheduled to sail north to the Scotian Banks. But the fog sat heavily over Cape Cod, and *Puritan* and *Mayflower* found themselves tied side by side in Provincetown, waiting for the weather to clear.

**Ford was the same length as *Puritan*, at 137.8 feet, but of a more conventional design.

⟨ *The* Henry Ford, *photographed in 1923, did not win the International Cup but did take the Lipton Cup in 1922.*

The skippers talked. Men exchanged glances over the two vessels, and then it happened: It was unavoidable; there would be a match race off Provincetown. The *Mayflower* had taken on all contenders last year while on the grounds, and now here was *Puritan*. For his part, Jeff Thomas loved speed and never shied away from a contest.

They waited for the weather to clear and a good breeze to fill in. When all was set, the two vessels left harbor and the race was on. As the crews set the sails and the skippers stood at the helm, *Puritan* pulled away. Her speed was unquestioned. She was the new queen of the American fleet. For the one and only time in her fishing career, *Mayflower* lost a match race on the grounds. America had a new champion, and as far as Gloucester was concerned, there was nothing the Trustees could do about it.

Sailing north, *Puritan* made Boothbay, Maine, in five hours, covering the distance at a remarkable 14½ knots, and that with only a normal breeze. "It was nothing for this great schooner to reel off 15 knots. She was a real flying fool. . . . She was at her best in heavy weather, but, 'any breeze would do.'"[57] *Puritan* arrived back in Gloucester on May 9, after first selling her 25,000-pound fare of halibut in Boston. Captain Thomas judged his new vessel to be a "fast and able" sailor. She had stood up to heavy weather as well as any schooner afloat, while her lofty spars had proved a huge asset in light winds. Privately, Thomas said she was a tricky vessel to sail, a true demon of speed.[58] She couldn't be held back. *Mayflower*, *Bluenose*, it made no difference: No one could take *Puritan*.

On her second trip, *Puritan* first docked at Matthews Wharf in Canso to take on ice and bait. Then she sailed to the Scotian Banks in search of halibut. As she did, the *Henry Ford* sat tied up in Gloucester awaiting a ruling from the insurance underwriters following the vessel's grounding in the Essex River. Not until May 27 did the crew from Marion Cooney's sail loft bend on her sails, and on June 2, with the insurance issues finally settled, Captain Clayton Morrissey threw off the lines at the Atlantic Supply Company wharf. *Henry Ford* was away on her maiden salt-bank trip to the Grand Banks. Morrissey stopped at Shelburne for dories and baited at the Magdalens. And although the schooner was technically ineligible for the fall's International Fishermen's Race, Morrissey was pleased to hear that the Trustees might be lenient.

The *Ford* left Gloucester on a Friday, the superstitious in the community calling for Captain Morrissey to avoid the day, but in his view "some of the biggest and best trips" were made on Fridays and he was "not a bit leery."

Puritan returned from her second trip on June 10, landing 40,000 pounds of fresh halibut and 11,000 pounds of salt fish. Thomas again remarked on the craft's great speed in the wind. He thought "he had a vessel that would show them a thing or two." In his words, there "ought to be some fun" in the fall. *Puritan* refit and on June 17 sailed on her third trip. One week later, on June 23, word came that *Ford* had landed her first fare at North Sydney, Cape Breton, some 2,000 pounds of halibut. Gloucester had two new racing fishermen.

One day later there came news that shocked the community: *Puritan* had been lost on the northwest bar of Sable Island. She went down at 7:30 in the evening in heavy fog, in a stiff southerly breeze. With heavy swells and treacherous breakers, the men were just about to make a sounding when she hit.[59] She was making 9 or 10 knots at the time, and Captain Thomas was unaware that he was anywhere near the Sable Island bar. The speed of his vessel had carried her ahead of the course he had set. In fact, just before she hit, the skipper went below for but a minute, telling the man at the wheel to call him immediately if anything came up. But, even as his feet touched the cabin floor, *Puritan* rammed into the bar. Thomas heard the giant thump, crack, and smash, and *Puritan* began to heel over. The men saw part of her keel float to the surface; her rudder was jammed up through the wheel box.

)((*The schooner* Puritan *under full sail (save for her fore topsail). Designed by Starling Burgess, she was built to race in 1922. Undoubtedly fast and admired by all who knew her, she never had the chance to prove herself fully as she was wrecked off Sable Island in July 1922; only one man was lost.*

Heavy seas washed over the vessel and all knew she was lost. When she struck, *Puritan* had been under mainsail, foresail, and both topsails, and in minutes the cabin floor was awash with water.

As Captain Thomas rushed up from the cabin, he first thought *Puritan* had been run down by a steamer. He believed they were 20 to 30 miles east of Sable Island, heading on a northeast course for the Grand Banks. The man at the wheel yelled to the skipper that she was ashore, but Captain Jeff still couldn't believe it, yelling back, "It can't be." But it was true. Now the focus was on saving the men, but the fog was dense, the sea rough, and the vessel was sinking. Men scrambled into dories, and as they left the schooner at least one dory capsized. They soon lost contact with one another as the fog shut in thicker than before. As the sea broke over the wreck of the *Puritan*, Jeff's brother Peter dragged him into a dory. He was lucky to be alive. His vessel was filling quickly and was fully awash within 20 minutes. Only one man, a Dane named Chris Johnson, was lost. It was his first trip on the *Puri-*

tan. He had only recently arrived in Gloucester from Yarmouth, Nova Scotia. A breaker struck his dory crosswise and turned it over, throwing him and two other men into the water. One grabbed the overturned dory, the second swam to the schooner, and both watched powerlessly as Johnson disappeared under the waves.

Captain Thomas and five others landed on Sable Island, where they spent several days before being taken off by the Coast Guard cutter *Tampa.* The others headed for the Nova Scotian coast and were picked up by Canadian vessels. With the loss, Captain Thomas had had his fill of racing vessels and returned to fishing.* Initial reports had Captain Thomas taking out the schooner *Elizabeth Howard*, being refit for Ben Pine and Marion Cooney, but with delays, he moved into a Gorton-Pew vessel, the schooner *Corinthian*. He would take her out on shacking and fresh-fish trips throughout the remainder of the year; and as he left on his first trip, the American race committee met to set the date for the September elimination race.

The chairman of the American committee appointed a subgroup to secure measurements and other information on the entrants for review by the Halifax Trustees. One new contestant would be *Elizabeth Howard*, an immense older vessel, now under contract to Captain Ben Pine and Marion Cooney. They had her painted white, and a gang of longshoremen were putting in long hours fitting her out for a summer trip. Piney (Captain Ben Pine) had been bitten by the racing bug—if he did not have *Puritan,* then *Elizabeth Howard* was his next best bet.

On August 3, Captain Albert Picco sailed the *Howard* north on a shacking trip, meeting the terms of the deed of gift, and in early September the Trustees announced that she was qualified for the International contest. As for the *Henry Ford*, she, too, was qualified. The Halifax Trustees officially waived the rule on the date for sailing for the fishing grounds. In their words, a series of mishaps had prevented the schooner from sailing on time, and they were willing to issue the waiver. They did not believe *Ford* was a serious contender for the cup, or at least not in the same league with *Puritan* and *Mayflower.* Of course, had they not so ruled they would have raised the ire of

---

*Jeff went back to the Gorton-Pew Fisheries. His son Gordon said that he always felt well treated by the firm (Jeff and Gordon Thomas Oral History Collection). Then in 1925, he moved to United Fisheries, taking out the *Falmouth* for about a year. He next skippered *Oretha F. Spinney*, Captain Lem Spinney having largely "retired" from the sea. The *Spinney* was a big knockabout, and Jeff was pleased with her, taking her out for several years. In fact he liked her so much that he commissioned the schooner *Adventure* following almost the same lines as those of the *Spinney.* She was named after a childhood drawing by Gordon, who recalled, "He recognized my drawings and I had a part in the vessel. Little did I think . . . what would turn out in the future. We had just selected the name of the last American dory trawler and . . . the biggest money maker of all dory trawlers, and the last Gloucester schooner in actual commission [here in 1981]."

the Gloucester committee. The terms of the deed of gift were not as invariant as they had appeared when applied to *Mayflower*. The Halifax Trustee Committee said they were awaiting a copy of the vessel plans from the Boston syndicate, and only with the plans in hand would they rule.

In mid-September the Trustees once more excluded *Mayflower*, and again their reasoning was twisted and self-serving. While admitting that she may be a fisherman, or at least one that might be suited to the fresh-fishing industry as carried out in Boston, in their view she was not up to the standards of a Lunenburg salt-banker. Remarkably—or perhaps not—they could "find no sufficient reason for altering last year's judgment." They focused this year's decision on *Mayflower*'s cargo-carrying capacity, and in that they found her wanting. Of course, this was once again an untruth. But at least they had given a reason that could be addressed. Burgess, the designer of *Mayflower*, set off for Halifax to prove that *Mayflower*'s cargo-carrying capacity was up to that of the Lunenburg schooners. His information proved beyond any doubt that *Mayflower* could carry 525,000 pounds of fish, the same capacity as *Bluenose*. He also offered to change any details in *Mayflower*'s rigging that the Trustees found not to conform to Nova Scotian standards. But the Trustees refused to lift the ban, insisting that, "in the information furnished by designer Burgess regarding the *Mayflower*, there is nothing to warrant them in changing their previous decision, barring the *Mayflower*."

Z. H. Wicker, the managing owner of *Bluenose*, was pleased with *Mayflower*'s exclusion, saying "the news was not unexpected to him, as he always said the Deed of Gift could not be altered. He did not consider the *Mayflower* was a vessel anything like the *Bluenose*, as she could not carry a large cargo of dry fish out to the West Indies and bring back a load of salt in the winter time, as the *Bluenose* does every winter." It was an amazing assertion, for Wicker raised *Bluenose*'s involvement as a simple cargo freighter against the record of *Mayflower*, a 12-month fisherman. Captain Angus Walters of the *Bluenose* said it best: "If the *Mayflower* had been allowed to enter there would have been no vessel from Lunenburg in the International Race." The message from Nova Scotia was clear: If *Mayflower* raced, *Bluenose* would lose, and thus she would not have participated.

After *Esperanto*, *Mayflower* and *Puritan* were unquestionably the two great vessels of the early 1920s. *Bluenose* was also great, but thanks to the actions of the Trustees her reputation was built on deception, cunning, and pretense. In 1922, there was to be one more opportunity to make it right, the match race agreed to earlier in the year. The *Mayflower* syndicate was still willing to put up the $5,000 in prize money. And should *Bluenose* win the International Trophy off Cape Ann, her owners and skipper had agreed to

the race. Now all waited to see whether the match race would come off, or if the Nova Scotians would find yet another excuse for avoiding a race with *Mayflower*.

The American elimination race occurred in October. Gloucester was getting much-needed national publicity for its fisheries, and the local Chamber of Commerce could not have been more pleased. They had surveyed their counterparts from all across the country and found that Gloucester fish was highly regarded, but people were not sufficiently acquainted with fish of any kind. In the Chamber's view, the International race was just the vehicle to carry the news of fish to every home in the land, providing the fishing industry with an opportunity to increase the sale of fish.

News of the impending trial quickly went forth to the nation. One of the contestants, *Elizabeth Howard*, arrived in Gloucester on October 4 from a Middle Bank trip. She landed 100,000 pounds of salt fish and 50,000 pounds of fresh fish. She unloaded, took off her dories and gear, and refit for the race. On the following day, Captain Angus Walters confirmed that *Bluenose* would race *Mayflower* if he won the International race, assuming Boston put up the purse. They were to race 40 miles to sea, with a five-hour time limit, with oversight by the American racing committee.

On October 24, the *Ford* clinched the elimination race, taking race two over the *Elizabeth Howard* and the other competitors, after taking race one when the *Howard*'s main-topmast broke. On that same day, *Bluenose* arrived in Gloucester. *Bluenose* and *Ford* were similar in size: 141 versus 137 feet long, and with about the same square footage of canvas: 9,771 square feet versus 9,522. On paper, they were two finely matched schooners. The table was now set, and in the first of the International races *Ford* did the unthinkable: She beat *Bluenose*, having a good lead at the finish. Might the two best schooners on the North Atlantic be American—*Ford* and *Mayflower*? But events soon turned ugly, as the first of a series of nonrace issues surfaced that would forever becloud what happened off Cape Ann. Strange events ruled the day, and in our telling we will first look at the races and then the story behind the races.

As the participants gathered that evening to speak of the day's race, the official judges declared they had "sensed" a light wind at the start and had tried to postpone the race at 10:30. But the two fishermen had good starts. They refused to turn back and in fact completed the race. The judges followed the craft for three legs, timing their progress as the schooners turned the course. But they pulled up before the last leg, assuming *Bluenose* and *Ford* could not finish within the time limit. They were wrong, and *Ford* took the race. The International race committee, however, called it a nonrace, since they had not been present to see it end, even though everyone knew who had won. Captain Walters knew better. *Ford* had won, and he said if *Ford* won the second race, the trophy belonged fairly to the Americans. Captain Morrissey, too, knew he had won the race, and was livid. Should he refuse to go on? Should he take Captain Walters at his word and simply say that one more race would determine the cup? Eventually, Captain Morrissey made a difficult decision: He bowed to the late-evening decision of the race committee, listening to the pleadings of the vessel owners and American officials.

Two days later, *Ford* again defeated *Bluenose*. Two races, two defeats, and both by the third-best American craft of the day. By all rights, *Ford* should now have won the trophy—Morrissey knew it, Walters knew it, all of Gloucester knew it. But Walters did not step aside; *Ford*'s two-race victory was not to be. The "official" count was one race for *Ford*. There was to be a third race, and now Morrissey was incensed. He had won and was prepared to end the series then and there. He told those close to him that he would take *Ford* out of racing and get back to fishing. Only after the intervention of the U.S. Secretary of the Navy and the renewed pleas of several deep-pocketed friends at Eastern Point was he prevailed upon not to overhaul *Ford* and go fishing. But he never forgave himself. He knew what he should have done. Walters had said it all: If the *Ford* took the second race, the cup was hers.

Bluenose won the next race, doing well in a heavy breeze. She then won the next, and last, when *Ford*'s fore-topmast snapped during the contest. With two consecutive triumphs, *Bluenose* was declared the winner. But word soon spread that the last two contests had themselves been seriously compromised by the actions of the Nova Scotians. The races had not been won at sea but on the docks. The Nova Scotians claimed that the *Ford*'s sail area exceeded that allowed in the deed of gift. How else could *Ford* have beaten *Bluenose*? Remeasurements had been called for, and Captain Morrissey agreed to accommodate the Canadians. The *Ford*'s mainsail, which he had used at sea during the fishing season, was said to be too large and was cut down. She went out to race. The sail was measured again, cut back again, and in the process was butchered.[60] It no longer fit on the main gaff. It was useless in a good breeze of wind, as had prevailed in the last two races off Cape Ann.

In a later conversation, Gordon Thomas and Harry Eustis (men who understood the race and the vessels) described how the race had been won:[61]

> *Harry*: She beat the *Bluenose* in two races. That was when they should have called the races off altogether. They had them cut the sails.
> *Gordon*: That foolish committee, did you ever hear of a fisherman having to cut the sails?

Harry: Not only once, but twice. They took so much cloth out of the mainsail.

Gordon: They cut from the garth down to the boom. Took strips out of her. Shortened up the mainsail, said she had too much mainsail according to the rules.

Harry: She was fishing with those sails for three or four months. Her sails were made to go fishing. It was a fisherman's race.

Gordon: Crazy . . . Cutting the sails. For God sake, fishermen aren't like that. They're not yachtsmen.

Nor did this end the strange behavior of the Nova Scotians. On the Thursday night after *Bluenose*'s victory, Captain Larkin and two of the principal owners of *Mayflower* met with the *Bluenose* group to arrange for the much-anticipated match race. They approached Mr. Wicker, managing owner of *Bluenose*, to discover the conditions on which he would agree to race *Mayflower*. But now it seemed the race was off. Wicker claimed there was not enough time, but privately, when pressed by John Kelley of the *New York World* and others, he said, "If I lose, where is my reputation?"

Later that evening, Wicker and Captain Walters met with members of the Gloucester committee and the *Mayflower* group in the back room of the Chamber of Commerce offices. Every possible argument was used to pressure the two men to honor their pledge. As Walters added condition after condition as the basis for the race, Captain Larkin for *Mayflower* agreed to them all. Finally, at midnight, Captain Walters said he would race *Mayflower*. There would be a five-hour time limit. The course would be determined by *Bluenose*, and there were to be no restrictions on sails or ballast. *Bluenose* was scheduled to go to Boston to haul out and take whatever time was needed to get the job done. It was agreed.

With the dawn of a new day, however, Captain Walters changed his mind, and on that afternoon *Bluenose* left for Halifax, slipping silently out of port. About an hour later, when Captain Larkin came back to Gloucester from Boston, he was fit to be tied. He had come for the *Bluenose*, but in his words, she "had dogged it for home at the first opportunity after one of the most colossal messes ever made of a sporting event." There was no match race, and *Mayflower* prepared to have her topmasts taken down and her canvas bent on in preparation for winter haddocking.

By November 16, *Bluenose* had hauled out at Lunenburg, where she remained until spring. In the words of Howard Chapelle, "The defeat of the *Ford* by the *Bluenose* was a noxious affair in which the sail plan of the *Ford* was repeatedly reduced by the race committee, to the point where her mainsail would not stand properly. . . . All of those [races] in which the Canadian

Captain Walters sailed were distinguished by a complete lack of sportsman-ship and by much bickering."

Yet racing remained on the minds of the Gloucester community. The fol-lowing year, 1923, was to be the three-hundredth anniversary of the city of Gloucester, and Sir Thomas Lipton, of America's Cup yachting fame, offered a cup for a fisherman's race. It was to become one of the major events in the year's celebration. There was also word that Gloucester would have a new racing schooner, the *Columbia*. Captain Ben Pine and Marion J. Cooney commissioned Burgess & Paine to design her. She was to have a length of about 108 feet at the waterline, just 3 feet more than that of *Puritan*, and an overall length of 123.9 feet. Her masts were somewhat shorter than *Puri-tan*'s, with longer booms and gaffs. But like *Puritan*, she was to be a practical fisherman—Ben Pine was determined to bring the cup back to Gloucester.[62]

But it was also a sad time. Word came that the *Mayflower* syndicate was fitting the great schooner with a 100-horsepower engine. She was to become just another auxiliary vessel. Her towering spars were shortened by 9 feet and her sail plan altered to suit. Never again would she challenge for the trophy.

CHAPTER 30

1923—Yet Another Great Gloucester Racing Schooner: "Piney's" *Columbia*

By early December 1922, work had begun on *Columbia*, and her owners had enlisted Captain Alden Geele to take her salt-banking. If the Trustees best understood vessels that went salt-banking, then *Columbia* would be a salt-banker. Piney was well aware of the uncertainties of this fishery—owners seldom recovered the cost of outfitting a vessel, let alone making a profit on the catch—but if *Bluenose* was a salt-banker, then so was *Columbia*.

Yet no matter *Columbia*'s fate, the fishermen of Gloucester had long known that their once-mighty salt-bank fleet was a relic of the past. For years, Gloucester's prosperity had rested on the success of its fresh-fishing fleet bringing in haddock, halibut, and mackerel, and these craft did well throughout the rest of the decade. Outfitters also added new powered draggers to the fleet and motorized almost all of the existing schooners. The dragger was a smaller, economical version of an otter trawler—diesel powered, about 85 feet and 100 tons, with a covered wheelhouse in the stern and an open deck in front to handle the fish as the nets came in over the side. New fish processing businesses opened, and the local community joined in an organized effort to stimulate regional and national interest in canned, frozen, and fresh fish.* What did not change, however, was the fishing season itself. In early January, haddockers were the first to leave—shore and offshore craft making for Georges and the Gulf of Maine. Meanwhile, at Billy Thomas's Gloucester Cold Storage plant, the first of the halibuters could be seen taking on bait and stowing ice as they prepared for trips to the Scotian and Grand banks.

For the early-season fleet, the weather remained a challenge. In one early February trip, the *Elizabeth Howard* rounded Eastern Point with her

*On April 11, 1923, the expanding, and increasingly successful, Italian fleet sailing south for mackerel included the *American*, Captain Thomas Scola; *Mary Scola*, Captain Sam Scola; *77c*, Captain Joseph Frontero; *Rosalie*, Captain Tony Batiste; *Felicia*, Captain Philip Curcuru; *Philomena*, Captain Vita Meone; and *Sea Bird*, Captain Jack Argusso. On July 6, Joseph Frontero received his new schooner from A. L. True of Amesbury, the largest fishing boat to join the Italian fleet. She was 50 feet long, 60 tons gross, and was fitted with a 55-horsepower engine. She had bunks for eight and could carry approximately 50,000 pounds of iced fish below deck, while her high sides permitted more to be carried above.

decks and rigging encased in ice. Water flooded into her hull through open seams, and a jury-rigged steering mechanism controlled her rudder as she eased into Harbor Cove. Her problems had begun on Quero Bank, when a gale swept in as she lay at anchor. As her anchor dragged, bolts pulled out of the steering gear and the helmsman lost control of the rudder. The wheel spun freely in his hands and only the jury rig that was quickly set up gave him some rudimentary control of the vessel. But as the *Howard* twisted uncontrollably in the churning waters, the seams on the starboard side opened, just forward of the foremast, and the pumps had to be worked around the clock to keep the water at bay. The schooner was in trouble and forced to make a perilous 80-hour return voyage to Gloucester with few fish in her hold.

Stories such as this would not be uncommon in the early winter of 1923, as gales swept the Banks. Nevertheless, by late February, the Gloucester outfitters had no choice but to fit out the rest of the fleet for late winter and early spring trips. Lent was about to start, and Nova Scotian fishermen soon crowded into Duncan Street. They unloaded mattresses and bags from wagons, took rooms at the boardinghouses, and headed to the waterfront in search of work.

In April, *Columbia* was launched in Essex at the Arthur D. Story yard and took on a Nova Scotian crew of handline fishermen. She was a big vessel: 123.9 feet long, 108 feet at the waterline, with a beam of 25.4 feet, a depth of 12.4 feet, and a displacement of 152 gross tons. The only larger vessels in the Gloucester fleet were the permanently mothballed otter trawlers *Seal* and *Walrus*. The *Columbia* was the new queen of the fleet. When her masts were stepped in Gloucester, the main stood at 93 feet, the fore-topmast at 85 feet, and with all of her sails in place, she could carry 8,800 square feet of canvas. On her maiden salt-bank trip, however, she went out with a staysail as her main driving sail, as the loft had yet to cut her mainsail. She was stocked for a three- to four-month trip.

Harry Eustis, one of the dorymen on the schooner *Catherine*, remembered seeing *Columbia* off the Nova Scotian coast.[63] He was struck by her unusual gray paint, and remembered thinking that she was not a particularly pretty vessel. In fact, to the men on the *Catherine*, "she looked like an old coastal. . . . She was light. They had no salt in her (and she was riding high). They were going to pick up salt down there [in Shelburne], and dories."

As *Columbia* rounded Eastern Point on her way out of Gloucester, *Ingomar* passed her coming up the harbor. She had off-loaded her 50,000-pound catch of halibut in Portland, and as the men came ashore they bore witness to an individual act of heroism by Captain Olsen, an act that forever set

)(*The racing schooner* Columbia *under full sail in the 1920s. Her lee rail is deep in the water as Captain Ben Pine mans the wheel and a crew member stands high on a cross tee.*

him apart from other skippers of his era.* One of Olsen's dorymen, Chris Nielsen, had lost his footing when his dory was being lowered on a flying set, and Captain Olsen dove fully clothed into the windswept sea to save him. The sea was rough and Nielsen and his mate had been about to set the last dory to be put over the side. Only Captain Olsen was on deck as Nielsen went overboard; all the other dories had been launched and were spreading out across the open sea. Nielsen and his dory mate, Augustus Johnson, had hoisted the dory clear of the rail, and the accident happened as they were

---

*On February 12, 1924, Captain Olsen received the William Penn Harding Award from the Massachusetts Humane Society, a medal, and a $50 prize for having effected the bravest rescue of a man from drowning in all of 1923. He had earlier been given the Helen Livingstone MacCormack Medal, for the same rescue, awarded annually to the fisherman making the most heroic rescue. In the 1924 award ceremony, Captain Olsen was called into the offices of the Gorton-Pew Vessel Company a little after 2 o'clock in the afternoon to receive his award. Mayor Maconnais represented the city, Ellery H. Clark represented the Massachusetts Humane Society, and there were two representatives of Gorton-Pew—John McLeod, manager; and Thomas J. Carroll, general manager. Captain Olsen had not been told beforehand that he was about to receive this prestigious award. Instead, he had been told to go to the office on a matter related to the fitting out of his schooner. Mr. Carroll then told Captain Olsen that he was to get an award for his actions the prior April. Mr. Carroll indicated that the award was richly deserved, and the act of heroism had not been surpassed in the history of the industry. Mayor Maconnais then lauded the fisherman, who was leaning against the windowpane, in terms that made Captain Olsen blush. He said the deed would go down in the history of the fishing industry and the city as being the nerviest act performed by a mariner.

lowering it into the sea. Nielsen was aft, holding the rope looped through the stern, keeping the stern tackle from slipping off. Johnson was forward holding the painter. As the dory hung over the side a wave struck the schooner; the water rushed upward and struck the descending dory. The bow tackle let loose and the dory was wrenched from Nielsen's grasp as he stood with one foot on the schooner and one in the dory. In that instant he was hurled into the water, loaded down with fishing gear, oilskin coveralls, and hip-high rubber boots. Captain Olsen was at the wheel thinking of how he and the cook would go after a swordfish or two once this last dory was away. But as he saw Nielsen disappear, Carl yelled to the cook, "Come up and bring her around, I've got to get Chris." In the blink of an eye, Olsen dove into the frigid sea, hitting the water an instant after Nielsen bobbed to the surface. When the cook reached the deck, Olsen was nowhere to be seen.

As the captain swam toward the doryman, he removed his oversized rubber boots and shouted encouragement: "Hang on Chris, I am coming." A strong swimmer, Olsen quickly closed the distance between himself and Nielsen, but the doryman was not a strong swimmer. He was not even a regular seaman; he was a rigger. Only the air in Nielsen's oilskins was keeping him afloat. His waterlogged rubber boots were pulling him down, and he was becoming numb in the cold water. But he was fighting to survive. Olsen continued to shout words of encouragement, closing the distance to his floundering mate. On reaching him, he helped him place one hand on his shoulder and had him climb on his back. He told him, "If you don't go home, I won't go home either."

Olsen held Nielsen close to his back and swam toward the approaching schooner, which by then had turned and was making for them. But then Olsen spied parts from the overturned dory on the crest of a rising wave, and left Nielsen to make for two floating oars. He reached them in an instant and pushed them back toward his friend, foot by foot, through the rising swell. Once alongside, he passed one oar to Chris and kept one for himself. The two tired men now kept afloat without having to extend themselves.

By now the dorymen on the last of the dories to have been set saw the men bobbing in the waves. Leaving their gear, they rowed frantically to their shipmates. The two waterlogged fishermen were soon hauled into the dory and then onto the schooner. The cook had coffee for Carl—for he was not a drinking man—and whiskey for Chris, who was quickly wrapped in six or seven warm layers of dry clothes and bedding.

Looking back on the event, one of the crew said it all: "That fellow Olsen can swim like a dog, and the way he swam after that pair of oars was wonderful." Nielsen was about to surrender to the ocean, but Captain Olsen had kept his word; they both returned to Gloucester. Both men were from Norway, and the 6-foot 3-inch Olsen had lived with Nielsen for years; he was like a father to him.

When a newspaper reporter approached Captain Olsen for comment, Olsen smiled, shook his head, and said, "Sure it was nothing; any man would do it." Olsen was a humble man, but as the story spread, all along the waterfront men spoke of this singular act of heroism. It was not an act that just any man would have performed, far from it. Only the most exceptional fisherman would have dared. As the full details of the incident spread, even the most grizzled seamen soon sang Captain Olsen's praise.

Gorton-Pew hauled Olsen's *Ingomar* onto the Parkhurst railways, where she was scraped and touched up, and within only a matter of days Captain Olsen again sailed her north for halibut, joining the many other halibuters in making repeated landings in Portland and Boston over the course of the 1923 season. This was also a time of year when the fresh haddockers, including craft such as the *Elk*, *Joffre*, *Natalie Hammond*, and *Elmer E. Gray*, began the switch to halibuting.

In mid-May, Captain Olsen came into Portland with 22,000 pounds of fresh halibut. He then made for Gloucester with 23,000 pounds of salt cod.* The halibuters were having a fine season, and all in Gloucester could see that the fisheries had turned around. But in June, the city turned its attention from charting the progress of its successful fishing fleet to the vessels that were to race in the fall's International Fishermen's Race. After the debacle of *Ford* and *Bluenose*, Gloucester was ready to even the score.

On June 27, the city's new challenger, *Columbia*, returned from her maiden salt-bank trip to the Sable Island Bank. Captain Geele came in four weeks early with a respectable 340,000-pound fare of salt cod. He said he never set foot on the deck of a finer craft, but having only a riding sail in place of the mainsail, he had been unable to test her full sailing abilities. Gorton-Pew bought *Columbia*'s fare for $12,693, an unspectacular price, and her owners barely broke even for the trip. But at least Gorton-Pew was again buying fish—the firm was turning the corner.

Columbia had returned much earlier than expected, and Captain Pine could not justify holding her in port for two months to wait for the Lipton Cup race in late August. So on July 5, *Columbia* left on a second salt-bank trip. She carried the mainsail from her sister schooner the *Elizabeth Howard*, which had been hauled out for repairs. As she left the inner harbor the

---

*Salt cod was still a Gloucester staple, and to increase demand a number of local firms test marketed a booklet explaining the proper way to prepare the fish. The test took place in several New York cities, with sponsoring firms including C. F. Wonson & Co.; Frank C. Pearce Co.; Gorton-Pew Fisheries Co.; C. F. Mattlage & Sons Co.; Frank E. Davis Co.; Davis Brothers Fisheries; and the Consumers Fish Co.

skipper set her fore- and mainsails, and as he did, the large crowds on Western Avenue cheered. As she rounded Eastern Point and headed north to the Scotian Banks, the Lipton Cup Committee announced that they expected as many as three schooners to compete—the new *Shamrock*, nearing completion at the Arthur D. Story yard, Ben Pine's *Elizabeth Howard*, and perhaps even the Boston schooner *Mayflower*. Meanwhile, they had no information on Captain Morrissey's plans for *Henry Ford*. He was still at sea, and the *Columbia* could race only if Captain Geele was able to return from his second salt-bank trip in record time.

But the *Columbia* did not make it back. In mid-July, the French trawler *La Champlain* ran into her off Sable Island. *Columbia*'s rigging, rails, and bowsprit were damaged, and Captain Geele had to go into St. Pierre for temporary repairs.

But *Henry Ford* did make it back in time, and when she did, something heartfelt happened: The people of Gloucester, sensing the injustice of previous International Fishermen's races, subscribed funds to replace *Ford*'s mutilated mainsail. Meanwhile, the *Shamrock* had been launched and towed to Gloucester to fit out for the anniversary race, and a week later *Elizabeth Howard* returned, unloaded her catch, and Ben Pine announced that she, too, was being readied for the race.

The Lipton Cup race was scheduled for August 28, but on that date Cape Ann sat under an impenetrable bank of fog. The fog hung heavy over the city, blanketing the wharves of the inner harbor, and as *Elizabeth Howard* pulled away from the Atlantic Supply Company wharf the paper said she looked "like a ghost as she loomed up in the fog and then faded from vision as she rounded the point at the Fort." Little headway was possible and the race was postponed. Two days later the committee tried again, but none of the competitors crossed the finish line within the allotted six-hour time limit. Finally, on the third try, *Ford* took the Lipton Cup. It was a close race; *Howard* trailed by only a minute, while the third-place *Shamrock* was a more distant 14 minutes and 58 seconds behind the winner.

So ended Gloucester's celebration of its three-hundredth anniversary.* The race had proved a satisfying warm-up for the Cape Ann elimination race

---

*As one more feature of this celebration, the city commissioned a permanent memorial to its fishing heritage, a statue to be erected on Western Avenue. After examining eight proposals, the design submitted by Leonard Craske was selected. The figure, entitled "Homeward Bound," was that of a typical Gloucesterman at the wheel of a schooner, bringing his vessel into the wind, but its erection was delayed until 1924. The statue was to be 15 feet high and cost $20,000, half provided by the state, half by the city. In July 1924, Craske came forward with a 3-foot model for the fishermen's memorial. It represented a major departure from the initial model, reflecting insights he gained from a trip to the Banks in the dead of winter on the schooner *Elizabeth Nunanm*, under Herbert Thompson. Two things were indelibly impressed upon him from the trip—"no longer will he believe a fishing schooner goes in the water because by actual experience he found that the water goes on the schooner; and second, never again will he kick about the price of fish; it is worth every darned cent you pay for it." During

for the International Cup. But that was a month away, and for now all the contestants refit to go back to sea. On *Ford*, riggers took down the fore-topmast and Captain Morrissey left on a fresh fishing trip. On *Howard*, Captain Gille fit out for a halibuting trip and left within the week.

As the *Howard* left, her sister ship, the *Columbia*, returned from her second trip. She had met the terms of the deed of gift for the International Race. She landed 225,000 pounds of salt cod, receiving a stock of $9,011 for two months' work. Captain Geele came ashore and did not go back to sea for almost two years. *Columbia* had been roughly used on the trip. Her chain plates were still twisted from the collision with the French trawler, and there was a dent in her hull under the fore-rigging. The red lead inserted in her seams at St. Pierre showed up clearly all along her port side. But she was home, and after her cargo was unloaded, Piney had her fit out for the elimination race.

Both *Ford* and *Howard* were still out on the Banks, although they were expected to return in time for the elimination race. But the longer the delay, the less likely the elimination race could take place. It seemed that *Columbia* might become the American challenger by default. On October 8, Piney had *Columbia* hauled out on the Parkhurst railways. Fishermen and skippers stood along the narrow pier on either side of the rail bed and up against the engine house, as the heavy chain slowly wrenched *Columbia* up the ways. Her shoe was to be repaired, her rudder replaced, and other work completed.

As *Columbia* sat on the railway, *Mayflower* came into Gloucester from Quero, and *Elsie* came down from Boston with fresh fish for the splitters. Neither would participate in the 1923 elimination race. *Elizabeth Howard* came down from Boston with 10,000 pounds of salt fish, but Ben Pine said he and his associates did not plan to fit her out for the race. *Columbia* was their vessel. On October 14, the American Race committee asked the Trustees to postpone the International match for two weeks, as the date of the American elimination race was uncertain. *Ford* had not yet returned. But on October 18, with *Ford* still not in port, the American committee ruled that *Columbia* would be chosen unless *Ford* arrived that day. At noon, *Ford* came up the harbor and anchored. She sat deep in the water, having 176,000

---

the trip, Craske stood a watch of 22 consecutive hours with the men at the wheel (who would have had 4-hour watches), "getting the various expressions, their moves, their stance, and every detail so impressed on his mind that he carried the picture back into the studio and created the model which it is believed will satisfy the most exacting critic." Thus, the Gloucester fishermen's statue is a composite of what he saw on the *Numanm*. At the same time, "fortunately for him, a heavy wind came up one day and gave him the desired opportunity to view Captain Herbert Thompson in oilskins and intense action at the wheel, striving to steady the frail craft. He made it a point to study the expressions and features of several skippers before he began." On August 24, 1924, 5,000 people packed the Western Avenue parkway to witness the unveiling of the fishermen's statue, followed by a service for lost fishermen. Captain John A. MacKinnon, a local master mariner, pulled the rope unveiling the figure of the helmsman, a permanent tribute to Gloucester fishermen of all time.

pounds of fresh fish and 60,000 pounds of salt fish on board. Morrissey had been delayed by bad weather during nearly the entire trip, lying over for days as a hurricane swept over the Nova Scotian coast.

The next day, Captain Morrissey said he was willing to race *Columbia*, but believed Piney should take *Columbia* to Halifax to sail against *Bluenose*. The American committee would hear none of it. All the pieces had fallen into place and the elimination race was on. Frank Pearce paid his men to work overtime to unload *Ford*, and the last fish came off during the morning of the next day.

Morrissey decided not to haul out *Ford*. Her bottom was not fouled badly enough to retard the speed; all that was needed was to put up the fore- and aft-topmasts and bend on the sails. The elimination race could take place in 24 hours. But regardless of who the challenger might be, Captain Morrissey said that "there will be no more sail slashing or cutting on my vessel. . . . They talked me into it last year, but if I had been myself, I would never have stood for it. I have a good sail now, and it's going to stay good, too."

Columbia was the favorite, but she would be tested not only by *Ford*, but also by the unexpected late entry of *Elizabeth Howard*, increasing the chances of the Pine group becoming the American challenger. On the day of the race, the Gloucester Auto Bus Company arranged for special trips to Eastern Point for those who wished to see the competition. The results were as expected: *Columbia* would represent the United States in the International Fishermen's Race.

Yet, even as *Columbia* prepared for the trip north, Captain Walters tried to change the rules. Knowing that *Bluenose* was best in heavy weather, they asked for a five-hour time limit. The American committee refused, and a compromise of a six-hour limit was agreed.

Columbia left for Halifax on October 25 and came in after a 38-hour trip. She went out immediately to become familiar with the 40-mile course. As she checked the course, Captain Walters had yet another trick up his sleeve. The *Bluenose* skipper now said that if *Columbia* won the race, he would insist on her being hauled out and measured. He and many others in Halifax expected that *Columbia* would not meet the new jury-rigged rules on sail layout. Piney would have none of it. He was willing to haul out his craft before the first race, but not after it. If she won, *Columbia* was going to stay in the water.

The first race was held two days later. *Bluenose* won by 50 seconds, the closest race on record in the International series. The next race was abandoned because of a lack of wind. Halifax was blanketed with thick fog, as were the waters over which the race was to have been contested. Later that day, *Columbia* claimed to have been fouled by *Bluenose* in the first race. According to Piney, *Bluenose* had forced an illegal passage in a shoal area close to shore. But Piney had not lodged a formal protest, nor had the race committee seen fit to take action. Without this, *Bluenose* would not have taken the lead, and there would have been no 50-second victory. It had been a purposeful, dangerous move, and in so doing *Bluenose* had blanketed *Columbia*, swiping her rigging with the mainboom.

In the second race, *Columbia* led *Bluenose* in light air but failed to finish the course in the six-hour time allowance. *Bluenose* was astern of *Columbia* when the race ended, after covering only 24 of the 40 miles. At a meeting later that day, the sailing committee, recognizing the injustice of the results from the first race, adopted rules to prevent the contenders fouling one another's rigging. They also prohibited passing to windward of buoys marking shoal waters.

In the next scheduled race, *Bluenose* led *Columbia* by two minutes over the 40-mile course. With a moderate wind of 15 miles per hour that increased as the race progressed, *Bluenose* crossed the finish line first and appeared to be the winner. But she had not won; she had fouled *Columbia*, just as she had done in race one. *Columbia* filed a formal protest and was declared the victor. A third race was needed to decide the victor. But now things took yet another bizarre turn. Captain Walters refused to race *Columbia*. He and his crew packed up and sailed home to Lunenburg. In the words of Harry Eustis, an old-time Gloucester doryman, Captain Walters was a "very controversial" man. When Walters was told he had fouled on race two, "Angus got his gang and went aboard and went home. He wouldn't race no more after that. . . . He wasn't a sport. . . . He wasn't very well liked. And he was stubborn. . . . He wasn't a sport. No way. He was master and lord of his vessel."[64]

Now all that remained for *Columbia* to win the trophy was for Captain Pine to bring his craft up on the line and sail around the course, but Piney was unwilling to win in such a manner. The racing series was declared unfinished and the International Trophy placed back in the hands of the Trustees.

Back in Gloucester, as *Columbia* refit for fishing, Piney spoke of the fiasco off Halifax: "We went 700 miles to race him and he ran away from us. No, sir, there'll be no racing for the *Columbia* this year. That's final. Tomorrow we start to get ready for fishing."* But even as *Columbia* was having her pig-iron ballast removed, Captain Angus Walters was at it again. *Bluenose*'s attorney threatened to go to court if the Trustees refused to award him the

---

*Before *Columbia* left, Captain Pine received word that the *Elizabeth Howard* had gone down and was lost east of Halifax, on Porter's Island. Captain John MacInnis and his crew were safe. She had been on her way to the Bay of Islands to load a cargo of herring.

trophy. He challenged the committee's right to award race two to *Columbia*, but the challenge went nowhere. The 1923 racing season had passed.

In 1924, the Trustees called off the race, due principally to the unsatisfactory ending of the 1923 race. In 1925, the Nova Scotian fishing industry having experienced a lean year, there was little enthusiasm for an International Fishermen's Race. But interest remained high in Gloucester. On September 30, *Columbia* was hauled from her berth at the Booth Fisheries Company in East Gloucester to be made ready should the race be held. Workmen scraped her spars and installed yet another new rudder, after which she made a trial run off Eastern Point.

In early October, the Trustees responded that if a race were to be held, it had to be off Halifax, with the earliest date being November 5. They asked if the American committee was willing to combine the elimination and international races off Halifax. *Bluenose* was getting ready to haul out in Lunenburg, and they expected her owners to be more likely to go to Halifax under this plan. Yet finances were still an issue. By October 20, it was clear that the Trustees could not schedule the International Race under the terms of the deed of gift; there would be no Fishermen's Race in 1925.

Nevertheless, the American committee had not given up. In March of the following year, they asked the Trustees to set preliminary dates for a fall race. In July, the Trustees announced an elimination race off Halifax, but provided no assurance that the winner would come to Gloucester in the fall. By mid-August, the Trustees all but gave up hope for an International Race in 1926.

Meanwhile, Captain Morrissey made a formal challenge for a race between *Ford* and *Columbia*. The challenge arose out of a friendly discussion over the merits of the two vessels, with Morrissey telling Piney "he was not so sure the *Columbia* was a better boat than the *Henry Ford*." Morrissey soon put the challenge in writing. Ben Pine accepted for *Columbia*, and both skippers agreed to a two-out-of-three format with the race managed by the American Fishermen's Race Committee. They put up a $500 side bet on the outcome of the race.

Nothing more had been heard from Nova Scotia about an International Race, although a dispatch from Captain Walters showed he still bore a grudge: *Bluenose* would not race at Gloucester until the money taken from him and his craft in the 1923 International Race was paid up. Captain Walters had a long memory. But no matter, in Gloucester enthusiasm was building for the contest between *Ford* and *Columbia*. The skippers agreed to sail with fishermen's rules, with only a few special instructions about passing rights when on a new course, passing and rounding marks, and so forth. There were to be

)(*Two of Gloucester's finest: the racing skippers Ben Pine and Clayton Morrissey, pictured in 1926.*

two cups and cash prizes—$1,500 for first, $1,000 for second, and assuming others could be coaxed to enter, $500 for third. And the date of the first race was moved. The committee did not wish to race on the first day of baseball's World Series. The new date was set for the second week of October.

The American committee invited three Nova Scotia craft to enter the race: *Bluenose, Haliogonian,* and *Mayotte.* Only *Haliogonian* expressed an interest, but her owners wired that she could not come south because of damaged sails and rigging. Thus, only *Ford* and *Columbia* met off Eastern Point. For the fishermen in Gloucester, these were the two premier fishing schooners in the North Atlantic. If Walters was unwilling to enter the race, it said much about the skipper and his craft. Harry Eustis captured the Gloucester feeling: She was fast, but "*Bluenose* was an overrated vessel."

The two races were sailed in moderate winds and smooth seas, with the people of Cape Ann looking on from the Back Shore to the Eastern Point Light. Gordon Thomas remembers arriving on a small jitney bus (like a station wagon). The day was cloudy, with a good breeze blowing, and Thomas got out of the bus at Rocky Neck and walked up Eastern Point Road, going left onto Grapevine Road.[65] He arrived at Bass Rocks just as *Henry Ford* passed by, "plowing into it, it was rougher than the devil." But *Columbia* took race one by 1 minute and 4 seconds, and then came back to win race two by 4 minutes and 48 seconds. *Columbia* was the fastest fishing schooner on the North Atlantic.

In the early summer of 1927, the American committee notified the Trustees that they would put forth the winner of a local elimination race, probably *Columbia*, to race *Bluenose* or any other Nova Scotian craft. Captain Pine announced that if *Columbia* won the local race, she would sail to Halifax with Clayton Morrissey at the wheel, enter the harbor with the American and British flags flying aloft, and say, "Here we are, come and race us." He went on to say, "We'll go down there and wait 10 days for them, and then race them for money, cup or nothing. If they want to race. And if they don't, then we'll sail back home satisfied that we have made them put up or shut up and decide the thing once more for all." According to Captain Pine, all he heard along the coast, whether in Newfoundland or Nova Scotia, was, "When are you going to finish the race?" He was disturbed that there were some who felt *Columbia* was holding back, and he wished to set the record straight.

As the American committee awaited word from the Trustees, there came news of the worst kind. Forty Lunenburg schooners had been caught in a major coastal hurricane on August 24.* There had been heavy damage all along the coast, with boats sunk in harbors from Yarmouth to St. Pierre Island. The Lunenburg schooners *Clayton Walters* and *Joyce Smith* had been lost near Sable Island, and others were unaccounted for.

Lunenburg was now much more concerned over the fate of its fleet than a contest for a racing trophy. Many Gloucester craft had also been on the grounds during the hurricane, and they now limped slowly back to port. On August 27, the Boston schooner *Yankee* and the Gloucester schooner *Arthur James*, both of which had been on Brown's Bank during the hurricane, came into Yarmouth to make repairs. Two days later came word that the skipper of the *Mayflower*, Captain Alvaro P. Quadros, had been swept from the deck while on Western Bank as a giant comber rose up and smothered the craft.

Another half-dozen Gloucester vessels soon staggered into other Nova Scotian ports, bearing visible evidence of their battle for survival during the height of the storm. The schooner *Edith C. Rose* was the first to make it home, coming in on August 30. She had been fishing on Western Bank as the weather began to look gloomy and the wind freshened. The storm hit with great intensity, and in the skipper's words, "it was every man for himself and trust to the kind providence of God for safe delivery." Then came word of the loss of the schooner *Marion McLoon* at Yankee Harbor. The storm had

---

*A hurricane had formed east of the Leeward Islands in the Caribbean on August 18 or earlier. By August 19 it had reached hurricane strength. When it made Nova Scotia it had maximum winds of 105 mph. The storm grazed Cape Cod but struck headlong into Nova Scotia. It made landfall near Yarmouth, Nova Scotia, and swept over the entire peninsula.

thrown her on the rocks, the men taking to the rigging, "praying, that the wind would stop." On September 13 there came the most ominous report yet—five (later corrected to two) battered dories and a pair of oars marked "Columbia" washed up on Sable Island. This was all, but it was enough to cause all to wonder whether *Columbia* had been lost in the hurricane. But Captain John Carrancho of the *Herbert Parker* had sighted and spoken with *Columbia* some 50 miles south of Sable Island on August 23. Carrancho did not believe *Columbia* could have been off Sable Island during the storm. Piney also reported that some 60 of *Columbia*'s old dories had been previously sold to different fishermen. Perhaps it had been one of those that had been found. He reminded the community that *Columbia* was not expected back until October 1. It was too soon to worry.

A week later, Captain Lem Spinney reported having sighted *Columbia* in the distance on Quero in early September. On September 22, the Halifax office of the Marine and Fisheries Department said *Columbia* had been reported fishing on Quero Bank on September 7. Yet, on October 1, *Columbia* had failed to return, and the U.S. cutter *Tampa* was ordered to Sable Island to search for her. *Columbia* and her 22-man Nova Scotian crew, including Captain Lewis Wharton of Liverpool, Nova Scotia, had left on July 3, fitted out for a little more than three months on the Banks. On July 15 she had put into Liverpool and had not been seen since.

The evidence since then was skimpy, but troubling. Among the floating debris found was a scrap of dory planking with "Columbia" stenciled in yellow paint, as well as the oars marked "Columbia." Debris had also washed up that would not have been found on a Canadian vessel, including containers of Gold Medal flour, New Jersey brand milk, and oak hatches. And Mrs. Wharton had not received a letter from the skipper in six weeks. This was unusual, as he was in the habit of sending a letter home on a passing ship every two weeks.

In Canada, one of Wharton's brothers-in-law remained hopeful: "If he didn't have a full cargo of fish, he wouldn't come in—he'd stay out till he got them. I'm not worried, and I won't be until after the 15th of October." But when the cutter *Jackson* returned to Halifax on October 7, she reported no sighting of the vessel. A few days later the commander of the *Tampa* said that he, too, failed to find any trace of *Columbia*. Later in the month, when all of *Columbia*'s supplies would have run out, Captain Pine made a brief announcement admitting the loss of *Columbia*, and on that same day, the haddocker *Mary Sears* brought in a dory bearing the name "Columbia," containing a bait knife with the initial "M" carved on the handle. Six of *Columbia*'s crew had a last name beginning with *M*, and all had been lost. The dory was smashed up, but it was clear that its painter had been cut off short at the knot in the bow, suggesting that an attempt had been made to escape in a period of dire stress. Her arrival was followed by a report that the skipper of the *R. H. McKenzie*, a Lunenburg vessel, had sighted *Columbia* just before the storm, not 50 miles south of Sable Island, but much closer, lying near the anchored schooners *Uda R. Corkum* and *Joyce Smith*, two of the four Lunenburg vessels lost in the storm.

Once the *Columbia* was given up as lost, Captain Pine began to receive letters written by Nova Scotian clergy asking if there was to be any help for the widows of the men who had been lost. Mrs. Arthur Firth from Shelburne, Nova Scotia, said:

> As my husband was in your vessel *Columbia* and I am left destitute with two small children and no means whatever for support. The men out of Lunenburg that was lost in Lunenburg vessel draws a compensation fund but lost souls understand there is none for us. So have been advice to write you to see if you could not give us some support. . . . I am a widow along in years and not able to work much. The ages of my children are 14 and 15. As you know, I had my husband and son both lost in your vessel. I will expect to hear from you soon.

Mrs. Fran Dedrick of Shelburne wrote in the same manner:

> I am writing to ask you if you intend to give me and the rest of the widows and fatherless children any help or support whatever. I am left alone with six small children, left destitute, no home, and homeless. As my husband was lost in your vessel the *Columbia* I think it is about time something was done to help us poor souls out, as all my help and support is taken from me and my little children.

Finally, Mrs. Rupert [Bertha] Bragg wrote, "If Mr. Bragg has any money down there will you please forward it to me as I am badly in need of it. I was left without anything and a few dollars would help me quite a bit, especially this winter."[66]

Something had to be done, and the American Fishermen's Race Committee stepped forward to lead a fund-raising effort to aid the families of the men lost on *Columbia*. They hoped to raise $20,000; the members of the race committee contributed $2,000 to start the campaign. Contributions of $100 each came from Clayton Morrissey, Ben Pine, Miss Rayne Adams (Captain Pine's associate in the Atlantic Supply Company, acting as bookkeeper, who would become the principal contact with the widows),* Alex-

*She later married Ben Pine.

ander Chisholm, and Marion Cooney. Another sizable contribution, of some $240, came from Captain Lemuel E. Firth and the members of the crew of the mackerel steamer *Three Sisters*. The need for the fund had been obvious, but the committee had waited for the formal announcement by *Columbia*'s owners that the vessel had been lost. Now that the announcement had come, the members of the committee came face-to-face with the serious side of the fishing business.

The campaign got off to a good start. Daily updates appeared in the local newspaper. On November 3, a contribution of $200 came in from John Hays Hammond, along with an announcement by the committee of various chairmen appointed to reach out to groups across the community.* Two weeks into the campaign, the fund hit $7,720, and the human side of the loss was reinforced when the local paper described the memorial service for the *Columbia*'s crew at the Trinity Church in Liverpool, Nova Scotia.[†]

Within weeks came word that many families were in need of immediate relief. The men left 74 dependents, including 57 children less than 12 years old. Almost five months had passed since the men had gone to sea, bills had piled up, rents had become due, and many families were in dire straits. An immediate response was needed, and the race committee appointed a group of nine to be responsible for the care and distribution of the funds.

On December 3, the fund had reached $20,096, but then something unexpected happened: It continued to grow. The fund was kept open until January 21, 1928, when it finally topped out at $35,766. Gloucester had shown its heart to the Nova Scotian fishermen. At Christmas, money and baskets flowed north to meet the immediate needs of the families, and before long notes came south thanking the committee for the checks, baskets of food, clothes, and payment of old debts.[67] The relief committee had listened with great compassion to the sad accounts and new pleas included in the thank you notes from the families. Mrs. Arthur Firth, of Shelburne, Nova Scotia, wrote:

> I received the box of clothes to day. Thanks for them. Some of them will come in good, but Dorothy was disappointed to think her raincoat was

---

*There were chairmen for the fisheries firms, merchants, fishing captains, the Portuguese fishermen, and the Italian fishermen. There were also chairmen for fraternal organizations, the Boston Fish Pier, laundries, insurance agencies, physicians, lawyers, automobile dealers and garages, coal and lumber dealers, summer hotels, restaurants, and winter hotels. Captain Matheson arranged to broadcast updates every morning on the local radio station, telling the story of the disaster and the progress of the canvass of funds. In addition, the women's organizations of the community looked for ways to help in the campaign.

[†]On January 3, 1928, the beam trawler *Venosta* reported bringing up *Columbia*'s hull in a drag. Because of rough weather, the hull broke away and sank immediately. The wreck was found at 43.24°N, 061.27.5°W. Her sails and booms were gone, and her bowsprit was broken off, while the hull was intact. The location was 40 miles west-southwest of Sable Island, or about 115 miles south-southwest of Halifax. She was in 40 fathoms of water.

not there. As she needs them so much and her overshoes. She asked for them when Captain Cameron was here, and he said they were coming. Can't you get her a coat and overshoes. She wants new ones. Size sixes, coat for a girl eighteen. Just find them for her and can't you send me a size of clothes for a boy fourteen. And send them down by Captain John McKenzie for I need them for the children to go to school and I haven't means two by them. I had to pay one dollar and a quarter to get the box from the station and there is only three things that is any good to us.

Other letters, such as that from Mrs. Belang and Mrs. Johnson of West Green Harbor, Nova Scotia, expressed satisfaction with the gifts: "Dear Miss Adams. We received our money and also the Xmas boxes. We have plenty of food now and if we need anything we can buy it. Now we want to thank you so much for all of our things as it was so kind and thoughtful of you all. Mrs. Johnson wants to thank you all too. Baby received his jacket and it fitted fine. We appreciate your kindness so much so we all want to wish you a Happy New Year."

These were poor people, and the Gloucester relief fund was to prove a valuable resource as they worked to get on with their lives. The full record of what happened is lost to history, but what is known is that they were helped through the kindness of the people of Gloucester.*

The *Columbia* was no more. She never again raced with the Nova Scotian craft *Bluenose*, and Captain Walters never received his trophy money from the race with *Columbia*.

---

*Captain Reuben Cameron, on behalf of the fund, visited the families as his ship berthed at Nova Scotian ports, and left the following account from July 1928. It describes the status of families one year after the loss of *Columbia*.

I called at Gunning Cove and made inquires regarding Mrs. Brodrick and find that she is getting along very well. . . . Regarding Mrs. Firth of Shelburne: called at her home and found she was acting as a nurse for some woman at the Poor Farm. Her home was closed and the two children were away. I learned from her that one girl reaches the age of 16 in September; she intends to go to the states in the fall. Her mother Mrs. Firth will not dispose of the house. All other families in the section seem to be getting along fine. . . . In the case of Foster McKay and Samuel Belang, members of the *Columbia* crew; both of these boys were taken by James N. Williams when they were infant babies and were just the same as their own children. Since they were lost Mr. Williams who is partly crippled has found it very hard for he and his wife to get along and it is evident from what reliable business men tell me, how this aged couple will have to go to the Poor Farm this fall unless something is done for them. These men think $20 per month will take care of them, and I think it would be a very good thing to do. If this recommendation is accepted I would not suggest any back allowance but simply start payments from this time. . . . I inquired about the Copp family at Eagle Head and found them to be doing very well. In Halifax, I found that the family of Albert Mayo had moved, but located a party that lives on the same street with them before they moved, and he reported that conditions were very satisfactory with them. Located the family of Leon White, whose address up to this time was unknown. They live at Belle Cote, Cape Breton. They have several sons living, two employed in the states and one in Western Canada, and one at home, who is about 45 years old and goes fishing there, and looks after the farm in general. They have a daughter about 30 years of age who lives at home. Their place looks nice, the house is well kept and I think they get along very well at present, and do not seem to be in need of anything.

CHAPTER 31

The Fisheries in the Mid- to Late 1920s, Beginning with the 1924 Turnaround

The age of the great International Fishermen's Races now slipped quietly into the pages of history. The race had served a purpose, but by 1924, having passed through the darkest days of the decade, Gloucester now had several strong and growing companies and a stable fleet. The mackerel had returned, and for some, rum-running had become a profitable sideline. For the first time since the war, a new year began with the business of fishing on a stable footing.

Mackerel fishermen would have a splendid year in 1924, while the fresh haddock and halibut fisheries would continue to prosper. New vessels, including several smaller, economically viable draggers entered the fleet. On the most active days, upwards of 30 otter trawlers and 50 schooners went out for fresh fish on the nearby banks, with another 25 to 30 craft on the Scotian and Grand banks seeking halibut.

Over the spring and summer of 1924, Gloucester's splitting prices rebounded as the salt-cod processors began to find a more stable national market. Good marketing of a convenient, low-cost, quality product began to find traction in the American marketplace. Fresh, iced cod came in from the Georges, Western, and Sable Island banks; and when Boston was glutted, the Gloucester splitting firms took in 25 million pounds of fish that otherwise had no marketable outlet, enabling the fresh-fish fleet to stay at sea on a year-round basis. New businesses were opened, such as the fish-processing Fort Company, which was organized with a capital stock of $100,000. They purchased the Fort property of the former Gloucester Fresh Fish Company and went into the fresh-fish business. Changes were made to adapt the plant for the new business, including installation of an ice-making plant capable of producing 40 tons of ice per day. Provisions were made to supply crushed or cake ice to the fleet at water's edge. Storage tanks were also installed for dispensing fuel oil and gasoline. Once again there was a Gorton-Pew connection, as William T. Gamage led the effort.

✕ *1921 Elimination Race off Eastern Point. Left to right:* Elsie *and* Arthur James *covering each other,* Philip P. Manta, Ralph Brown, *and* Elsie G. Silva.

Of all the positive changes, however, the most important was the survival and later flourishing of Gorton-Pew Fisheries. The move into receivership proved important to its survival and eventual emergence as the dominant fish-processing and -marketing firm in the United States. As a first step, the receivers, Henry J. Guild and Arthur J. Santry, arranged to sell the company to A. Stanley North of Boston. The book price was $500,000, and with the sale the company was isolated from its creditors. North then transferred the property of the company to Gorton-Pew Fisheries Company Limited, a new corporation formed under the laws of the Commonwealth of Massachusetts. Property so transferred consisted of the rights, title, and interest of the Gorton-Pew Fisheries Company, including all real estate, personal property, and vessels, as well as its business and goodwill. As one condition of the sale, the limited partners agreed to keep Gorton-Pew intact; the firm was not to be broken up and sold for its pieces, nor was it to move out of Gloucester. It was welcome news, as many hundreds of people—dockworkers, fish handlers, and fishermen—were dependent on the company for their livelihood: Gloucester without Gorton-Pew was inconceivable.

The company was back in business. They maintained the remaining vessels in their fleet: 21 craft in 1924, 24 in 1925, and about 20 a year for the rest of the decade.* The company no longer had to sell its fleet to raise capital. Gorton-Pew also became a major buyer of foreign and domestic cod to produce its extensive array of salt-cod products for the American market. In the summer, they took in splitting fares from Boston. Boats either sailed to Gloucester with the fish, or the fish was trucked from Boston to Gorton-Pew's Gloucester plants. Over the year, the company took in foreign fares from Canada, Iceland, and beyond. In the winter of 1924, a Danish craft brought in a record 1.3 million pounds of Icelandic cod. The fare was split between Gorton-Pew and the Frank C. Pearce Company; a check for nearly $17,000 was paid into the Custom House treasury in duty. But it was only the beginning. Later in the decade, there were several instances when a single steamer brought in upwards of 3 to 3.5 million pounds of cod from Iceland or Europe for Gorton-Pew.

Having emerged from the dark days of the postwar depression, the firms of Gloucester began to look for space to grow their businesses. As they did, Gorton-Pew proved well placed to sell its excess waterfront wharves and fish-processing buildings at a profit. In 1924, land near the Eastern Avenue end of Main Street was sold to the Hart Garage Company. For the first time in years, the old-time waterfront landmark no longer provided wharf or dock space, or buildings for curing fish. In September 1925, another wharf parcel was sold for $80,000 to the newly formed Producers' Fish Company, an Italian-run firm that caught and marketed fish, as well as supplying ice and oil to fishing vessels. Located in the Fort section of the waterfront, on Harbor Cove, the 229-foot Producers' Fish Company wharf extended 169 feet inland. The new company built a concrete block building to take in and package fresh fish, while outfitting another building (known as a cooler) in which to store ice. One of the principals of the new firm was Gorton-Pew executive Thomas J. Carroll.

Gorton-Pew also sold one of its unused plots of waterfront land to Clarence Birdseye, a man who was to revolutionize America's food-processing

---

*In January 1925, Gorton-Pew did sell off *Walrus* and *Seal*. They had been tied up since 1920 at the Cunningham & Thompson branch of the company, and were now sold to Newfoundland interests. The world record beam trawler fare was landed in the *Seal* by Captain Carl Olsen, some 475,800 pounds. In April 1925, the boilers and supply tanks of the craft were filled and a fire started in their boilers. They were being moved to dry dock in Boston in order to complete the necessary overhauling prior to being sent to their new home port: They were to be used as freighters, moving pulpwood from St. John's, Newfoundland, to Bangor, Maine. The two steamers are estimated to have cost close to $250,000 each; their sale price was in the vicinity of $20,000 per vessel.

industry.* The property was situated on Commercial Street, facing Harbor Cove to the east and the outer harbor to the west, with the large cement building on the southern side of the property becoming home to Birdseye's General Seafoods Corporation—a company that would revolutionize the way fish and other foods were processed in America and across the globe. Over the prior two years, Birdseye had carried on research in New York on new ways to quick-freeze fish, and now with the financial backing of Wetmore Hodges, William Gamage, Basset Jones, I. L. Rice, and J. J. Barry, it was time to move from the experimental stage to the production phase of this venture. Under Birdseye's new process, fish were flash frozen at a low temperature, resulting in little damage to the product; the fish were brought to −50°F so quickly that virtually no cell-destroying ice crystals formed.

Although a food revolution had begun, in the 1920s, with limited installation of freezers in grocery stores or homes, much of the promise lay in the future. Nevertheless, the General Seafoods Corporation was a growing concern. They had found a niche, and in January 1928, the company bought the assets of the Whitman, Ward and Lee Corporation of Boston, including two steam trawlers and a fresh-seafood business. They now had fresh fish to complement their "frosted" specialties. By this time, Birdseye had also upgraded its Gloucester facilities[†] and installed their patented equipment in plants in other ports, including one in Halifax, Nova Scotia. Then, in May 1929, the Postum Company bought out General Seafoods. While the principal plant and research division remained in Gloucester for years, Postum soon expanded the company's operations beyond its Gloucester roots. They applied the quick-freezing process to the preservation of meat, poultry, vegetables, and fruit.[§]

---

*Other changes along the waterfront continued to be noted in mid-decade. On January 7, 1926, O'Hara Brothers of Boston purchased the Davis Brothers Fisheries property at 44-63 Rogers Street, comprising about an acre and a half of wharf land. The firm operated 11 stores on the fish pier in South Boston, and a large fleet of vessels as well. O'Hara Brothers intended to carry on the business of the Davis plant as before, but on a larger scale. They went extensively into the smoking and canning of fish. They planned to have their fleet come to Gloucester to split and the fleet fit out of Gloucester in the summer season, when the craft brought a large part of their catch for splitting purposes. On March 12, 1926, a newly organized Gloucester firm, the Independent Fisheries Company, bought out the former Andrew Leighton wharf and buildings. The incorporators of the new firm were A. Harold Brown, Captain Robert Wharton, and Captain Wallace Bruce. They engaged in the outfitting of vessels and the buying of fish. The schooner *Joffre* was one local craft that moved to the new wharf, as she was operated by Independent Fisheries and went seining under Captain Bruce. The wharf was replanked and the buildings renovated.

[†]They installed a new pump house on an adjoining wharf to secure a continuous source of filtered water for their plants.

[§]General Seafoods continued to expand its Gloucester operations, but before long moved its principal business to Boston. Other buildings in the former Cunningham & Thompson property on Commercial Street were purchased. Their wharf holdings were expanded so that multiple vessels could be taken out at any one time, and they placed an order with the Fore River plant of the Bethlehem Ship Building Corporation for three beam trawlers at a cost of $400,000, and with J. F. James & Sons in Essex for two wooden draggers.

Thus, these two companies, Gorton-Pew and General Seafoods, re-established the importance of Gloucester as a fish-processing center. Canned and frozen fish were in demand, and Gloucester's substantial wharf space and processing facilities helped to ensure that the community remained an important marketing center in the 1920s and beyond.

In the mid- and late 1920s, Gloucester's fishermen focused their energies on mackerel, haddock, cod, and halibut. The fresh haddock and halibut fisheries prospered, and the mackerel schools had returned. With a stable fleet, the men led reasonably comfortable lives, and the Italians moved aggressively into the fresh fisheries. When the children of a doryman like Steve Olsson or a skipper like Clayton Morrissey graduated from high school, they were able to further their education in business schools and colleges, buy cars, and find jobs unrelated to the fishing industry. This was a prosperous time; families built their savings, and the fishermen had options in how to live their lives. A man could return to Newfoundland and fish out of St. Mary's Bay, while another could contemplate commissioning a new schooner.

As had been the case so many times in the past, when the mackerel returned, prosperity was soon to follow. By early spring, the mackerelers were once again fitting out for trips south. The air was filled with the familiar

Unbaited trawl lines in buckets are set out on board a schooner prior to departure for the fishing grounds, 1928.

smell of paint on newly spruced-up vessels and tar on miles of new seine lines. Gloucester's industry was awakening,[68] and as mackerel poured in, hundreds of jobs were created and orders were placed for new vessels. The decade had begun with just 15 vessels in the southern mackerel fleet; before its end, there would be more than 60. The 1924 southern catch proved the best in years: 2.7 million pounds by early June, against 1 million pounds in 1923, and in 1925 the catch doubled again.

With this rebirth, shipbuilding in Essex boomed, with each spring seeing new vessels crammed into every inch of space at the Story and James yard. Established Gloucester skippers and outfitters were at the forefront of those commissioning new craft: the 60-ton *Almon D. Mallock* by Captain Ben Pine; the *Babe Sears* by Captain Joseph Sears; the 60-ton *Evelina M. Goulart* by Captain Manual Goulart; the 140-foot *Eleanor Nickerson* by Captain Enos Nickerson; and the 107-foot *Adventure* by Captain Jeffrey Thomas.* Gloucester's Italian community also took advantage of the booming mackerel fishery, building up its fleet of smaller 45- to 60-foot craft. They placed orders with yards in Essex, Newburyport, Amesbury, and Maine; the hulls of many of the new vessels were "transitional" in style, suitable for dory fishing or dragging.

The southern season remained strong through 1928, but in 1929 the mackerel again disappeared. Early in the season, the southern fishery was the lightest in five years. Fish appeared in small pods and were often too wild and difficult to catch. The prospects for the year seemed grim. But then, in late June, the Cape Shore fleet had several fine weeks. Reports came in of good schools on Georges, and in early fall, the shores off Cape Ann were literally alive with small mackerel. On one afternoon in September, the Back Shore from the Eastern Point breakwater to Thatcher's Island was a veritable harvest ground. One million pounds of mackerel were landed in a single day. Smaller boats even seined fish inside the breakwater as spectators looked on from the shore. Fares ran from 12,000 to 100,000 pounds, and by the end of 1929, the total catch was 42.5 million pounds—the best since 1926.

The halibut catch remained stable, although there was a slight drop in tonnage and vessels in the middle of the decade. Captains Carl Olsen, Charles Colson, Lem Spinney, and Robert Porper remained among the fleet's leaders, and in May 1924, Captain Olsen received a record stock of $15,036 for 86,000 pounds of halibut taken at Hawkins' Spot on St. Pierre's Bank. In February 1925, he set another record, receiving $16,250 for a 90,000-

---

*To read more on the *Adventure*, see Joseph E. Garland, *Adventure* (Camden, ME: Down East Books, 1985).

pound catch. One month later, he again "set the tongues along the water-front a-going," when he received $16,928 for 79,880 pounds of halibut.

Carl Olsen and the other halibuters were a force to be reckoned with, the fleet having continued success throughout the spring and summer. But then, on a foggy June morning in mid-decade, disaster struck and the community once more came face-to-face with lost lives. The vessel was the schooner *Rex*, and on that June morning she sat anchored on Quero Bank, nine early risers talking quietly on deck as they looked for the morning sun to burn off the fog. They were cutting herring, baiting trawl lines, and telling jokes. Without warning, a Cunard liner emerged from the fog, and the men looked up to see her bearing down on *Rex*'s port quarter. The liner blew her whistle, but it was too late. Her gray form closed quickly on the stationary vessel. The men had hardly a moment to think. They yelled to their crewmates below, and engineer Albert Roberts in his bunk and Captain Thomas O. Downie standing by the stove filling his pipe heard the shouts from above. The skipper moved for the gangway and Roberts followed, reaching the deck just as the bow of the steamer cut into the *Rex*. For those on deck, it was every man for himself as they jumped over the railing into the calm sea below. In an instant the *Rex* had disappeared. The survivors, all in a bunch, grabbed anything they could get hold of, dodging blocks of ice that were coming to the surface from the sinking vessel.

The Cunard liner *Tusconia* came to an immediate stop; passengers stood along the rails and lifeboats were lowered, but for 14 men and boys, including Captain Downie and Albert Roberts, it was too late. Most of the lost were trapped in their bunks, with no chance of escape. The cook, Austin Firth, and his 10-year-old son, Charles, had been in the forecastle, the cook preparing breakfast while the boy lay asleep in his dad's berth. Others had just begun to stir, to put on their gear, to get ready for a hard day's work. But it had all happened so suddenly. There was no warning, and the survivors doubted the lost men even knew what had happened. They felt especially sorry for the boy; he was a good-natured little fellow, always in the dory with Oscar Williams, another of the lost men. One of the crew lamented that "little Charlie Firth, enjoying his first trip on the water, was enthused with the prospects of going back home and telling his playmates how big fish were caught, and what a wonderful time he had with his dad out on the fishing banks." Now he was gone.

The loss of life, with boys and men taken so suddenly to their graves, shocked the community. It was the only source of conversation along the waterfront, and a committee was formed to raise funds to help the widows and orphans of the crew. Many on Cape Ann contributed, and trustees were

346346346326326326326326326346346346346346346346326326346326346346326346346346326326326346326326326346346326346326346346326346326346326326326326326326346326326346326346346326346346326346346326346346326346346326326346326346346346346346346346346346346346346326326346346346326326346346346346346346346346346346346346326326326326326346346346346346326326346346346346326326326346326346346346346346346326326326326326326326326326326326346326346346326326346346326326346346346326346346346346346326346346346326326326346346346326346326326346346326346346346346346346326326326326326326326326326326326326326326326346346346346346346346326326326326326346346346346346346346346326326346326346326346346346346346346346346326326326326326326346346346346346346346346346326326326326326346326346326326326326346346346346346346326326326326326346346346346346346346346346326326346326326346326326346346346346346346346346346326326326346346346346346346346326326346346326346346346346346326346326346346346326

appointed to look into the needs of the families. The community eventually raised $9,200 to help bring relief to those left behind, and the trustees did all in their power to ease the pain.*

Notwithstanding the loss of the *Rex*, the fresh-halibut fishery continued to prosper. In the upcoming year Captain Olsen skippered the *Progress*. He had a one-half interest in the craft and with her he remained first on the list of the highliners until the end of the 1928 season, when he purchased a controlling interest in the *Oretha F. Spinney* from Captain Lem Spinney. On his first trip on the *Spinney*, he fished at Hawkins' Spot and returned with 60,000 pounds of fresh halibut.

Even as the halibut fleet was producing fine stocks, Gloucester's fresh-haddock fleet continued to send more vessels to sea. Winter haddocking brought fine prices in Boston, while summer splitting prices improved. Boston's weekly fresh-fish catch was often enormous. For a two-week period in February 1924, more than 9 million pounds of cod, haddock, and other ground fish came into the port, although with such enormous supplies much had to be sold to the Gloucester splitters. But as Lent began, the catch increased and the prices rose.

Finally, at the end of the year, vast amounts of fresh fish would once more come in as the entire fleet returned to celebrate the Christmas holiday.

---

*This was also a time when the Gloucester fishing community was called on to take note of the strange passing of several distinguished skippers. Joseph Bonia drowned while serving on a Nova Scotian schooner, working on a merchant trading vessel, and Alden Geele and Fred Thompson both took their own lives under the most unexpected of circumstances.

For Captain Bonia, the time was March 1924, the vessel was the Nova Scotian schooner *Independence*, and he and his crewmates were sailing from Ardrossan, Scotland, to Antwerp, Belgium. The craft simply disappeared, never to be seen again, although on April 4 in that year a vessel bound to Mallanear picked up and landed a lifeboat from the *Independence*. The *Independence* was a staunch vessel, having previously sailed against *Bluenose* in the Halifax elimination races, and like so many Lunenburg craft, she was engaged in winter freighting at the time of her loss. She sailed from Halifax on January 2, and when she reached Ardrossan, Scotland, many of her crew decided to return to their homes in Nova Scotia, but Captain Bonia signed on a Scottish crew, loaded a cargo of liquor, and sailed on March 19 for Antwerp. She was never seen again.

For Captain Geele, the time was April 10, 1926, and he was to take his life while aboard the schooner *Thomas S. Gorton*, the last all-sail vessel to fish out of Gloucester—at Shelburne, Nova Scotia. He was found dead in his bunk with a bullet hole in the side of his head and a 32-caliber Colt lying on the floor by his side. Captain Geele left a note saying he could not stand the pain in his back from which he had been suffering. (*Halifax Bulletin*, April 13, 1926.)

Geele had sailed from Gloucester on a dory handline trip, and put into Shelburne for shelter prior to leaving for the Banks. He was a native of Shelburne and about 60 years of age. He had a long and distinguished career commanding Gloucester vessels and was one of the greatest salt-bankers ever to sail out of Gloucester. At the height of his career, he brought in an average of 800,000 pounds of salt cod each year, and it was Geele who first commanded *Columbia*. He relinquished command when *Columbia* returned for the race, and had remained ashore in 1924 and 1925.

For Captain Fred J. Thompson, the time was November 5, 1928, when he took his life in a swamp about 50 yards from the Riverdale Willows, on Washington Street. He was last seen by his son Fred on Saturday afternoon, when the latter took his father to the camp of Charles Young, for a supposed visit. But he did not stop there. The son said his father appeared to be in good spirits. Captain Thompson had come to Gloucester from Norway 42 years earlier and had followed the fisheries ever since. It was believed that he was in a state of despondency.

On December 23, 1924, 21 craft brought nearly 1 million pounds of fresh fish into Boston, and then for two days, December 24 and 25, the docks were bare. The schooners had motored back to Gloucester and were tied up at the wharves or anchored in the channels as their crews spent Christmas with their families. On December 26, after storing gear, food, ice, and bait, many of them left for the Banks again. The new year's fishing had begun, with the offshore haddockers leaving for Georges and the Scotian Banks. Some family members came out to Eastern Point to bid adieu to vessels such as the *Natalie Hammond, Elk, Heja, Thomas S. Gorton,* and *Silva* as they set off.

But haddockers were still subject to the whims of the weather, and in an early January trip the *Oretha F. Spinney* came into Boston encased in nearly two feet of ice. Captain Jeff Thomas said spray had coated her deck and rigging, instantly forming a solid block of ice. She had nearly as much weight in ice as she did fish in her hold. But despite the difficulties at sea, what stood out were the fine catches brought in and the good prices paid at the Boston docks. On February 9, 1925, 43 vessels landed a one-day winter record catch of 1.5 million pounds, followed by 1.2 million pounds the next day. Such days became common, as Boston set records from one year to the next. In 1927, the port recorded 4,310 arrivals and landings of 171,370,000 pounds of fish. Yet, as the decade unfolded, the schooner-based dory fishermen were no longer bringing in the lion's share of the catch. That honor fell to the expanding fleet of smaller otter trawlers, the draggers: By mid-decade, they landed 60 percent of Boston's fresh catch—dorymen having moved from setting trawl in small boats to setting large nets from the decks of draggers.

The new generation of otter trawlers was smaller and more economical than the large vessels of the war years. The 1926 fleet still included a few of the old-style beam trawlers, but there were many more of the smaller draggers, even smaller baby draggers, built for speed.

One of the more unexpected positive turns for the fishermen was the running of illegal alcohol from offshore mother ships into the harbors, bays, and beaches of Cape Ann. The national prohibition law outlawing the manufacture, sale, or transportation of intoxicating liquors went into effect on January 16, 1920. There was money to be made, the Massachusetts coast and harbors made rum-running easy, and many a Massachusetts and Nova Scotian fisherman turned to the new industry. But there were dangers. The first recorded seizure of liquor from a Gloucester "rumrunner" occurred on March 25, 1920, when 60 bottles of brandy were confiscated from the schooner *Monarch* as she came in after having delivered a load of fish to Greece.

Key to the trade was the presence off the New England coast of a string of merchant craft and schooners loaded to the gunwales with the choicest of liquors. Many came from Lunenburg, including the onetime Gloucester vessels *Arethusa*, *Henry L. Marshall*, and *Seal*. They sat 15 to 30 miles off the coast from Chatham to Cape Ann, and it was to these larger ships that smaller, faster Gloucester craft made evening runs to load up and run back to shore. Everyone knew the names of the offshore craft; the local paper identified them by name and even by location.

In February 1923, as Gloucester and federal authorities sought to increase the visibility of their supposed vigilance, the local police descended on the Fort section of the city with search warrants: They seized 227 bottles of whiskey found in a shed. A few months later, the steamer *Joppaite* of Gloucester, carrying between 350 and 500 cases of whiskey, was seized by a seagoing revenue cutter. The skipper said he "did not know where he got the liquor, what he was going to do with it, or where it was supposed to be delivered." However, the boat was 15 miles from shore when seized, fit out as a mackereler, with a mackerel seine-boat in tow and a man posted as if on the watch for mackerel.

In 1924, the local police, in conjunction with their federal counterparts, moved to catch rumrunners, one man at a time, confiscating liquor in a storeroom, on a dock, in a warehouse, and on a vessel. The example of Howard Blackburn says much about the effort. He was the best-known fisherman in Gloucester, having lost his hands many years earlier while rowing for the safety of the Newfoundland shore in the dead of winter. Before prohibition, he had been the owner-operator of a famous Gloucester saloon,* an ideal, high-visibility target for the enforcers. On March 2, 1924, federal agents paid an unexpected call on Blackburn, seizing large quantities of moonshine. A tank containing moonshine was found concealed in the wall of a rear room: The liquor was being distributed to purchasers through a piece of hose used as a siphon. For some weeks an agent, passing as a retired sea captain, had frequented Blackburn's place and identified the operation. The confiscating agents arrived in town in a large touring car and left once their mission was accomplished. An example had been set. But it was a mundane arrest. It did little to change the underlying cycle of demand and reward, and as a major port city, Gloucester would see many more arrests and daring chases as vessels brought liquor into Cape Ann. Lillian Lund Files tells a story of the rumrunners of those days:

---

*To read more on Howard Blackburn, see Joseph E. Garland, *Lone Voyager: Extraordinary Adventures of Howard Blackburn, Hero Fisherman of Gloucester* (Boston: Little, Brown and Company, 1963) (as well as reprints by others).

Alvin Lund's divorced wife Ida ran a Swedish boarding on Main Street, and one of their daughters, Gerda, married the "rum runner" Carl Cameron. There came a day when a fishing boat Carl had commissioned came into Gloucester with liquor from Canada, but they got caught by the Coast Guard. The liquor was confiscated and the whole crew was sent up to Boston to jail. "This was against the law." Well, Carl Cameron, who was the head of the thing, and had a little restaurant, a diner, on Main Street, went to Boston and bailed out all the Gloucester guys. "They come back on the train to the depot in Gloucester and Carl had a big brass band welcoming them."[69]

At about the time of the Blackburn arrest, rumrunners coming in on a fast boat were forced to make a mad dash to outrun Gloucester's special duty squad. Under cover of darkness, the rumrunners had entered the inner harbor and tied up at the New England Fish Company. They were soon detected. Local officers in a Coast Guard launch crept up on the men and called for them to remain where they were, as they were to be boarded. But the men on the dock cut the ropes and slowly backed the boat away. As they did, the police unholstered their revolvers and opened fire on the craft's gasoline tank, but to no avail. A chase followed out into the harbor into the blackness of the night. The boat rounded the Fort and made for the Blynman Canal and Annisquam River on the western side of the harbor. Once in the canal, the boat ran aground in the rear of a tenement house, where police later confiscated 13 cases of liquor. Upon investigation, the officers found their bullets had penetrated the gasoline tank, the fuel had leaked out, and the craft had had to be beached. The men escaped on foot and no arrests were made.

Just before Christmas in 1925, the local police captured the small, 23-ton Gloucester schooner *Mineola*, loaded to her scuppers with holiday liquor. Her capture came in the evening, shortly after she made port. On board the officers found cases of Old Smuggler whiskey and champagne. The vessel was registered to Manuel Machado, and the seized liquor was worth $25,000. As she had come into the inner harbor, an officer on patrol heard the putt-putt of a motor through the haze. Hurrying to the old Leighton wharf, the officer saw the boat tie up and a crewman jump to the wharf. The officer rushed to get a warrant, but on his return with an augmented boarding party, he found no one aboard. After smashing open the hatches, they found case after case of Old Smuggler whiskey—135 cases of whiskey and 33 cases of champagne were seized.

While colorful, and probably profitable, stories of rumrunning are incidental to Gloucester's true success as a fishing port and processing center during the 1920s. Her fishing fleet had come back (albeit with more powered draggers), processing plants prospered, and the future seemed bright. And while there was the grief of unexpected loss at sea, the people of Cape Ann rallied to support the widows and orphans, and the community came together.

CHAPTER 32

The Great Depression

Then it happened: In the early fall of 1929, the American stock market crashed. In Washington, over the next three and a half years, President Herbert Hoover* and his Republican stalwarts failed to appreciate the full gravity of the situation, wedded as they were to a self-correcting business cycle model. For these men, a return to prosperity was always just around the corner. The Republicans, in general, retained great faith in President Hoover, seeing him as a man of sound views and great wisdom. The editorialist in the *Gloucester Daily Times* said that "seven years more of Hoover will raise the standard of living in this country."

But the Republican pundits misread the future. On the eve of Hoover's reelection campaign in 1932, one in four Americans was out of work. Wages had fallen by 20 percent,[70] banks were failing at the rate of about six a day, and New England fishermen had seen a 30 percent decline in their catch. The Hoover policies had not brought prosperity to America, and in 1932 the people had the final say. Hoover would be a one-term President, with the responsibility for running the country now falling to the Democrat Franklin Roosevelt. Nationally, Roosevelt carried 57 percent of the vote, Hoover 39 percent, while in the Electoral College vote Roosevelt carried 89 percent of the total. Roosevelt also carried Massachusetts by 110,000 votes, but Gloucester remained solidly Republican, concerned that a Democrat would not support protective tariffs on the fisheries. But the Democrats now had control of the White House and both houses of Congress—they had a 20-vote margin in the Senate and controlled two-thirds of the House.

In Gloucester, by 1929, only two pure sailing craft remained in the fleet, and they too were soon relegated to the role of transport vessels. For the briefest of moments, the Gorton-Pew Vessel Company considered outfitting them for the salt-bank fisheries. Despite their age, the schooners had been well maintained and were well suited for long voyages to the far eastern grounds.

---

*On November 7, 1928, Herbert Hoover won in a landslide, carrying 38 states. In Gloucester, it was Hoover 5,093, Al Smith 3,262.

But it was not to be. The federal government lowered tariffs on imported bulk salt fish, leaving Gloucester craft at a competitive disadvantage with salt-bankers from Nova Scotia. It would now cost more to mount a salt-bank trip out of Gloucester than out of Canadian or Provincial ports. But the tariff ruling did not, in itself, end Gorton-Pew's interest in the venture.* Because Nova Scotia had been hard hit by the depression, many maritime fishermen were more than willing to sail on such a salt-bank trip. Yet, their willingness was soon irrelevant. With a worsening depression, the U.S. Department of Labor prohibited seasonal entry of workers from the Provinces. Immigration all but stopped, and Gorton-Pew was forced to give up the idea of resuming the local salt-bank fishery. The last two of Gloucester's nonmotorized schooners moved into transportation, going port to port to pick up Canadian salt fish for the trip south to Gloucester. As for the other old-time schooners, all of which now had diesel engines, some went out with dragging rigs, while others fit out to go dory trawling for haddock, cod, and halibut. They went with shortened fore- and mainmasts, and only a bare minimum of sails for maneuvering on the grounds—a foresail, jibs, and a riding sail.

From 1931 to 1935, the number of vessels of 20 or more tons dropped by 22 percent, from 171 to 134 craft. Yet, few changed hands. There was little demand for old Gloucester schooners; vessels were not replaced as they were lost, but even so, in the days leading up to the stock market crash, a number of large, majestic schooners could still be seen fitting out at the docks or leaving for the grounds. They represented a bygone era, and in a wave of nostalgia the people of Cape Ann came together to support one last race off Eastern Point. For one more time, topmasts would go up, sails would be bent on, and schooners would compete for the title of fastest of the fleet. It might be the last time such a race could be mounted, and in the fall of 1929, four schooners sailed out of the harbor to race over a 40-mile triangular course.

But events took an unexpected turn. A group of affluent summer residents joined with a number of fishing masters to commission a new fishing schooner, the *Gertrude L. Thebaud*. Under the leadership of Ben Pine, they saw her as following in the wakes of *Mayflower*, *Puritan*, *Ford*, and *Columbia*—going to sea before the first of May, fishing over the summer, and then competing with the best of the Nova Scotian fleet in a revival of the International Fishermen's Race in the fall. Gloucester reconstituted the American committee and forwarded the challenge of *Thebaud* to the Nova Scotians. The Gloucester committee would put up the prize money. All the Halifax

*In June 1930, President Hoover signed a tariff bill that was generally beneficial to the Gloucester fisheries, with the exception of salt cod. Gloucester's local Congressman, A. Piatt Andrew, was instrumental in protecting the fishing industry.

Trustees had to do was reconstitute their committee and send *Bluenose* south, but they lacked the will to act. The Trustees would never fulfill their part of the bargain, but the owners of *Bluenose* took up the challenge. They would come south in 1930 to engage in a special invitational international match race with *Thebaud* off Cape Ann.

In Gloucester, the men were too eager to wait, and in 1929 the people of Cape Ann were treated to a race among four local schooners: the *Elsie*, *Mary* (renamed the *Arthur D. Story* by her owner Captain Ben Pine), *Progress*, and *Thomas S. Gorton*. Only the *Elsie* had previously raced (unsuccessfully) for the International Fisherman's Cup, but all were real fishermen, averaging 19 years on the Banks, and as they fit out it soon looked like old times around the waterfront. Topmasts were going up, rigging was being fixed, and upper and lower sails were flapping gently in the breeze. Schooners slipped down the harbor under full sail, stretching their billowing canvas and testing their new rigging. Finally, on the last day of August, they were ready to go, but light winds pushed the race back by two days. Then, on a day when the winds never really picked up, the smallest of the four, the *Progress*, won the race. She crossed the finish line 10 minutes ahead of the *Arthur D. Story*, overtaking the larger vessel over the last six miles of the course.

On the *Gorton*, Harry Eustis remembered the contest as "a real race . . . a holiday race," one without enough wind to push the bigger vessels over the course.[71] The *Gorton* was an immense vessel and without a lot of wind she fell behind the fleet. Meanwhile, on the *Progress*, the skipper smartly set a huge amount of canvas, picking up what little wind he could.*

The *Gertrude L. Thebaud* was built over the winter of 1929–1930. She was a good-sized schooner, built along the general lines of *Columbia* at 115.8 feet and 137 gross tons. She was to have a 400-horsepower auxiliary engine and would go to sea with the backup sails from the ill-fated schooner *Columbia*. She would be the last full-rigged fishing schooner ever commissioned for the Gloucester fleet and was named after the wife of her primary backer, but in line with the changing times her sails would be seldom used except in racing. She slid into the Essex River almost four years after the last of the great commercial offshore schooners had been launched—*Adventure*, *Mary Sears*, and *Eleanor Nickerson*[72]—and no matter why she came into being, *Thebaud* was a true fisherman; she would go to the Banks, set trawl, and bring in a catch.

---

*On June 2, 1930, the schooner *Progress* sank off Nantucket Shoals, the victim of an explosion. The crew escaped safely. The explosion sent tongues of flames leaping high in the crosstrees and running like a flash throughout every rope and plank, consuming the vessel in a few minutes.

By late August, the race between *Thebaud* and *Bluenose* was assured, and both sides hoped it would remind the American public of the fishermen and their problems. While the fisheries were still doing reasonably well, unemployment was on the rise, and the future was uncertain; all saw the race as a splendid opportunity to spread the story of Gloucester and the North American fisheries.*

The race was scheduled for October, and in mid-September *Thebaud* had her first trial under full sail. Two weeks later, *Bluenose* dropped anchor off Ten Pound Island and the excitement built. The rules were simple: It was to be a fishermen's contest. Each race would begin at 10 o'clock in the morning and would have a six-hour time limit. There were to be no intrusive rules on sails, crew, or ballast. The two boats would sail off Cape Ann, with the skippers deciding how to fit out their craft and contest the race. The winner would receive the Lipton trophy, not the Dennis trophy controlled by the Halifax International racing committee.

On October 7, Cape Ann threw a "royal welcome" for the crew of *Bluenose*. Thousands lined the streets. Cheers rose as Captain Walters and his men passed in review; the Nova Scotians responding with waves and broad smiles. The people followed the parade to the wharf, filling every nook and cranny to get a glimpse of the visitor. Local fishermen were mingled in the crowd, having stayed over to see the race.

Soon all the preparations had been made, and the vessels were brought on line for the first race. It would be a best-out-of-three series, and in the first race, on a day with heavy seas and a stiff northwest breeze, *Thebaud* crossed the start line 1 minute and 40 seconds ahead of *Bluenose*. It was a perfect day for racing, and by the end of the contest, *Thebaud* carried the race by 15 minutes and 40 seconds. The wind averaged 15 knots on the windward legs, and the two craft averaged a remarkable 12.3 knots over the course.

In the next race, with hardly a breath of air to fill the sails, the contest was called off with *Thebaud* in the lead by more than four minutes. Three days later saw no wind on the course, and the contest was again called off. On the next day, they went out on the course and the race would be called off with *Bluenose* in the lead; the schooners had sailed once around the 15-mile triangle, but the wind then rose to 40 knots and with reduced visibility neither boat could find the next mark. It was a blow to Captain Walters, for he had about a 1-mile lead on *Thebaud* when the race was called. In fact, the lead had actually been twice that, but Captain Walters had lost time as he

---

*Advertisements were placed in trains and in the North Station, while Gloucester's Legion post sent notices to 10,500 other posts across the country. News stories and photos appeared in at least 50 newspapers.

sailed back and forth in a fruitless search for the buoy. Back in port, Walters made known his displeasure in having the contest called, and the committee session was "almost as stormy as the gale."

Finally, in what proved to be the deciding race, in an 18-knot northwest wind, *Thebaud* crossed the starting line about a half-boat length behind *Bluenose*. On the second leg, *Bluenose* gained a lead of 8 minutes. But *Thebaud* caught up with and ultimately passed *Bluenose*, finishing 11 minutes ahead of the Nova Scotian craft. Thus did *Thebaud* win the $3,000 cash prize, as well as the Sir Thomas Lipton Trophy, "emblematic of the North Atlantic sailing championship for fishing schooners."* It had been a fine match contest with little rancor—the way it was supposed to be.

Unfortunately, and predictably, the competition failed to sustain the fisheries. Catch levels began to fall and unemployment rose. Families faced a bleak winter, and the city fathers looked for ways to help those out of work, hoping to tide them over until the much-anticipated spring resurgence in the fisheries. A committee was formed to identify available public and private seasonal jobs, and the unemployed were called to sign up for work. People from all walks of life sought help—painters, plumbers, fish workers, and many others. The city quickly hired 200 seasonal workers, including Steve Olsson. For almost two months, Steve would join others laying pipe in ditches, and his daughter Gerda said Steve was "disgusted with some of the guys who loafed on-the-job." Men wandered off and did not do a day's work, and it troubled him as so many others he knew were begging for work. More jobs were needed, and Gloucester's mayor pitched in by working with the City Council to extend a sewer project on Chestnut Street, have City Hall repainted, and begin a bond drive with the aim of "buying a day's labor" from the unemployed. A $5 city bond entitled the buyer to one day's work, and within a few weeks $1,200 worth of bonds had been sold, translating into 240 man-days of work. Of that total, one-third had been turned back to the committee to be used as they saw fit, and men were hired to clean beaches and roads. Yet, despite these efforts, many of Gloucester's unemployed ended up on the "dole," costing the city about $45,000 in 1930, or $1 on the tax rate.

---

*On September 16, 1931, Halifax accepted a challenge race between *Bluenose* and the *Gertrude L. Thebaud*. In preparation, *Thebaud* had a series of trial races with the *Elsie*. Her standing rigging was overhauled and tightened up, masts scraped down and slushed, and a good part of the running rigging replaced. In the first of the October trial races against *Elsie*, *Thebaud* won by only 2 minutes and 27 seconds—much too close for comfort. *Thebaud* seemed sluggish for so large a vessel. In a second trial race, *Thebaud* won by a more convincing 7 minutes and 54 seconds. Once at Halifax, *Thebaud* lost the first race by 3 miles. It was claimed that she was too heavily ballasted for the conditions. *Bluenose* then won the second race, and the cup, by 1 mile in a 16-knot fresh breeze. There was a final race between these two vessels in 1938, with *Bluenose* winning the series three victories to two. She took the last race in rather still winds, when a little puff picked up her sails and carried her to victory.

In the spring, with the mackerel fleet preparing to head south, Gloucester's unemployment committee disbanded. The chairman said that those who had been offered temporary employment would gradually find steady work. He believed the crisis of unemployment was over, and more avenues for work could be expected to gradually open. The fitting out of vessels for the south had afforded extra work along the waterfront, and more would surely follow. Work on the highway and water department projects was also in full swing, and the chairman believed private individuals would now step forward to hire the unemployed to work on their properties. Workers would be needed to open summer estates, erect new buildings, and alter existing buildings. In his view, "very soon now there should be no difficulty for those in need of finding work."

But this optimism failed to foresee the continued decline in the fisheries, with lower catch levels and a falling stock. New England fish landings in March were lower than those of the same month in 1930: 29.1 million pounds versus 31.2 million pounds, with an even steeper drop in the price paid, from $1,371,475 to $980,830.

By May, when full employment had been expected, the city relief rolls were climbing. Yet the economic pundits of the day, as well as the administration in Washington, held out hope for better business ahead. They expected to see gradual improvement, believing that the normal seasonal gains of spring had only been delayed. Surely the summer would bring a quickening in trade. In President Hoover's view, it was not the province of legislatures to cure a worldwide depression. That task must be done by individuals, and the rebound in the economy could be expected over the summer. But there was no summer rebound in the fisheries, and people were tiring of calls for patience, hard work, and common sense. By mid-October, 700 banks outside New England had failed. They were typically smaller banks, but in New England a single large bank had failed.

In Gloucester, after four years, the schooner *Columbia* fund had finally expended all of the $35,766 raised to help the families of those who had been lost. The fund had provided help to 35 adults and 53 children, and the families were now on their own.

By November, Gloucester's unemployment was at an all-time high. The Gloucester National Bank, whose roots went back to 1796, closed its doors in December. They did so to protect against an expected ruinous run following the closure of the Federal National Bank of Boston and its five branches and subsidiaries. The Gloucester bank reorganized and reopened in March the following year. In Boston, fish landings were down by a staggering 83 million pounds.

By January 1932, Gloucester's mayor, responding to the unfortunate circumstances of so many families, called for those who were working to contribute to a fund for the less fortunate. He hoped to raise $75,000. Some 585 people had registered for work, and the mayor questioned whether the city could meet the situation through its Public Welfare and Soldiers' Relief Departments. To seek help, 40 members of the Gloucester Unemployment Relief Committee canvassed the city, contacting upwards of 700 businesses and their employees. One month into the campaign, with the fund at only $18,000, there seemed no end to those seeking work. By late February, a committee of eight began to check out the applicants and fix the work assignments for the seasonal jobs that had been identified. They included replacing cement-lined water pipes, installing sewer pipes, and building a new tennis court at the entrance to Stage Fort Park. The assignments were for three, four, five, or six days a week, depending on the needs of the family. The relief fund eventually topped out at $43,000, with donations from about 1,700 people.

By mid-March, 450 people were working under the auspices of the Unemployment Relief Committee, more than 200 of whom had been taken off the Public Welfare Department list. They were receiving an average weekly salary of $15. As in 1931, the committee hoped that with the spring upturn of the fisheries most of the unemployed would find work. But if this was unrealistic in 1931, it was a delusion in 1932.

In May, the price paid for fresh fish in Boston slumped even further, with demand falling far behind supply. As new lows were set for haddock and mackerel, many craft were forced to sell their fish to the splitters. By June, the New England fish catch was 6 million pounds below the low of the previous year. As the price of pork and beef declined, so did the price of fish.

As Roosevelt swept into office, Gloucester faced the worst winter of the depression. Houses that had been occupied by paying tenants were vacant. The ranks of the unemployed climbed, and few fishermen had been able to save for the slow winter months. Governor Joseph B. Ely warned that vessel owners and fishermen had earned next to nothing during the year, but no one seemed to know what to do.

Gorton-Pew decided to concentrate on fitting out 10 fishing craft for the handline haddock fishery, traditionally the most profitable of the winter fresh fisheries. The firm had little need of more fish at its Gloucester plants, and every dragger was hauled out, stripped of its gear, and readied for handline haddocking. In itself this provided one month's employment to carpenters, painters, caulkers, riggers, and longshoremen. Everything else had failed over the year, but this seemed the best course for the firm—bringing a catch of the best fresh fish into Boston.

Gloucester's monthly Welfare Department costs reached a staggering $39,332 in December—$33,701 for public welfare and $4,831 for soldiers' relief. It represented an increase of $21,000 over December 1931. For the year, the cost of public welfare totaled $259,919, with costs increasing month by month. By year's end, 1,084 families were receiving help. The situation was intolerable; the Welfare Department signed on four volunteer investigators to look for cheats and tightened the rules for rental support. The maximum payment of $3 a week toward rent would no longer be given until the tenant was at least three months in arrears. Finally, the department called on everyone in the city to consider themselves an informal member of the investigating committee. People were asked to turn in those neighbors whom they suspected of violating the welfare rules. One month into the review, the committee reported that 78 cases had been dropped from the rolls, saving the city $573 a week. Twenty-four men were dropped because they were among the lucky few to receive employment on a new city sewer construction, but over the same period, 16 cases were added. The committee reported that some Gloucester residents had written in about their neighbors, but the results were not what many had expected:

> We hear reports of this or that one living the "life of Riley" so to speak, and supposedly getting welfare aid. We investigate and find that they never were on the rolls at all. . . . When we checked their names with our list, we have found in some instances not a single name on the welfare rolls, only going to show that many of our people think that certain families are on our rolls, when in reality they may be in need and still have not resorted to welfare aid and we make this explanation for the benefit of many such cases.

By the end of the review, a little less than a quarter of the total cases (256) had been dropped, saving the city $1,665 a week. With the hunt for the "undeserving" now past, Gloucester hoped for better times as the people looked forward to the inauguration of President Roosevelt. And Roosevelt lived up to their expectations. New programs proliferated under the New Deal, and Gloucester's mackerel seiners found themselves being supported by one of them.

CHAPTER 33

The Mackerel Fishery in the Depression

In the years leading up to the Great Depression, the mackerel fishery had been prosperous. Catch levels had been high, new motorized vessels had come on line, and the fishing supported more than 800 men.* Six months into the depression, in 1930, 55 vessels, or roughly half of Gloucester's entire fleet, once more prepared for the trip south. Thirty sailed under Italian masters from the west side of Harbor Cove—continuing a 20-year tradition of fishing in motorized vessels out of the Fort, while 25 sailed under Provincial and American-born skippers from the east side of the Cove. There was still a sense of optimism as machinists, dockworkers, riggers, fishermen, and families stored supplies and checked gear.

Several Italian vessels had been on the southern grounds all winter, dragging for sea bass and 1- to 3-pound scup; and rather than returning to Gloucester, their mackerel nets and seine-boats were brought south by other Italian craft. In the spring, the fleet fished off the coast of New Jersey and Maryland, landing much of their catch at Cape May, New Jersey. Over the summer and through early fall, they fished on Georges and in the Gulf of Maine, and all in all they had a reasonably good year.

But in 1931 the fishery began to change. Record stocks of fish sat unsold in New England freezers and warehouses. Iceland and Norway produced an overabundance of both cod and mackerel, driving prices even lower. Worse yet, in Boston, landings of fresh fish fell, and counter to the experience of earlier years, so did prices. But for a while the mackerel fishery remained profitable. On April 9, the steamer *Rita*, under Captain Frank Favoloro, brought the first 30 barrels of southern mackerel into Cape May, and as other craft followed, much of the fish was sent overland to Boston, where it brought a respectable 6 cents per pound. Vessels made good catches on dark, moonless nights, and Gloucester was buoyed by the early success. But as the season progressed, the mackerel catch fell considerably behind that of the

---

*But by 1932, Gloucester's fleet had all but ceased to grow. Only 2 new vessels were added in 1932, none in 1933, 3 in 1934, and none in 1935. The Gloucester fleet, which at one time had numbered upwards of 500 schooners, fell from a total of 181 craft in 1929 to 134 in 1935.

previous year, and most of the Italian craft were left no option but to fit out for winter dragging for scup and sea bass off Virginia, spending months away from their families. Most had failed even to cover their outfitting costs, and with liens on their craft, further bank loans were out of the question. Men had averaged only $250 for an entire season's work.

It was a rare day that winter when a vessel entered Harbor Cove or the giant sails of a schooner fluttered in the wind as she prepared to go to sea. Not until March did the waterfront again come alive, as the Italian fleet returned from Virginia to refit for the southern mackerel fishery. By 1933 there were many troubling signs: The price received for a pound of fish had dropped from 6 cents in 1931 to 2.7 cents in 1932; many vessels no longer carried marine insurance; and the average vessel owner had cleared only $1,178 in 1932. Factoring in depreciation, this amounted to a staggering loss of $4,700 per vessel.[73] Something had to be done. Gloucester's Captain Ben Pine suggested sending a message to the new administration in Washington. Franklin Roosevelt had just been sworn in as the thirty-second President of the United States and was about to begin the most remarkable first 100 days of any President in the history of the country. There would be bank holidays, bank-deposit insurance was introduced, the gold standard was abandoned, and an "alphabet" of new programs began. Gloucester may not have voted for Roosevelt, but now he seemed the only hope.

Piney wished to get the President's attention, perhaps even making the needs of the fishermen one of his priorities. To this end, the veteran owner called on the community to send his vessel, the *Gertrude L. Thebaud*, to Washington. The *Thebaud* was Gloucester's best-known vessel, and he hoped the press would be intrigued by so unusual a trip, made by so famous a racer. The City Council saw the wisdom in the proposal, and put up $1,500 to fit out *Thebaud* for the trip to the capital. *Thebaud*'s hold was steam cleaned and her forecastle and cabin spruced up to house the old-time skippers who were to make up the crew. But Piney refused to repaint the vessel: She should appear just as she did when coming in from a trip, her sides weathered from the sea and chewed up from the pounding of dories as the men unloaded their catch.

As she set canvas on April 20, *Thebaud* carried a distinguished group of elder master mariners: Ben Pine, Clayton Morrissey (retired), Joe Mesquita (retired), William L. Nickerson (vice president of the Master Mariners'), "Charlie" Stewart, John Brymer, and Oscar Johnson. Captain James Abbott was the master for the voyage. The trip south was largely uneventful, or at least it was until *Thebaud* had sailed halfway up the Potomac. For here, in one of those twists of fate, she passed the Presidential launch, *Sequoia*, as it came down the river. On board *Sequoia* were President Franklin Roosevelt, his guest, Prime Minister Ramsay MacDonald of Great Britain, and members of their families.

Piney, recognizing the drama of the moment, called on the Gloucestermen to line the rail, and the aged fishermen raised a hearty cheer for the new leader of the nation. Noting the salute, President and Mrs. Roosevelt returned the cheers of the fishermen with lusty waves of their own. Captain Abbott dipped *Thebaud*'s flag and personally pulled the cord on her horn 21 times. With this sign of respect, the President and others cheered as the two vessels parted in midstream. But that did not end the story. Later that day, when *Sequoia* returned to the Navy yard, the Presidential party visited *Thebaud*. The President inspected the vessel from stem to stern, shaking the hand of every skipper on the schooner. He singled out Captain Pine, saying he knew his name from the schooner races. He also recognized the name of Captain Mesquita from the President's earlier days in the Navy Department during the Great War, when the *Francis O'Hara* was sunk by the Germans.

The story of the chance meeting was picked up by the press and appeared in papers across the country. They said there was something likable about the old Gloucester skippers who had made the trip to Washington, with their endearing "childlike faith in their venture." Yet, despite the publicity and the direct exchange with the President, nothing changed. There was sympathy for the fishermen, and some Senators signed on to their cause, but nothing of substance came out of the romantic trip of Gloucester's last great sailing vessel to the nation's capital. The mackerelers, haddockers, and halibuters were left to fend as best they could in a complex, changing world. The federal government did not step in to support them.

Yet, as *Thebaud* returned to Gloucester, all of this was unknown, and she was met some 10 miles out at sea by an escort of Coast Guard craft and fishing vessels. Horns rang out and people cheered as she sailed up the harbor, and later throngs climbed over the vessel at Piney's Atlantic Supply Company wharf. Piney repainted his craft, and Captain Abbott soon had her back in the haddock fishery.

One of the most respected of Gloucester's old-time skippers, Captain Clayton Morrissey, was about to come out of retirement to try his hand in the mackerel fishery. His new 40-foot boat, the *Nimbus*,* was built by his son-in-law, Ralph Nelson. She was launched at Bear Skin Neck, Rockport, and by mid-May was off Block Island in search of mackerel. But 1933 was a bad year to enter the fishery. If overabundance in 1932 had been a problem, in

---

*Named in memory of a boat owned by his father when Captain Morrissey was a boy. It was the craft on which he had his first experience fishing.

1933 it was a disaster. The first signs of trouble came on April 22, when 22 craft brought in fares of small fish that could not be sold. Benjamin Curcuru of the Producers' Fish Company radioed a message over a public broadcast station advising the fleet to take no more small fish.

On April 25, 33 craft brought in more than 1 million pounds to Boston, but demand was light and prices were down. By early May, the southern catch was already 15 percent ahead of the prior year, and when 20 seiners landed in Boston on May 7, buyers offered only 1½ to 2 cents per pound. The following day the price dropped to 1 cent per pound, and two days later it fell to only .75 cents per pound. Yet, even at such low prices, skippers were hard-pressed to sell their fish. There was just too much mackerel and too little demand. Few people had money, and those who did were pounding on their bank doors to get it out. Seiners returned to Gloucester and tied up for several days, if not a week or more. There was little sense in wasting money on fuel for engines and supplies for men.

Over the next few days, Gloucester mackerelers held the first of a series of open meetings. Owners and skippers came together to work out a plan to control supply and to set a price floor. Everyone was in port: The Fort and inner harbor presented a literal forest of masts as the men met at the rooms of the Master Mariners' Association to plan their future. The first plan called for no mackerel to be sold for less than 2 cents per pound, with each craft limited to a maximum of 25,000 pounds per trip. If a skipper had more than 25,000, he had to get rid of the surplus in whatever manner he chose, excluding selling the fish. He could dump the catch at sea or give it to another craft that had less than 25,000 pounds, or he could even give it to the poor who frequented the Gloucester docks. With the agreement in place, the fleet went back to sea. On May 23, the agreement was tested when 25 seiners brought fish into Boston. Little on shore had changed, and by noon only two fares had been sold.

A message went out on the fish report show carried on radio station WHDH advising the seiners to forgo trying to bring mackerel to Boston. They were to make for Gloucester to take part in the second of what became an ongoing series of mackerel meetings. One craft, the *Jackie B*, under Captain Stephen Post, tied up at Gloucester's Davis Brothers wharf and gave away all its fish. Children filled pails, bags, and boxes, while adults strung fish on strings.

The next plan called for seiners that were able to sell their catch during the week to pool their receipts. But with fresh mackerel selling at only .50 cents per pound there was little to share; the plan lasted less than a week.

)(*A mackerel schooner at Fort Wharf in Harbor Cove. Two seine-boats—with purse-seines piled high in the stern—sit ready to leave, but on this summer's day finely dressed children and adults climb over the vessels.*

The third plan reduced the maximum catch per trip to 20,000 pounds and set a minimum selling price of 3 cents per pound for large and medium, and 2 cents per pound for small. Craft were divided into three squadrons, the first third sailing the day after the meeting, the next third a day later, and the last the day after that. On June 5, 31 seiners brought in 542,000 pounds of fresh mackerel, but again supply exceeded demand. By nightfall many vessels still had fish in their holds. This plan, too, had failed. On the following day, the seiner *Alden*, under Captain Percy Faith, pulled into Gloucester's Independent Fisheries wharf and gave fish to the needy families of the city. Across Harbor Cove, the Italian seining fleet had come home, tying up until the skippers worked out yet another plan.

Two days later the skippers met, but the news from ongoing discussions with the Boston buyers was not good. The buyers insisted they could not make a profit if they had to pay 2 to 3 cents per pound, for it meant the retail price was more than people could, or would, pay. Fish prices were set relative to the price of meat, and meat prices were low. Gloucester retail stores were selling lamb for 8 cents per pound, pork at 10 cents, rib roast at 15 cents, and hamburger at 8 cents. But the fishermen could not operate at a lower price and, instead, aimed to further restrict supply. The fleet was divided into two equal squadrons of 29 vessels each. The first squadron sailed on Sunday and fished continuously until the following Friday. All of their fish had to be sold by nightfall of that day. They then had to tie up for the next week as the second group went to sea. The asking price per pound was to remain at 3 cents for large and 2 cents for small. The maximum for any one trip remained at 20,000 pounds, and all boats in each squadron were to pool their receipts for the week. On June 15, the first squadron landed 250,000 pounds in Boston, and received an acceptable 3.1 cents per pound for large and 2.5 cents for small.* But within a week, supplies again had exceeded demand, and craft held over from one day to the next, some coming to Gloucester to tie up rather than making yet another losing trip.

On June 26 the plan was tweaked; stocks would now be pooled across both weeks. The modification seemed to work. Over the next few weeks, with lower supplies, Boston buyers bought fish at the minimum levels demanded by the fishermen. However, in late July the entire agreement seemed in jeopardy. A few of the smaller mackerel boats signaled a willingness to accept 1 cent per pound in Gloucester. An excited group of Italian skippers swept out of the Fort. The exchange at the nearby wharf was peaceful, but the message was clear: Gloucester's mackerelers were standing together. The price line would be held. Yet by mid-August, even as the elaborate system for restricting supply appeared to be working, some fishermen were receiving as little as $6 every two weeks, while some owners felt that they would have been better off if they had just tied up for the season.

And so the season limped to a close. Prices never improved; many vessels failed to cover the cost of oil, food, and supplies; and the men had averaged only $300 for the season. By mid-September, many of the Italian seiners had changed over to dragging. They headed south to set their drags off Virginia and North Carolina for scup and sea bass, and to the relief of all, the winter fishery proved a success—shares ranged from $200 to $800 per man.

The next two years saw no improvement in the mackerel fishery. Prices remained low, even as the seiners made halfhearted attempts to develop rules

---

*On June 16, there was to be the first general meeting in Gloucester to consider organizing the fishermen into an official industry group that could then apply to the federal government for recognition under the new National Recovery Act (NRA) of 1933. The program seemed tailor-made for the mackerelers, for the act provided that in cases where the majority of an industry group formulated rules for its salvation, the government stood in back of them in having the minority obey those rules. The seiners moved to get this authority, but the federal bureaucrats first dragged their feet so long that nothing happened, and then the Supreme Court was to rule the program to be unconstitutional.

to govern their actions. Maximum catch levels were established, but neither minimum prices nor the order of sailing was governed. The few rules enacted had little or no effect. In Boston, many fares went unsold, and fish prices went as low as 0.65 cents per pound. Nothing worked, and by the fall, craft again had to refit for dragging on the southern grounds. This time, however, many of the non-Italian seiners joined in; indeed, a good many stayed south into May, rather than returning home to engage in the uncertain spring mackerel fishery.

CHAPTER 34

The Halibut Fishery in the Depression

In 1930, some 20 schooners were engaged in the halibut fishery, and through the next few years, 22 of Gloucester's most experienced skippers would go out after halibut—scarcity, combined with a high demand for halibut, helped to sustain a reasonable market. Captains Charles Colson, Archie McLeod, and Simon Theriault would make 30 to 40 trips each; Captain Carl Olsen would make more than 60. But theirs is the story of the ending of an era: Craft were lost, skippers and dorymen left the trawl fishery for draggers, and the schooners that remained went to sea with larger and larger crews, as ex-skippers signed on as dorymen. By 1935, only 6 halibut vessels remained.

Early in the depression, the fleet averaged 2.5 million pounds of fresh halibut a year. In later years, the catch dropped to 1.1 million pounds, and with the collapse of the economy it became more difficult to sell the catch. Captains Olsen and McLeod were always the first to fit out for early trips to the Grand Banks. In 1930, among the dorymen were Charles Daley* and Steve Olsson. Steve had left the *Natalie Hammond* as Captain Colson's health declined, while Charles Daley returned to Gloucester following the collapse of the St. Mary's Bay fisheries.† These two men now came together as dorymen on the jewel of the Gloucester fleet, the *Oretha F. Spinney* under Captain Carl Olsen.

The 1931 season opened on January 16, when the *Oretha F. Spinney* left for the Grand Banks. Gone for three weeks, she returned with 53,000 pounds of fresh halibut. Captain Olsen received an $8,600 stock, while each man received the exceptional share of $200. The *Pilgrim*, under Captain Christopher Gibbs, was the next to return, but she had had a trying time at sea and landed only 20,000 pounds of halibut. While fishing on Green Bank

---

*During this period, Charles Daley also made trips as a doryman on the *Jean Smith* and *Imperator*.

†Until the St. Mary's Bay fisheries collapsed, Charles had made a reasonable living as a salt-cod fisherman. He had gone to sea in his own boat, a black, 30-foot open boat with a heavy diesel engine. The locals called her a Branch skiff, named after the place at which she was purchased. The skiff had a small house but no cabin, and Charles went out with a hired hand. The skiff towed a dory astern, and once on the grounds Charles set the anchor, and the two men laid trawl. When he returned to St. Joseph's, his wife, Lucy, helped unload and salt the fish. Charles anchored near his house, rowing a dory back and forth to the small wharf at the foot of his property.

in the worst kind of weather, a huge wave had swept across her deck, taking two men into the sea. One regained the vessel on his own; the other was saved when another man dove in to rescue him.

By mid-March, as winter haddockers began to convert to halibuting, Captain Olsen returned from his second trip with 50,000 pounds of halibut, but there also came news of the loss of one of the fleet. While fishing on Burgeo Bank, the seams on the *Angie L. Marshall* had opened up under the constant pounding of the sea on the hull. Captain Robert Porper* ordered the men to work the pumps, but they could not keep up with the on-rushing sea. The *Marshall* went down, and although the men were saved, they lost all of their personal belongings.

By summer, halibut had become scarce on the Grand Banks, and vessels brought in meager fares from the Peak of Brown's and Georges. Yet, prices were down. Men were lucky to receive an $80 share for a 50,000-pound fare. For the first time in years, a full catch of fresh halibut had to be brought to Gloucester for flitching. By October, the vessel owners had pulled their craft out of halibuting. As they moved into other fisheries, several were lost in November, and once a vessel went down it was never replaced. The *Edith and Elinor* was rammed off Yarmouth, Nova Scotia, and went down in only four minutes with the loss of six men. A few weeks later, the *Squanto* piled up at Flat Rock in the Bay of St. George. Her crew was saved.

In 1932 Captain Charles Colson became too ill to take the *Hammond* to sea, and the vessel sat on the sidelines for more than a year; she never again went out for halibut. Captain Colson died a year later of mouth cancer, brought on by his habit of chewing tobacco. He was 59.

In May, another old-time Gloucester schooner, the *Louise B. Marshall*, was lost, but there were to be questions about how she went down. A member of the crew said he saw the skipper, John Marshall, take kerosene to the pilothouse when the other men were away in their dories. The captain asked him to go to "mug-up" in the forecastle, and it was not long before smoke came pouring out of the ventilator. However, the federal court later ruled that Captain Marshall had no motive to set the fire, and he was released.

In June another halibuter, the *Azores*, was lost when a fire broke out in the engine room and the batteries blew up. The men watched from their dories as the schooner burned. The age of the Gloucester dorymen was drawing ever closer to its end. Yet a few craft still engaged in the fishery, and in 1933, Carl Olsen left for the grounds in mid-January, returning three weeks

---

*Captain Porper made one more halibut trip in 1931, three in 1932 on the schooner *Raymonde*, and then left this fishery.

)((*Captains Jeff Thomas and Ben Pine (from right) with other Master Mariners, dressed to kill, c. 1930.*

later with a fine fare. Each of his men received a $131 share. Ten days later, Captain Archie McLeod shifted the schooner *Catherine* from haddocking to halibuting. The *Catherine* made Boston on March 20 with 93,000 pounds, the largest halibut fare in many years. But prices were low; white halibut sold for only 14 cents per pound and gray for 8 cents, thus each man's share was only $117. McLeod described the trip as uneventful, outside of high winds and choppy seas. But it was also lonely. In five weeks on the Grand Banks he spoke with just one other vessel—no halibuter, no salt-banker, no shacker. The Banks had become silent.*

As the season continued, craft landed fares in Boston, returned to Gloucester to refit, and sailed back to the grounds. At the end of April, Captain Olsen tried his luck on Georges, while Captain McLeod headed east to the Grand Banks. The halibuters continued to sail alone on the vast seas.[†] When Captain "Strings" Giffin came in on *Hesperus* after a 17-day trip on Georges, he told a reporter that "he thought the war must still be on for he saw nary a sail or a craft of any description during the whole halibut trip. The ocean was all his own as far as he could see." Even though he was alone, the fishing had been poor, and he quickly refit to get back to the grounds.

Low fares and prices still bedeviled the halibuters. Even the once-mighty halibut fishery had at last been beaten down by the new realities of depression-age America. Captain Olsen did not escape: He found poor fishing on Georges, steamed north to Quero, and on May 16 landed only 20,000 pounds of halibut. Were it not for a supplemental catch of fresh and salt fish, the men would have had little to show for their efforts. As it was, they received only $44 each.

On his next trip, Captain Olsen had the pleasure of once more welcoming his son Einar on board for a summer's fishing. Carl did not want his children to go into fishing, and neither of his boys ever made a living in the industry, but summer fishing was different.[74] Einar went out with Carl every summer. Such trips began when Einar was eight and ended in 1935 when he graduated from high school. Einar was the son of a first, failed marriage. Carl had legal custody of the boy, but when he remarried, his new wife,

---

*Meanwhile, the U.S. House of Representatives had followed the Senate in voting to repeal prohibition, and on April 7, even before the states officially joined Congress in repealing prohibition, President Roosevelt signed legislation allowing the manufacture, sale, and consumption of 3.2 percent beer and wine. Now it was up to the states, and the Massachusetts legislature and Governor quickly added their approval. Gloucester soon set up a new beer licensing board and quickly granted 20 licenses to restaurants and other establishments. On April 12, 1933, people swooped into Gloucester restaurants to be among the first to taste 3.2 percent beer.

[†]Captain Eric Carlson skippered the *Pollyana*, and sailed with a largely Scandinavian crew, including John Lund from Gullholmen. Among the other craft, Captain Archie McLeod took out *Catherine*, Captain Jeff Thomas took out *Adventure*, and Captain Olsen remained on the *Oretha F. Spinney*. Captain Simon Theriault now sailed on the *American*, having repurchased the vessel from Nova Scotia to replace the *Edith and Elinor*.

)(*Toward the end of the dory-fishing era, men bait up trawl lines on the deck of a schooner, c. 1935.*

Florence, did not want to share her home with the child of the first marriage, and Florence was the boss.* Carl arranged for Einar to live as a boarder with a Finnish family in Lanesville. Thus, from the age of eight to eighteen Einar lived with two Lanesville families—one during the school year, the other in the summer on those few days when he was home between trips. Only three houses separated Einar's summer and winter homes in Lanesville's Folly

*Carl Jr. gives a good description of his mother—she was a take-charge type of person. "She was the chief cook and bottle washer. She was the disciplinarian. She was the one that presented good times. You didn't see him very often, and when you did see him he was too busy because in those days he settled his own trips at a speci-fied place, and made the payoff there. My mother had worked for Gorton-Pew in the vessels office and that was the job she had, in settling the boats. That is how he met her. At that time when she was working there they had . . . (many) vessels under the flag of Gorton-Pew, sailing out of Gloucester. Carl understood who ran the family, at no time was there any conflicts of that kind or otherwise. However, there was no question who the boss was. That went without saying. I had no relationship at that time with half-brother Einar. Did not see him as a kid. Mother had no fears of Carl going to sea as a fisherman. Nothing that ever came to voice. In fact in growing up I never thought of a thing like that. He went to sea that was all there was to it."

Cove area.* Carl arranged to have the *Spinney* in Gloucester to take his son on board as soon as school let out.†

Einar first served as a cabin boy—a catchee—staying on board the vessel, assisting wherever he could. He helped the men launch and retrieve their dories, and when the dories were away, he fished off the sides or did other odd jobs until everyone returned. On occasion he was allowed to take the wheel. But as a young boy, he sometimes "messed up," and his father then waited a while before again giving Einar the helm. By 1932, however, Einar was an experienced sailor, having spent many summers on the Grand Banks with his father, and it was about this time that Einar said Steve Olsson spoke up for him. As the captain's son, Einar was getting less than a full share for a summer's work, and Steve, seeing that the boy was doing a man's work, went to Carl and told him to pay his son full wages. Steve was 60, the senior and most respected member of the crew, and others soon joined in with the chant: "He deserves full wages." Carl acquiesced, and from that time on, he paid Einar full wages. Steve and Einar had always gotten along well. When Einar was younger Steve watched out for him, and Steve was the first to take him out on a dory. Carl knew whom he could trust with his son.

Over the summer of 1932, Einar recounted two harrowing experiences as they fished on the northern grounds. In the first, Captain Olsen brought the *Spinney* through a narrow strait with the tide running against him and a driving storm hovering over the vessel. It was touch and go. Carl had ordered everybody into the dories. Then, with the engine running, he took the schooner through himself. If the boat was to sink, he would be the only one to go down with the ship; his men would be all right, no matter his fate. But they all made it through, dories and schooner, and before long Einar and the rest of the crew were back on board. In a second episode, Einar had a run-in with a swordfish that he never reported to his father. He was alone in the dory, and when he got the swordfish into the boat it began to flop around and knocked the peg out of the back of the dory. The water bubbled up all around Einar and the now-reviving fish. Einar was able to knock out the swordfish, replace the plug, and bail out the dory. He didn't tell Carl what had happened—he knew his father would not look kindly on the fish knocking out the plug. But

---

*The family was the Shermans, and they put Carl's son through college, and not just him, but other children in the neighborhood. They had no children of their own and found great satisfaction in helping to pay the tuition for the Finnish kids living in the Lanesville area. Mr. Sherman worked for a steel company that was later bought out by U.S. Steel.

†As the *Spinney* fished over the summer, Gloucester voted overwhelmingly to support the repeal of prohibition—2,122 for repeal to 444 against—and Massachusetts became the eleventh successive state to vote to repeal prohibition. Repeal carried the state by a margin of 4½ to 1.

Einar was proud of his win and brought home the sword as a memento of the fight.

In these hard times, many old-time skippers were concerned with the future of their children. Some, like Jeff Thomas, kept their sons ashore, while others permitted theirs to go to sea over the summer. One skipper following this model was Captain Archie McLeod, who, like Captain Olsen, took his son out over the school break. For Archie "Sonny" McLeod Jr., 1932 was to be his second halibuting season, and his buddy, William Maki, joined him in the adventure.

On shore, one of Gloucester's old-time retired skippers, Captain Lem Spinney, who still held a minority share on the *Oretha F. Spinney*, was considering a return to halibuting. He commissioned a boat in Rockport and told his cronies at the Master Mariners' that he was going back to sea. These old men still came together at the Association rooms to talk of life and wonder about the future. But their ranks were shrinking. In the summer of 1933, three chairs sat empty—the men who regularly occupied them, Captain Frank Hall, Captain Joe Mesquita, and ex-postmaster Charles Cressey, lay sick at home. Captain Hall, known for greeting all who came into the rooms with a hearty "Good morning, men," seemed destined not to make it through the summer.

These were somber times, and even as the old men talked of days gone by, in early August, the *Oretha F. Spinney* came into Portland with only a small fare. Captain Olsen told the other skippers that the high run of tides on the full moon had made halibut fishing practically impossible. He had little to show for the trip, and he immediately turned around and sailed again for the Grand Banks. He hoped that the heavy prevailing winds would give him a chance, but the weather did not cooperate. Fish remained scarce and halibuters experienced some of the worst August weather imaginable. Captain McLeod was one of the few who had a successful trip, bringing 55,000 pounds into Portland, but the market remained soft and the prices low. As one of his men said, "For four weeks of hard labor, we can only say we contributed our energy to the depression." They had sailed as far as the Funks, between Newfoundland and Greenland, some 1,400 miles from Gloucester, the longest trip in years, finding field ice and large bergs, but they received little for their trouble.

On August 22, the *Spinney* was back in Portland. Each crewman, including Einar, received a $76 share for a two-week trip, and Captain Olsen went right back to Georges. Even with low shares, the halibuters were better off than many in Gloucester. Mayor Parker appealed to the community to rally behind a drive to clothe the poor. He spoke of the thousands who were in

)(*Fishermen Kit Hines and John Dall at the focsle table on the schooner Corinthian, c. 1935.*

dire need of clothing, stockings, shoes, and rubber boots, asking people to collect old shoes and leave them at one of the fire stations. He pleaded on behalf of the many children who had only the clothes on their backs as they returned to school.

On September 12, as fall approached, Captain Carl Olsen returned to Gloucester a day or so before the fall school term began. It was fine for his son Einar to spend the summer working in dories with Steve Olsson and the other members of the crew, but Carl was determined that his two sons should have a different life from the one he had led; Einar would go to college. Einar said good-bye as he returned to Lanesville to begin the school year. After a summer's hard work, Einar had a fine set of new clothes on his back. There would be no public handouts for the Olsens.

The halibuters now began to shift over to haddocking. The *Oretha F. Spinney* was the last to hold out, bringing in a fine fare from the Peak of Brown's in early October: 40,000 pounds of halibut, 35 swordfish, 10,000 pounds of fresh cod, and 13,000 pounds of salt cod. It was a bountiful trip, and each man received a $150 share. Captain Olsen then tied up the *Spinney* until the new year, when he would resume halibuting.

As Carl Olsen and the other skippers prepared to spend time ashore, Gloucester lost one of her great skippers, Captain Joseph Mesquita. After an illness of several weeks, he died at his Prospect Street home at the age of 74. He had been in fishing since arriving in Gloucester as a young lad, but had been retired for about 10 years.

Then in December the fleet lost another vessel. The *Ellen T. Marshall* burned and sank some 30 miles off Cape Sable. She was the second of the Marshall vessels to be lost to fire within a year. Neither Albert Hines, the skipper, nor the crew could explain how the fire began. But strong winds had sent the flames licking over the ship, and Captain Hines ordered the men to take to their dories. Poorly clothed, they had pulled their oars for hours through blinding sleet and snow as they made for Seal Island. All but three of them made it. One of the lost was the skipper's youngest son, Ivan John Hines, whose dory was swamped as they made for the island.

In the same month, Jeff Thomas's *Adventure* had a close call off Sheet Harbor, Nova Scotia. She had been running in a heavy snowstorm with winds from the southwest and large seas. While trying to make shelter she grounded on rocks. Captain Thomas was forced to beach her to prevent her from sinking. As the water rushed in at the rate of one inch per minute, the

)(*Raising first dory off the stack prior to launch.*

men pumped furiously to keep her afloat long enough to be beached. When the craft hit rocks, she heeled over, and some of the men started to put the dories over. But Captain Thomas was not about to give up his ship.[75] He went to the railing and yelled out, "By God we're going to save this vessel, she is not in that bad a shape, come back aboard, we are going to pump." The men responded, tossing overboard anything to lighten the vessel, including 40,000 pounds of fresh haddock and cod, and 6,000 pounds of bait. They pumped through the night and into the next day, when the *Adventure* was safely beached.

Captain Thomas arranged for a big lighter to tow the *Adventure* 40 miles west to Halifax for repairs. The men pumped all the way and were successful in keeping her afloat, but she had suffered considerable damage. Her rudder was gone, part of her keel was torn away by the pounding while aground, and then there was the loss of the fish. Captain Thomas and the men spent Christmas at Sheet Harbor: They had no family gatherings, no Christmas tree, no turkey; they feasted on haddock chowder. As the days dragged on, the weather was terrible, they could hardly walk on the streets, and Captain Thomas was becoming despondent. He could not put his crew to work and they had no money. They would not speak to him, but there was nothing the skipper could do until the insurance company settled the claim. He had put up all the money he carried as master to guarantee payment for the repairs being made. It was a horrible time. His wife, Lulu, had recently passed away, and now his men had turned on him. He began drinking, so much so that when he did eventually return to Gloucester he swore off the bottle. In late January, with the repairs complete and the insurance company having settled, the *Adventure* set out once more for the haddock grounds.

Back in Gloucester, two days after Christmas, Captain Carl Olsen fit out the *Spinney* for a January 8 trip to Hawkins' Spot, on St. Pierre's Bank. In addition to loading the normal stores and personal possessions of the crew, a 100-fathom section of cable was brought on board to replace a worn section in the main, 400-fathom cable. But before the *Spinney* left, there came word of the loss of yet another halibuter. On the *Catherine*, Captain McLeod had run into the severe December weather and gone into Liverpool, Nova Scotia, for repairs. He had then gone back to the fishing grounds. The weather continued to be abysmal, but McLeod found a few good days and had somehow managed to land 25,000 pounds of haddock and cod. But again the weather turned, and his ship was iced from stem to stern and up her rigging. She would have to go to port to chip off the ice. As they made for land, McLeod could hardly see his hand before his face. Navigating through fields of drift ice, approaching the northern entrance to the harbor, *Catherine* piled into

)(*Dory fishing always required hard manual labor in often difficult conditions. Clifford Goodick baits up during a snow squall aboard the schooner Corinthian, c. 1935.*

Bald Rock Shoal. McLeod said the vessel "trembled as she smashed against the adamant obstruction, and within a half hour keeled over on her side. . . . Hardly had the last dory been filled than the *Catherine* toppled over, and the galley stove in which a roaring fire was burning, because of the intense cold, upset. The flames licked . . . at the vessel, and finally reached the oil tanks, causing explosion after explosion to rend the air." Among those who escaped was "Muggins," a Scotch collie dog given to Captain McLeod at Liverpool while the *Catherine* had been on the ways.

Once ashore, Captain McLeod telephoned his wife in Gloucester. She reported, "It was as if he had lost his best friend, for the *Catherine* was a most important factor in his life." In Gloucester, the loss was deeply felt; there was real sadness at the loss of the "largest hooker of them all." For 18 years, the *Catherine* had been a reminder of the greatness that had been Gloucester. There were fewer and fewer of these ancient schooners in the fleet, and each loss was felt across the waterfront, for no new vessels were commissioned to replace them.

Even so, Captain McLeod was not about to leave the fisheries. He was a successful halibuter, and it was only a question of where he would alight, for as long as halibut could be found in harvestable quantities, dorymen would be called on to bring in the catch. There was talk that McLeod might buy the *Natalie Hammond*, but that was not to be. Rather, later that winter he took out Ben Pine's *Arthur D. Story* on a halibuting trip. He carried as many of his old crew as wanted to go with him, and he was back in the fray.

In mid-January 1934, however, Captain Carl Olsen would once again be the first to leave for the far eastern halibut grounds. There would be seven halibuters in the fleet, down from nine the year before. During the year they were to make 36 trips, against 57 in 1933, and as one Gloucester wag said, "That's no fleet at all, but that is the best Gloucester can offer today." Whether it was or was not a fleet mattered little to Captain Olsen, for he was still in the game. The weather off Nova Scotia was some of the worst on record, with few days on which to launch dories. Not until February 12 did Captain Olsen come into Boston. But what a fare he brought back: some 70,000 pounds of halibut that he sold for 21 cents, 16 cents, and 11 cents per pound, resulting in an $8,447 stock, and a $186 share for each of the men. It was a great trip, and one on which Captain Olsen had met some of the meanest weather in his career. "Snow, snow, and more snow, day in and day out was the order of the banks, and on the way home. With bitter cold weather, and the temperature down to the zero mark, frozen fingers and frozen hands were common with the crew who were reminded of how tough winter fishing can be." It took the *Spinney* 5 days to make the 800-mile trip to the Banks, but the weather was so horrible that Olsen did not put out a dory for the first 10 days. On hearing of the trials and tribulations of the *Spinney*, Captain Jeff Thomas on the *Adventure* decided to make a few more haddocking trips before shifting to halibuting.

Captain Olsen left on his second trip on February 19, and within the week Captain McLeod took out the *Arthur D. Story* on her first halibuting trip of 1934. McLeod stopped at Boothbay, Maine, for ice and bait. There was barely a scrap of bait at Gloucester's Cape Ann Cold Storage facility—all the previous year's mackerel supply had long since been sold—but the firm still had a fair supply of bluebacks for bait, and the alewife season was soon to start.

At the end of February, Captain Thomas returned in the *Adventure* from his first trip of the year, landing 53,000 pounds of ground fish. But prices were low, and each of the crew received a share of only $12. They immediately refitted and sailed back to the Scotian Banks. Then, on March 26 came news that Captain Jeff Thomas had died. On March 24, while on Western

)(*A dory is prepared to go over the side. The men are alert lest something go amiss in lowering the boat into the sea.*

Bank, the *Adventure* had become heavily iced up, and Captain Thomas, a heavy man at 260 pounds, went out to chop the ice from the rigging with a big wooden mallet. He must have felt something coming on, for he dropped the mallet and staggered to the pilothouse, where he suffered a heart attack.[76] Death had been instantaneous, although the men on board gave him what first aid they could, putting spirits of ammonium underneath his nostrils in an attempt to revive him. But he showed no sign of life. He was gone. Three of his men carried Jeff below, placing his body on top of the iced codfish to keep it cool. He was 59. The men landed his body in Halifax, and it was brought back to Boston by steamer and thence by train to Gloucester.

When word of Thomas's death spread, the American and Canadian vessels in Shelburne Harbor lowered their flags to half-mast. When the *Adventure** returned to Boston, Jeff's son, Gordon, sailed her back to Gloucester—for one trip, Gordon was a Gloucester skipper. In addition to Gordon, Captain Thomas was survived by two daughters and a brother (Captain Peter

---

*Captain Leo Hines took command of the *Adventure* and made a great record on her as a haddocker. For more on *Adventure* and Captain Hines, see Joseph Garland. *Adventure, Queen of the Windjammers.* (Camden, ME: Down East Books, 1985).

)(*The men approach the schooner with a full load of halibut. The dory has little freeboard showing, a good catch indeed.*

Thomas of Newton). His funeral was a huge affair. A great man had passed over, and all mourned his passing. But more, the Gloucester community recognized that with each such loss the city was burying a treasured piece of its past and fame.*

With the somber loss behind the community, the rest of the year passed without major incident among the halibuters. Carl Olsen made trip after trip, before pulling out for a winter's rest on October 16. He was once again highliner of the fleet. The *Spinney* stocked $62,129 for the year, with each of her crew receiving a fine share of $1,295. With fewer craft in the fleet, halibut prices had remained steady in 1934. For that year, at least, it was good to be a halibuter.

As for Captain Archie McLeod, he made only two trips on the *Arthur D. Story*. On the first, he brought in 60,000 pounds of halibut, but the stock and share were lower than hoped. He came in just as Lent was coming to an end, and prices had dropped. He received only 12 cents for white halibut, and 7½ cents for gray. On her next trip, the *Story* put into North Sydney with engine troubles. Captain McLeod shipped 30,000 pounds of halibut back to Boston, and after repairs had been made, returned to the grounds.

---

*Jeff Thomas's funeral was out of his home, with burial at the Oak Grove Cemetery. Skippers and shipmates attended the last rites in great numbers, with Rev. Clarence J. Cowing of the Universalist Church the officiating clergyman.

)X(*Men of the schooner* Sylvania—*Captain Jeff Thomas to right, arm resting
on overturned trawl bucket, wearing his street clothes. Seated second from
left in a white shirt, smoking a pipe, is Jeff's brother Peter.*

But once again the *Story* broke down and she returned to Gloucester for
repairs.* Captain Pine then offered McLeod the *Gertrude L. Thebaud* and
the skipper accepted. He took her halibuting for the rest of the season, making six trips until pulling her out on November 6. In 1935, Captain McLeod
planned to move to another Ben Pine vessel, the *Raymonde*. The vessel had been
idle for two years, but was only five years old, and Piney was having a new 180-
horsepower Cooper-Bessmer diesel engine installed. She would be a fine halibuter.

Then came the decision by retired skipper Lemuel Spinney to commission his new craft. The *Clara and Hester* (named after Captain Spinney's two
daughters) was built in Rockport and measured 66 feet. She was launched in
late August, and he planned to take her to sea with six dories and a crew of
13. Early in October, Captain Spinney tested the compass and engine on a

*At about the same time, the schooner *Pilgrim*, which had been halibuting under Captain Cecil Moulton, went
ashore on Cape Breton and was a total loss. The crew of 23 were saved. There was no insurance on the craft.

trial spin around the harbor, but a week later he injured his fingers and was out of commission for the rest of the year. And so did the 1934 halibut season come to a close. Great men had been lost, and vessels too; Captain Spinney was out of commission with an injury, and Captains Olsen and McLeod still sat atop the leader board of the small fresh halibut fleet.

CHAPTER 35

The End of an Era

In March 1935, Charles Daley had not been home to St. Joseph's for more than two years. As a U.S. citizen, he had secured a berth on the *Spinney* and regularly sent money home to Lucy and the children—they ranged in age from 3 to 17 years. He loved his family, but there was no going home, and Charles was looking forward to the spring, for he had saved enough money for Lucy to come south from St. Joseph's: The two of them talked of the family settling in Boston or Gloucester, with Charles assuming command of a merchant ship later in the year.

In these hard times Charles had not only been sending home his wages, but he had arranged with relatives and friends to gather new and old clothes, pack them up in barrels, and send them to his loved ones in Newfoundland. Such packages were not uncommon, and the family referred to them as "Barrels from Boston": They were two to three feet across and stood three feet high. Mercedes Daley Lee vividly remembered them:

> My father used to send nice clothes home for us children in a barrel. Good things, like nice little dresses. We had the best! But the barrel had to go through the customs, and he would also place some old clothes, or worn clothes. Some of them would be good. New clothes in among the old so the customs agents would not charge him for bringing in the new clothes. There would be an old pair of pants that would belong to some one—God knows who. We would have great fun opening the barrel—dressing up and all. Old hats would come too. I must say that most of it was well made use of. Many of the things were made over after we received the barrel. We would rip them up and make little coats. It was my mother's sister who had gone down from Newfoundland to Boston (Maud Doggin). Aunt Maud they used to call her, and when in Boston my father stayed with her. She helped Charles get the barrel ready to be sent up from the States. My father would also go to visit my mother's sister living in Brooklyn. She was good, she too would help get a barrel together.[77]

But no matter how much money or how many goods Charles sent home to St. Joseph's, Lucy still had to work hard to maintain the family. For several years she ran a store attached to the house. Here she sold candy, soups, and other foodstuff, as well as supplies for fishermen. But hers was not the only small store, and Lucy made little from the effort. As for the children, they knew and loved their mother, but the younger children never really knew their father; during the 1930s, he was rarely at home. But then, out of the blue, there would come a telegram saying Charles was on his way home for a winter break. Lucy's prayers had again been answered: Charles was safe.

Once home there was so much for Charles to do, and so little time to do it. The children went out on sleigh rides with their father, listened to his stories, and saw his kindness in helping the less fortunate in the village. But what stood out most was his devotion to them and their mother. Times being what they were, he would soon have to go away to support them, but there was never a question of his love or his kindness. It was difficult to save enough money to come back to St. Joseph's, but in 1935 Charles had at last saved enough for Lucy to come to Boston. He had even sent money for Lucy to buy a new plaid suit for the trip; it hung on the back of the door as Lucy waited for spring and the much-anticipated reunion with her beloved husband.

For Steve Olsson, 1934 and early 1935 had not been much different from any other time since his marriage in 1908. He was away for two or three weeks at a time, then back home for a few days with Hilda and the three of his five children who still lived in Gloucester at the family home. The two older children, Romaine and Gerda, lived at the Franklin Square House in Boston's South End—Gerda was working, and Romaine was in school.* His third daughter, Harriet, worked at a dry-cleaning store in downtown Glouces-ter (having graduated from high school in 1933); his son Jack worked at the LePage glue factory in West Gloucester; and his youngest daughter, Olive, was still in high school.

Gerda last saw her father as she was returning to Boston after visiting her family in East Gloucester.[78] She was waiting for the bus:

> Papa, Mama, and I were looking for the bus to go down the neck [Rocky Neck, at the end of its East Gloucester circuit]. It was wintertime, and I

---

*Even though they were living in Boston, Steve remained in active communication with his two girls. Visits home and letters took the place of a hug and a chat. One surviving letter from 1934 reveals the nature of their exchange."I got home two days ago, mother has been working for the Knols', two days now I shall go out. Till the morning I was in Boston over the night, but I could not see mother because of the clothes she was doing for Mrs. Knols. I came from Rockport now, now Aunt Bertha and Svea are well, how do you have it in Boston in this heat. Although it is cooler now here I see in Romaine's letter that you had a hot time. Now you must try to read Swedish as well as you can [that was the only real language in which Hilda could write]. Mother talked to [name of a man] . . . so he shall teach Jack to drive automobile. I hope you will be home by Memorial day when I will be next home. But maybe I have been home and out again. Harriet will go to Boston on Monday. Now you must write and let us know how you are. . . . Mother will write when she has done her work for Mrs Knols."

didn't want to stand on the bottom of the hill any longer than I had to. The minute we saw it go by the house, to go down the neck and come back, of course I had to say goodbye. I kissed them both. Never dreamed I would never see Papa again.

Romaine remembered last seeing her father in January 1935, when she, too, came home to visit her parents. For Romaine, as for Gerda, it was the bus trip that stood out in her mind.

I was about to return to Boston, beginning with the bus trip from East Gloucester to the Gloucester train station. The bus passed the bottom of Mt. Pleasant Avenue, went to Rocky Neck, turned around, and wound its way up East Main Street on its return trip through Gloucester. As was the custom, my mother, father and I were looking for the bus out of the bed-room window. Once it had gone by, I would walk to the bottom of the hill in time to get the bus on its return trip into Gloucester. I can remember the bus going by, and then kissing my father goodbye, never dreaming that it would be for the last time I would ever see him.

Jack last saw his father in February, when Steve left on his final trip.

I don't remember that his leaving was in any way unusual. Father said his goodbyes at our Mt. Pleasant Avenue home. As always, he was dressed in normal street clothes. He wouldn't change into his fishing gear until he got on the boat. We didn't make anything out of his leaving, we didn't see it as any "big deal." I remember that we acknowledged each other; we kind of nodded one to the other. We seldom embraced. We had our own way of expressing ourselves; we understood one another. Or in the words of Jack's sisters, "Steve and Jack were just not that close."

Although Harriet had seen Steve as he left on his last fishing trip, she could remember no details. A neighbor, however, said she was looking out of her window and saw Steve go out the door. He stood in the lane next to his house, at his front stoop, looked at the house for the longest time, and then walked off to get the bus.

Harriet knew that her father

didn't want to go to sea—it was a job. He just mentioned it, he never actu-ally complained. . . . It must have been awfully hard out there in the cold, cold nights. That is all he knew. He didn't say much about the sea. He just said he loved the countryside better. I never heard him complain about going out to sea. He did tell us that he was going to move to a dragger later that year, but who knows. He said his ship mates called him "pop," even though he was 63 years old. I would get so mad. To me he did not seem old. Usually your parents seem older, but he didn't. I would say to Papa,

"How can they call you that? Like you were an old man." But Papa said, "I am the oldest on the boat."

As Steve left his home, Hilda watched him walk down Mt. Pleasant Avenue. She went out on the front stoop to wave good-bye, and as she did, she heard the bells from the carillon of Our Lady of Good Voyage Church drifting softly over East Gloucester. In later days, as Hilda thought back on this image, it struck her that she had never before heard their sound from her East Gloucester home.

So the 1935 halibut season had opened, and with the coming of January the two dorymen, Charles Daley and Steve Olsson, once more joined Captain Carl Olsen on the *Oretha F. Spinney*. The fishing started slowly as howling winds battered any schooner unlucky enough to be out on the North Atlantic. Many dories were smashed, decks were swept, and rigging was soon encased in ice. Only Captains Olsen and McLeod scheduled January halibuting trips. The *Spinney* rounded Eastern Point on January 23. One day later, the worst blizzard since 1926 engulfed Cape Ann. A northeaster left 19 inches of snow, with drifts up to 20 feet in height. Six days later the temperature on Cape Ann dropped to −12°F.* On February 12, the *Spinney* made Boston with more than 60,000 pounds of fresh halibut. The stock was $7,300; each man received a fine share of $154, and the trip was the talk of the waterfront. Perhaps 1935 would be better than 1934. Four days later the *Spinney* left on her second trip of the new year. It was cloudy, cold, with a falling temperature and moderate west to northwest winds.

On March 6, Captain Pine received a telegram from Captain McLeod at St. Pierre and Miquelon. The weather to the east had been unusually severe. In three weeks McLeod had made only one set. On March 4, moderate gales moved from the west to northwest, accompanied by snow and low visibility. Blizzard conditions extended across all of Newfoundland. The storm over Newfoundland at last passed out to the Atlantic on March 6, but lingering fresh northerly winds and snow continued on the south coast. The winds on March 7 were moderate northwest to north, with moderate to good visibility. An area of high pressure and cold temperature extended from the New England states northwest to the Hudson Bay. As the weather improved, Carl Olsen called on his men to launch their dories on Hawkins' Spot. The

---

*A letter from Steve to Hilda let the family know he was all right—"Dear Hilda and children, I would like to let you know that we are alive, although it looked dark for us in this dangerous storm that we went into this night and today we arrived at 10 o'clock in Edgartown. We were away from land and it blew hurricane and thunder, so it looked like it was very deadly threatening. We were scared. I cannot explain in this card how we had it, or felt, so [who would have thought] I would go toward death yesterday when I left the house. Everybody believed that he would go to Provincetown, but no, rapidly a change. I shall write when we get time."

seas were rough, but not so rough as to cause the men to refuse the skipper's order to set trawl. In fact, like the skipper, they considered it a good winter fishing day, even though there was mist on the grounds, making it difficult, but not impossible, for the men to see one another after leaving the *Spinney*.

Charles Daley and Steve Olsson soon set their trawl, landed a partial catch, and appeared to have been in the process of bringing in the rest of their catch when they were lost. Unseen by the skipper, or by any of the other dorymen spread out across Hawkins' Spot, Charles and Steve's dory was upset and the two men disappeared under the waves. The next day, the overturned dory was found about one and one-half miles from where the two men had set their trawls. All its gear, except the oars, was still on board. One mitten, later determined to belong to Steve Olsson, was found caught on a trawl hook—a mute witness to the tragedy that had unfolded as the two men worked alone in their dory. It was the only physical remains ever found of the two lost dorymen.

Charles had apparently been working in the bow, pulling the trawl line from the seafloor below. Steve was working in the stern of the dory, coiling the trawl line that floated on the water behind the boat, and removing whatever fish had been brought up. They had been at work all day, having gone over the side early in the morning. Captain Olsen had swung by during the early afternoon to take off the fish Charles and Steve had already retrieved. The skipper then sailed off to pick up the catch from the remainder of his dories. As he left, he saw Daley and Olsson return to their trawls. When the *Spinney* later returned to pick up the dory and call it a day, Captain Olsen found only the buoy line at the spot where Charles and Steve had last been seen. There was no dory in sight.

Olsen began a wide search for his lost men, steaming back and forth over the grounds. But it was late in the day, and as the sun set the dory still had not been found. A double watch was set on the deck of the *Spinney*, the men straining their eyes across the empty sea in search of the two missing men. It was not until noon the next day that the overturned dory was found.

When the dory was taken on board the *Spinney*, the painter was found made fast to the plug strap, suggesting the men had been alive for some time after the dory overturned. The skipper surmised that the dory must have been hit by a wave, which capsized the boat and threw the men into the frigid waters of the North Atlantic. It also seemed likely that they fought for their lives. They made fast the painter line to the plug strap and clung on in hopes of the *Spinney* reappearing on the horizon. But in the icy cold water, and knowing that the *Spinney* was running down to pick up her full line of dories, the men must have realized there was little real chance they would survive. It would not be long before the cold water took its toll and they lost their grip. Only Steve's mitten remained. How long they struggled, what they thought, or how they felt would never be known.

Some among the crew suggested that one of the men might have caught his hand in a trawl hook and that a wave hit the dory just as he was leaning over to release the hook. Both men were clad in oilskins, hip-high rubber boots, and the usual heavy outfit of a doryman; they had little chance to battle the icy waters and the long swells, even if they had been able to hold on to the plug line and crawl onto the overturned dory.

Captain Olsen and his men hauled in the trawl line that Charles and Steve had set, thinking that one or both of their bodies might be caught in it. But none was found. The skipper steamed back and forth across the grounds all that day, but when another gray dawn came and still the sea was empty, it was time to head home. The *Spinney*'s foghorn was sounded one last time, but there was no answer, and at dusk on the second day the *Spinney* turned for Gloucester.

News of the tragedy first spread when the *Spinney* came into Boston at 1 P.M. on March 13. It was said that the "waterfront was saddened on learning of the fate of the two men . . . for they represented the highest in the tradition of the sea." One day later, newspapers in Boston, Gloucester, and St. John's printed the first accounts of the loss. The *St. John's Daily News* noted that Charles Daley was from Newfoundland, reporting only the bare facts of his loss and the *Spinney*'s search for the missing men. It was not long before the news reached the Daleys in St. Joseph's and the Olsson's in East Gloucester.

In St. Joseph's, the first visible sign of the tragedy was the appearance of a mailboat coming up St. Mary's Bay with its flag at half-mast, a sign whose meaning was clear to all. There was a telegram on board for someone in the village, and it bore the most chilling news. Meanwhile the postmistress, Mrs. Elsie McCormick (a good friend of Lucy Daley's), had just opened the local telegraph office and received the more immediate news of the loss. It fell to Elsie to carry the news to Lucy Daley. Not wanting to go alone, she phoned the parish priest, Father John Enright (who had come to St. Joseph's in 1919 from Limerick, Ireland), saying she was sure Lucy would want him to be there when the news was heard. The two then walked together along the shore road toward the home of Lucy Daley, first stopping at the home of Christine (Christy) Daley, who lived next door to Lucy. Christy was the wife of Charles's first cousin, Maurice Daley. Christy and the postmistress then led the way to Lucy's, the priest holding back. He was a "strict old Irishman," a man incapable of comforting those in pain, a man who truly "hated to break the news."

As they approached the Daley home, Lucy was standing in the threshold. Before Christy could speak, Lucy said, "It's Charles, isn't it?" Then, without saying a word in response, Mrs. McCormick passed the unopened telegram to Lucy. As Lucy read the contents she started to scream, and the postmistress held her in her arms, saying ever so gently, "Now, now, Lucy, remember you have little children depending on you." As for the priest, he could find no comforting words for the sorrowful widow, and soon turned on his heels and retraced his steps to the safety of his church.

Later, Lucy told Christy she had had a premonition that Charles was gone the previous night. She awoke, went into the room where her twin boys were sleeping, and awoke John and Alban (who were then about 13), telling them she feared something had happened to Charles. Lucy said she had been troubled for the past week, since the very day on which Charles had died. For on that day, Moore Daley, a cousin of Charles's, reported seeing an apparition as she emerged from the woods on old Daley's Road (no more than one-half mile south of Lucy's home). The apparition was none other than Charles Daley; he went across the end of the marsh in his oil-clothes—the clothes that he was wearing the day he drowned on St. Pierre's Bank.*

Mercedes Daley Lee, Charles's youngest daughter, a girl of seven at the time of her father's death, remembered well the day the telegram arrived. "It was a real stormy, cold, cold March day—I will always remember the stormy day when I came home from school, with so many people standing around in our home. Helen McCormick had come to school to get me, for she was a good friend of mom, and they let all of us out of school." However, neither Helen nor anyone at the school told Mercedes that her father had drowned, and when Mercedes entered the house she found her mother and everybody else crying. Mercedes was struck by her mother's grief. She had hardly ever seen her mother cry. Lucy and the children may have said the rosary every day for Charles's safe return, but now to actually lose him, to know that Charles would never return, Lucy could not be reconciled to the new reality. Mercedes remembered next going upstairs. "I wanted to cry like everybody

---

*Of course, at this time one also has to remember that there was a strong sense of faith and belief in this Irish-Catholic community. The spirits were real, as was a strong sense of superstition. Joe Dobbins (in a taped interview in 2003) remembers the St. Joseph's of the 1920s and 1930s as a religious community, in which there was a lot of superstition. Many people who traveled through the woods kept a piece of bread in their pocket to keep them from the fairies, and Father Enright believed in ghosts. But there was no denying the fact that this was the day Charles was lost on the Grand Banks.

else, but I didn't." What stood out for Mercedes was the plaid suit Charles had sent home for his wife's coming trip. "It was to sit on back of the door for years, that was her suit she was going to wear to go down and see Charles."

Mercedes also recalled that it was not long before one very physical remnant of Charles's last voyage was to return to the family, his "Labrador Box." It was about 24 inches wide, 24 inches deep, and 3 feet to 3 feet 6 inches in length, and made of pine. It was a place for a sailor to keep his belongings—clothes, tobacco, rum, and personal reading matter. The crew of the *Spinney* packed Charles's box and sent it to Lucy. The box remained in their house for years—Lucy would not part with it or its contents. Inside were the mittens Charles had worn, his protective wristbands—or pups—his oilskins and Waltham watch. "Mom had given it to him, and it was engraved on the back." The trunk became a sacred object, and Mercedes often peeked into it, trying to remember the man she never really knew. As time went on, Mercedes said, it seemed funny that she really didn't miss him. "That is so strange to say, but I really didn't know him. I was so young and he was gone so much. I was so young. All I can remember are people talking about my father. They always described him as a 'good' man. Every time I heard anything it was something good. He was a good worker."

In East Gloucester, Steve's daughter Harriet was the first of the five children to learn of his loss. She was working at the dry cleaners on Main Street, alone on the counter, when two policemen came in. Harriet thought:

> "Oh God, what did I do now? I was so nervous when they came in." They asked, "Are you Harriet Olsson?" I said, "Yes." Then they said, "Well, you have to go home." I said, "What?" They said, "Your mother needs you." They could not tell me that Papa was lost. All I could think about was that my mother was sick. I never thought about my father, not once. We were so used to Papa going to sea, coming and going. So, I said, "Well, OK, but I've got to lock up, I've got to wait for my partner to come." I was going to call her up at her house, and then I would get a bus when it came. They then said, "Don't wait for a bus, take a cab." I've often wondered why they didn't just take me home. Those days I guess they didn't do things the way they should. But next door was a store where this young fellow sold cards and stuff, and he was a real nice young man. He had seen the police come in and he called them over and said, "What's going on? What happened?" And, they told him, and they went. He then came in the store, and he said, "Gee, I hear the police say your mother needs you." And I said, "Yes, I wonder what's the matter?" I felt so worried. I said, "I've got to get home somehow." He said, "I'll take you home, I'll close up my store." So he took me home. Afterwards, he told me that the whole way to 143, he felt like

dying. I kept saying, "I wonder what's the matter with my mother?" Over and over and over. And he didn't have the heart to tell me.

So we got to Mt. Pleasant Avenue and there was a car out front. And this is how blank I felt, I could see that it was a police car but I thought it was a nurse or doctor's car. I really didn't take it in. I banged on the door. This guy opened it up 2 inches and said, "Who are you?" And I said, "I was told to come home, I live here." I could hear my mother crying out on the small sun porch off the dining and living rooms. That guy then told me what had happened. Right then he told me, I didn't get farther than the door. Then he said, "Is there anything I can do for you?" I just looked at him and I said, "No." And he said, "Will you and your mother be all right?" I said, "Yeah [softly]." I didn't know what I was saying. He said, "OK, so I'll leave now." He was glad to leave, because it made him nervous, and I don't blame him.

There was a neighbor in with my mother. My mother was on the small sun porch with her, and Mama was crying. Then we started to make phone calls. Gerda and Romaine came back from Boston on the train. At that time, Jack was in the bus coming home from LePage's where he worked in West Gloucester. He told us afterwards that it was queer that people were looking at him and were talking very low. Then they would look at him again. He thought, "I wonder what they are doing? What did I do?" And, all the time, they knew what had happened because the paper had come out with the story of the drowning of Steve Olsson and Charles Daley. They knew, and Jack couldn't figure out why they were looking at him. This was late in the afternoon. It was early March. It was cold and windy.

When I told Jack our father was drowned, Jack ran all through the house almost out of his mind. He thought the body was there, and he was saying "Where is he? Where is he?" He thought they had recovered the body. He was lying on the bed. Crying and sobbing. He was about 17 or 18 then. He really went right to pieces.

Then Olive came home. She had gone to a friend's house instead of coming right home after school—just like Olive. She came in, and many people were there by then, including Harry Cluett [the man who later married Gerda]. So everyone just looked at Olive. Poor thing, she was looking at us and we back at her, and Olive thought that she was getting these looks because she should have come home right after school. So Harry said, "Does she know?" And we shook our heads. And the sad part was that it was Olive's birthday a few days before and she had saved a piece of cake for Papa, for we all knew he was due home soon. She went right up stairs and none of us knew what to do. So Harry said, "I'll go up to talk to her," and that's who told her. We knew she would break down, that was Olive. She really broke down. She lost it to a point. I don't remember how Mama handled the rest of the day. We were all there. I think she just quietly went about what she was doing. I suppose my aunt Bertha came and was with her.

Meanwhile, Gerda was working at the Hotel Continental in Cambridge, and the telephone rang and it was Harry Cluett. He said, "Come home as quick as you can. Get hold of Romaine and come home quickly." She said, "Why, what's the matter? What's happened?" Gerda first thought something might have happened to her brother Jack or one of the kids. Harry said, "I am not going to talk about it now. You just come."

And then Gerda right away knew who it was.

I had a nightmare several days earlier, a week earlier. It was on March 7th, the day Papa drowned . . . and in the nightmare it seemed that some awful looking man came into my room and started to do something and I just woke up a wreck. I was so scared. So I was saying the next morning when I was going to work, I used to go with this girl who lived at the Franklin Sq. House. She said, "What's the matter? You don't seem to be yourself. You're quite nervous and downhearted." I said, "Well I had a terrible nightmare last night and I can't seem to get over it."

I must have called Romaine. I told her to meet me down at the North Station to get the train home. All the way home Romaine was saying, "What do you think happened?" I said, "I don't know," but I really knew inside of me what it was. So Romaine was saying, "Do you think something happened to Jack? To Harriet or to Olive?" When we got home, we took the bus to East Gloucester. We always took the bus. We got off at the foot of Mt. Pleasant and walked up the few feet to the house. We could hear Jack upstairs crying his eyes out. He was lying on the bed upstairs. So, I guess we knew then what happened. It was pretty horrible.

Then when Romaine and I got back to the Franklin Square House I could hear Romaine crying every night. She was two rooms down from me. I never went to her or tried to cheer her up.

There were brief memorial services, for Charles in the Catholic Church at St. Joseph's, and for Steve at the Episcopal Church in Gloucester. The Gloucester remembrance was a low-key affair, the priest mentioning the loss of the two men at the end of the 10:45 service. He also offered a remembrance of Steve's 87-year-old mother who had passed away while he was at sea; a remarkable coincidence. Steve loved his mother, and Hilda and the children had dreaded having to tell him of her death. Steve had not seen his mother for 42 years, but he wrote to her after each of his trips, and she wrote back to him. He had harbored a hope that one day he would again visit Gullholmen, once more see his mother. It was not to be.

Epilogue

The loss of Charles Daley and Steve Olsson shook the fishing community, but more was to come that year. As early as mid-March there was concern for the late return of the *Arthur D. Story*. She had left Gloucester, bound for Newfoundland, on January 30. Once there she had loaded 1,120 barrels of salt herring from various ports along the south coast. On March 1 she sailed from Belleoram, only to be driven back by heavy weather. She tried again on March 3, and there was speculation that she was caught in the ice fields, which extended for miles along the fishing banks. Her seven-man crew was well stocked with food, and as Gloucestermen often did, they could obtain water from the snow atop the ice. But, with the severe gales and cold weather of early March, there was genuine concern for the vessel and her crew.

On March 18, when Captain Albert Williams brought the schooner *Imperator* in from a halibut trip to the same grounds, he reported that since leaving on February 14, he had experienced heavy winds and bad weather. He was forced into St. Pierre and Miquelon on March 7. Some of his dories had been smashed in the gales, with four lost in the gale of March 4 and 5.

Also coming in at this time was the schooner *American*. Captain Simon P. Theriault described the weather as being as bad as he had ever survived; had it not been for the parting of his cable his craft would have gone to the bottom. As it was, his vessel was at the mercy of the "hurricane" for 44 hours, the wind carrying them along as if the schooner was a piece of kindling. Captain Theriault felt that if Captain Nickerson had all sails set, he had a chance.

Then came a glimmer of hope. At the end of the month, a Norwegian steamer reported a vessel caught in the ice off Cape Race, 150 miles east of Sable Island, near Hawkins' Spot, or perhaps just west of Hawkins' on the westerly end of St. Pierre's Bank.

But no more was heard, and on April 9 the owners of the *Arthur D. Story* officially gave up hope for her safe return. It was believed that she was caught in the hurricane of March 4 and 5, and was in the shoals of St. Pierre's Bank.

The annual fishermen's memorial service that July was especially poignant. It began with a procession from the downtown quarters of the Fishermen's Institute to the causeway where the fishermen's memorial statue looked toward the outer harbor. The procession included families and dignitaries, Boy Scouts and Girl Scouts, ministers and bands. At the fishermen's memorial, George E. Russell, chaplain of the Fishermen's Institute, called out the names of each of the 13 men lost that year—Ralph Capucco, Charles Daley, Ralph Fiander, Morris Fitzgerald, Steve Olsson, and the other eight who had been lost at sea. Captain John F. McKinnon, the retired president of the Master Mariners' Association, placed a wreath at the foot of the statue, while Mayor George H. Newel dropped another into the water. The people then walked the few hundred yards to Marine Park, at the Blynman Canal cut of the Annisquam River. There they sang hymns, pastors said prayers for the lost men, and the sound of a bugle floated over the water. There was a second reading of the names of the lost, and as each was read a young child came to the rail to scatter flowers in honor of the fallen fisherman. Hundreds were present. The youngest children to throw flowers were the four-year-old boy triplets of Captain and Mrs. Wylie C. Rudolph; the boys cast their flowers in memory of the late Philip "Flip" McCue. When the name Steve Olsson was read out, it was Olive Olson, Steve's 16-year-old daughter, the youngest of his five children, who "stepped forward to cast her floral tribute to the waters in his memory"—as she would do every year for the rest of her life.

As the Olssons and many others in the community remembered the men who had been lost, the few remaining halibuters continued their familiar cycle of trips to the Banks. By year's end, Captain McLeod had made 7 trips, while Captain Carl Olsen had made 11. Olsen was again the highliner, stocking $48,397, with each of his crew receiving a share of $925—a significant drop from 1934. The *Spinney* had been active for 229 days; her longest trip lasting 27 days, her shortest 15 days. Carl Olsen had left early in the season, found the best of grounds, and made more trips than any of the other halibuters. As Billy Shields noted, "Carl Olsen was the best halibuter. We used to make two trips while Archie McLeod made one. You couldn't beat Carl in halibut, he was a 'square head,' a Norwegian. . . . He was a good fisherman. Couldn't beat him."[79] But the men working under Captain McLeod developed a true bond with their gentle skipper. Thus, when the *Raymonde* ran into a piece of field ice, ramming a giant hole in the bow below the waterline, he was able to prevail upon his crew to stay and fight against what appeared to be insurmountable odds. The men worked hour after hour against the onrushing water, some manning pumps, others using whatever they could find to help plug the hole, and the *Raymonde* was saved.

 The schooner Oretha F. Spinney in 1935. Her bowsprit was a later addition fitted for her starring role in the Hollywood movie Captains Courageous.

Toward the end of 1935, representatives of Metro-Goldwyn-Mayer (MGM) came looking to purchase a schooner to use in the Gloucester-based movie *Captains Courageous*. The ship they ultimately purchased to play the *We're Here* was the *Oretha F. Spinney*.* Captain Olsen sold her for $25,000, on the very day that Everett James, the owner of the yard where she was built, died.† The *Spinney*'s sale was yet another blow to Gloucester's fishing community. As the highliner of the halibut fleet, she had been the finest deep-sea fishing vessel remaining in Gloucester, and she was never replaced. Before sailing to California, she was refit to resemble a 1895 schooner. Workmen knocked out the forward stump, inserted a long bowsprit, and added the topmasts and rigging for summer sailing.

For the long voyage to the West Coast, 6,000 gallons of fuel oil were stowed in her tanks, and Captain Olsen's trawl gear went below. Then, on January 8, at a time when Captain Olsen should have been fitting out for winter halibuting, the *Spinney* left Gloucester for the last time. Many mourned her departure; the remaining fleet sounded their whistles in salute as she left. The master of the *Spinney* for the 33-day trip to San Pedro, California, was Captain James A. Hershey of Yarmouth, Nova Scotia; the crew consisted of veteran North Atlantic fishermen. Ironically, the boatswain was Olaf Olson, a man who had served on the *Spinney* with Steve Olsson and Charles Daley; he had been born in Steve's home village of Gullholmen, Sweden.§

As for Captain Carl Olsen, he remained a Gloucester halibut skipper, although he now took out the vessels of others. Folks around the waterfront had speculated as to which firm would be fortunate enough to land the halibut highliner, and in mid-January Captain Olsen made his decision. He took out Ben Pine's schooner *Raymonde*. Piney had offered Carl the choice of any of his fleet, including the larger *Gertude L. Thebaud*, but the *Thebaud* was being skippered by Archie McLeod, and Olsen refused to bump his friend

---

*The movie was released in 1937, and Rudyard Kipling's classic adventure story of a young, rich lad who fell from an ocean liner and was rescued by a Portuguese fishermen from Gloucester enjoyed great success. Spencer Tracy won the Oscar for best actor for his role in the movie. The film is available on DVD and VCR and offers insight into life on board the schooner from which Charles Daley and Steve Olsson were lost.

†Everett James was 72 years of age. He built vessels in Essex from 1912—when his father brought him on as a partner—until he launched his last vessel in 1932. In this time, J. F. James & Son launched 72 vessels—48 of which were under Everett's sole direction.

§After the movie was filmed, the *Spinney* was tied to a dock in Long Beach, California, and lay forgotten and neglected. In 1941, the actor Sterling Hayden took notice, and in his book *Wanderer*, published in 1963, he describes his first contact with the *Spinney*. "I vaulted a fence, ducked a watchman, and made my way down the dock. She was a big Gloucesterman, the *Oretha F. Spinney*. . . . Her spars were tall. I paced her deck and I fondled her wheel; I sprang the lock on the forward scuttle with a rusted spike and sat for a time in the fo'c'sle. The stub of a candle lay on the galley pump. You could smell the old Banks aroma of oil and gurry mixed with tar and a touch of rot. . . . I rolled into a starboard bunk and I dozed. . . . I just had to have the vessel. Right now!" Have her he would, first taking her to the Caribbean and later selling her to new owners in Venezuela.

from the vessel.* He appreciated good, honest skippers, and there was none better than Captain Archie McLeod.

Near the end of October 1936, Captain Olsen, coming in from his eleventh and last trip of the year, was again the highliner of Gloucester's small fleet of halibuters. By 1937, there were only four halibuters in the fleet—*Raymonde*, *Thebaud*, *American* (Simon Theriault), and *Imperator* (Albert Williams). The lure of the halibut fishery had passed, and with it Gloucester had lost the last of its dorymen. Few men could stand the rigor of the work, or the long trips with comparatively poor reward, particularly as other more lucrative fisheries were coming to the fore in the late 1930s. It had always been a perilous fishery, and now there was little chance of getting fair pay for a week's work. Yet, Carl Olsen never regretted his decision to stay in halibuting or to take out Piney's *Raymonde*.†

As Gloucester's halibut fishery sputtered along, the small cadre of skippers brought in an ever-declining catch—dropping to less than 1 million pounds per year toward the end of the decade. Carl Olsen remained in the *Raymonde* until 1942, when he had the distinction of being the last of Gloucester's dory halibuters. While others went into dragging, he continued going to sea with a select crew of Nova Scotian and Scandinavian dorymen. He had some success, but his personal life had changed for the worse. As Carl Jr. said, "Oh Lord, did he like to drink. . . . They tell me that there were times when they came in that they had a favorite bar room and they would close the bar room except to the crew. That was the style of the day, we had more bar rooms than we had fisheries. Didn't impact on his fishing, cause when you went out you sobered up by the time you got to the fishing grounds." Captain Billy Shields says much the same of the man and his fishing skills: "Carl was the king of the harbor. I made a lot of money with him catching halibut, but . . . Carl Olsen was a drunkard on land. He was a good fisherman. You couldn't beat him when he was sober. Carl Olsen was the best halibuter. He wasn't dumb, he was just a drunkard, it was too bad."§

---

*Carl Olsen Jr. describes the relationship with Piney: "My father wasn't a friendly man, and Ben Pine was not a friend. Ben was a business associate. They were making money in the *Raymonde*. It was not unheard of that they would come in from a trip and tie her up and Ben would say, 'You know Carl the dories they don't look too good, I think they have seen better days. I got a set of dories coming for you.' And they would put a set of dories on the boat. Pine would put the old dories on his boats that were not making money. That was the times."

†Piney now had the finest rosters of highline fishermen in the port: Carl Olsen on the *Raymonde*; Archie McLeod on the *Gertrude L. Thebaud*; Mike Clark on the market fisherman *Ruth and Margaret*; Frank Foote on the southern dragger and mackerel seiner *Old Glory*; and Cecil Moulton on the fresh fisherman *Cape Ann*.

§Carl was far from unique in having a drinking problem. Though Gloucester reformers had tried, they had had little effect. When ashore, many of the men drank, and that was it. Harry Eustis, in the Jeff and Gordon Thomas tape series, tells the following humorous story that illustrates the point. It seems that as one of the fleet was getting ready to go to sea this fisherman, a little guy named Smithy, comes up to the skipper and the following conversation ensues. He comes up to the skipper (Big Brad) and says, "Captain." "Yes." He looks down, "Yes

Steve Olsen's description of his grandfather Carl in the 1940s is of a man who stood about 6 feet 3 inches, weighed about 340 pounds, and although not totally bald, was losing his hair. He had a big, red, bloated nose, as a result of a cancer contracted after he quit fishing. He also had unbelievably large hands—he was known for them. His son Carl Jr. said that no matter whether at sea or at home, Captain Olsen was a "son of a gun to work for." He always asked much of himself and much of others. Carl Jr. never feared his father; Carl never struck him. He went on to say, "I had complete confidence in the man. He had very strict standards."

In 1942, Carl Olsen became the last of the Gloucester halibuters.

> The government set the price of fish through the OPA [Office of Price Administration], and the OPA set the price on halibut, which was what they were fishing for, going to the Grand Banks. They set the price on halibut on the Pacific Coast. The Pacific halibut was smaller and a different fish, and traditionally sold at a lesser price. And when they set the price, he took most of his fish out in Boston, and he told the people of Boston don't get the idea that you can have a windfall because we can't operate at that price. Either you work out a method with the government to get a substantially different price or that is the end. And that was the end. That was the last halibuter to leave the shores. He went out on other vessels after that, but that was the end of the halibuting.

When Captain Olsen retired, he went to work for the fish companies in town, taking any job that helped him to get the "quarters of time" required for him to receive Social Security; he had not contributed to the system when he was fishing. He was also to derive income from his three-family home on Western Avenue, near Stage Fort Park. Captain Olsen died in 1965 at the age of 78;* Captain Archie McLeod (MacLeod) died two years later at the age of 84.† They had lived long beyond the age of the halibuters, dorymen, and

---

Smithy what is it?" He says, "Captain do I have enough time to go up town (meaning in Gloucester from the pier) to get a pint." He didn't say a pint of whiskey; he just said a pint. Big Brad looked at him. He got a little aggravated and said, "No, you don't have time. We are getting ready to go out to sea. Get back to work." So little Smithy takes five to ten steps away from the skipper. Stops, comes back. Comes in back of the skipper and says, "Captain." He looks down at him and says, "Yes Smithy, what is it?" "Do I have time to go up town and get a half pint?"

*The death notice, appearing in the *Gloucester Daily Times*, said, "He was the last and one of the greatest of the halibut fishermen who once made Gloucester the world's leading halibut port. He was a hard-driving captain of the old school, when Gloucester men baited long trawls and hauled them in hand over hand in two-man dories."

†Archie was a highliner in haddocking and halibuting, and his death notice indicated that "for 18 years the combination of Archie . . . and his schooner *Catherine* (launched in 1915) was practically unbeatable in fishing competition. He was born in St. Peter's, Cape Breton, Canada, May 11, 1883, son of Captain Norman and Catherine (MacCuish) MacLeod. He arrived in Gloucester as a youth of 16 and went fishing. He became a skipper 10 years later on the *Electric Flash*. Among the craft on which he was master were the *Agnes*, *John Hays Hammond*, *Hortense*, *Louise R. Silva*, *Georgianna*, *Bay State*, *Arthur D. Story*, *Gertrude L. Thebaud*, *Dawn*, and *Marjorie Parker*. After Archie retired as captain, he went as hand and was watchman on vessels in port. For the last two years he has been a patient at the Huntress Public Medical Center in Gloucester."

schooners, and the Gloucester of their old age bore little resemblance to the Gloucester of their youth.

By early summer 1936, the future of Gloucester's fisheries hung in the balance. It did not look good. Summer boarders driving along the main road around the harbor no longer had any reason to object to the "fish smell" coming from the wharves. The days of plenty when salt fish filled the wharves were in the past. There were even those who missed the odor. Yet, there were also the first signs of an economic revival that would sustain the Gloucester fisheries over the next two decades. Captain Benjamin Curcuru, manager of the Producers' Fish Company, placed an order in Maine for two 90-foot fishing vessels at $15,000 each. Curcuru held down the cost by using second-hand engines. A few months earlier, the Producers' Fish Company in the Fort, managed by Captain Curcuru, added a 2-million-pound capacity to its cold storage plant. At the same time, the Gorton-Pew cold storage plant on Rogers Street increased its capacity by 25 percent—from 4 million pounds to 5.3 million pounds. Both firms believed that with their increased capacity, more craft would come to Gloucester.

Gorton-Pew Fisheries did, indeed, see a substantial growth in its business, hiring an average of 836 men and women each week during the peak summer season. There was work on mackerel, whiting, herring, and most recently, redfish (or rosefish, a bottom-feeding member of the rockfish family, growing to 20 inches in length). Fish such as whiting and redfish had had little economic value in the past, but with a reduction in the fresh haddock catch and the expansion of firms with frozen filleting operations, they now became central to the prosperity of the Gloucester fisheries. Gorton-Pew and others developed a truly national market for their frozen fillets, and by 1936 the market had moved into the Midwest and South. Meanwhile the newly emerging redfish landings in Gloucester translated into real jobs on the wharves, with a 90,000-pound fare requiring $500 in labor just to handle it, and then keeping others busy filleting and freezing.

During the early and mid-1930s, Gloucester firms had laid the foundation for this new fishery, expanding their freezing and fresh-fish filleting processing capacity, following the lead of Clarence Birdseye. In the late 1930s, Gloucester and Boston craft came to Gloucester in increasing frequency. Some came when Boston fresh-fish prices were low, but for others Gloucester became the primary port for off-loading their catch. Gorton-Pew invested in an up-to-date quick-freezing and filleting plant, and thus became increasingly aggressive in buying fresh fish. The plant sat beside the Gloucester Cold Storage facility on Rogers Street, and would become quite busy when a ship pulled into port. As fish were off-loaded they were carried to the top of the processing building and then run through several filleting machines and packaged. They then went to the quick-freezing room, coming out neatly packed and frozen, ready for shipment.

The Gloucester fisheries began to turn around in the late 1930s. Frozen fish could reach New York in 10 hours, Chicago in 60 hours, and Denver in 110 hours. Redfish landings came in at 29.4 million pounds in 1938 and rose to 50 million pounds in 1939. Large refrigerated trucks carried filleted fish in attractive packages as far west as Denver. By 1939, 60 of Gloucester's 132 fishing craft, all draggers, participated in this fishery. With the explosion in redfish, fishermen began to earn larger and larger shares, up to $2,700 by 1939.

While the fisheries of Gloucester began to rebound, the widows of Charles Daley and Steve Olsson led difficult lives. Hilda continued to take in laundry over the summer of 1935 from the Fairview Hotel in East Glouces-ter. As one of her children was to say: "All Mama did was work in the sum-mer time doing that lousy laundry. I always hated it so much. . . . She worked hard. In fact, she worked too hard. This man [Joe Manning] said a couple of years ago, 'My mother-in-law and your mother were the hardest workers I ever knew.' That statement did not make me happy. I would much rather have him remember mama in those years as somebody who was more carefree. I can remember going to the beach, or sometimes when I had a job I had to walk past the beach up Farrington Avenue and I would see all of these people lying on the beach, enjoying themselves. But there would be mama home, ironing—washing and ironing. I would get really upset. Why wasn't she down the beach having fun?"* Hilda owned her own house, but she needed money to live.

Working in Boston, Gerda made barely enough money to survive. Gerda was taking in about $18 a week, while Romaine was going to Bryant and Stratton secretarial school. Each week Gerda paid $6.75 for her own rent and $6.75 for Romaine's rent at the Franklin Square House. The rest of her meager wages went for food and the odd "chocolate bar," leaving a dime or two to go to the movies. There was nothing left to send back to their mother, no matter how much they loved her.

---

*When alive in their late 80s, providing insights on the lives of their parents, my aunts Gerda (Olson) Cluett and Romaine Olson said they both felt sad to have to report that their mother (my grandmother) had to do other people's laundry. They asked that I leave this fact out of the story. Yet, I find this part of my grandmother's story indicative of her strength, resolve, and commitment to her family. Dear Aunts Gerda and Romaine, please forgive your loving nephew.

Four years after Steve's death, Hilda began to receive public support funds as the widow of a Gloucester fisherman.* Steve had not believed in life insurance. He told his children, "If I take life insurance, I'll die." He hated to hear anyone talk about insurance. His brother, who was in the insurance business in Sweden, wrote to him several times suggesting that Steve take out a policy, but to no avail.† His children continued to push Steve, at least to the point where it was obvious to them that this gentle man was thinking, "You want me to die!" Then they backed off. But he may finally have seen the sense of it. For that last winter he said, "When I come home from the next trip I will sign the papers." But he didn't come home from that trip, and there was no insurance to help Hilda carry on. There was no money, and even Steve's daughter Harriet, who lived at home and worked at the laundry, was bringing home little more than $10 a week.

The fall after Steve's death, Hilda found herself with no option but to go to Sewickley, Pennsylvania, to work for one of the summer families, the Bakers. She was having a hard time keeping up with the taxes on her home and meeting the needs of her children. The Bakers had a big house on Eastern Point, and when Hilda returned to Gloucester she continued to cook and clean for them in the summer season. Hilda stayed at her Mt. Pleasant home until 1940, when she sailed alone on a Grace Line steamer to Panama to join her daughter Gerda, Gerda's husband Harry, and their son Steve. The captain of the boat came from East Gloucester, Captain Swenson, and Hilda knew him. He took her under his wing. From then until the end of her long life, she lived with one or the other of three of her children, Gerda, Harriet, and Romaine. She was always a positive force in the lives of her children and the grandchildren she helped raise. We all loved her.

As for Lucy Daley in St. Joseph's, following Charles's death, his family, too, was hard-pressed to meet the needs of everyday life. Their one saving grace was the rural nature of the village and the presence of food from the

---

*Disbursements of cash to beneficiaries by Gloucester Fishermen's and Seamen's Widows' and Orphans' Aid Society: 1935, $3,592 (fund balance $74,665.46); 1936, $2,616; 1937, $2,103. Disbursements ranged from $6 to $24; almost all were for $24. Hilda Olsson got from fund: 1939, $24 (must be for month); 1940, $40; 1941, $12; 1942, $11; 1944, $25; 1945, $40; 1947, $38; 1948, $42; 1949, $38; 1950, $41; 1952, $43; 1954, $47.

†Letter from Albert Olsson, Steve's brother, in the late 1920s: "Surprised Steve that you didn't have such insurance. When you are younger the premiums are cheaper and you are more often healthier. According to our tariffs the highest age to get insurance is 55 years, which you are next birthday. But after the premium would be much higher. . . .The company accept even if you are more than 60 years but of course the premium is high. In that case you have to write to the company and write down how much the premium will be. If you will I can make myself informed about this, so write to me soon, or as soon as you can. An American citizen can have insurance in Sweden. Your kids are still young, cannot one of them take some kind of insurance? So the insurance will be low. Get back to me about this at the same time that you write about your own insurance."

garden, chickens, and pigs. There was no support from the Provincial government, for in 1934 it had gone bankrupt and Newfoundland had reverted back to British colonial status, becoming a Crown colony in Charles's last year of life. Charlie Daley left school at 12 in order to go to sea to help the family. By age 16, he was on his way to Europe with but $2 in his pocket. Three other sons of Charles Daley also went fishing. They went out with their uncle James Willie on the *Shamrock* to help feed the family. James Willie, brother of Charles Daley, was a top Newfoundland captain and a positive force in the lives of Charles's children; he lived to be 94.

Members of the family differ in their remembrance of whether they received aid from the States following Charles's death. Mercedes says the American Fishermen's Union was "good" to the family—providing help for probably a year—although she does not know exactly how much they provided. One of her grandchildren said Lucy received the equivalent of a pension from Gorton-Pew—$25 every three months, and "that's probably why she kept on referencing Gorton-Pew and how everybody remembers it." But it was just a "story." Lucy received no help from Gorton-Pew. The company paid no pensions, and even if they had, Charles had last worked for the company many, many years earlier. There was no pension for Lucy. According to Lucy's son Al, the years after Charles's death were hard. "She never got any assistance or nothing, except for a couple of months, she got assistance of $15 for three months or something. This was from the government."

After Lucy's son John left home, he sent money back to help his mother and siblings. He wasn't paid much on his first job, but whatever he could spare was sent back to his mother. John went to Gander, Newfoundland, to work at the new airport for 15 cents per hour. His first job was bringing water to the men on the base. His twin brother, Alban, joined him at Gander. From there they went to Goose Bay and became heavy equipment operators. Alban later went to work in the mill and became a millwright. Once they had these jobs, they always sent money home to their mother. As Mercedes said, "They were really, really good." For the Daley children who remained behind in St. Joseph's, the money from John and Alban made the difference between surviving and not surviving. There were seven children, and the twins gladly took up the mantle of helping to support the family. The next brother, Charlie, got into the fishery business. "He put his heart and soul into it."

In addition to the different streams of monetary support, and they were helpful, Mercedes believed the family survived because of her mother. "I will tell you why we survived. My mother had a vegetable garden. She kept a pig. We had a cow, and we had milk. The pig would be killed, and the hams would be smoked. I remember seeing the hams being smoked in the barrel. That is how we survived. We might only have porridge in the morning, but we had it. My mother kept us together."

So the story ends, with the widows of two of the last of the dorymen struggling to make ends meet. Where once there had been hundreds of schooners and thousands of dorymen, there were now but a handful of aging men setting trawl from dories. The era had ended, and what an era it had been: what men, what vessels, what a community. The life was hard, but there was much to love and honor in the men from America, Nova Scotia, Newfoundland, Sweden, Norway, Portugal, the Azores, Italy, and elsewhere. They were dorymen. They were Gloucestermen. They worked "alone at sea," and the community had been enriched by their presence.

Notes

1. Christine Leigh Heyrman, *Commerce and Culture: The Maritime Communities of Colonial Massachusetts, 1690–1750* (New York: W. W. Norton & Co., 1986), p. 46.
2. Dana A. Story, *The Shipbuilders of Essex* (Gloucester, MA: Ten Pound Island Book Co., 1997), pp. 9–10.
3. Daniel Vickers, *Farmers & Fishermen: Two Centuries of Work in Essex County Massachusetts, 1630–1830* (Chapel Hill, NC: University of North Carolina Press, 1994).
4. Ibid., p. 200.
5. Heyrman, *Commerce and Culture*, p. 60.
6. Ronald N. Tagney, *The World Turned Upside Down* (West Newbury, MA: Essex County History, 1989), p. 343.
7. Secured by John Adams under the 1783 Treaty of Paris that ended America's war with Great Britain.
8. Lemuel Gott, *History of the Town of Rockport* (Rockport, MA: Rockport Review Office, 1888).
9. Captain Sylvannus Smith, *Fisheries of Cape Ann* (Gloucester, MA: Press of Gloucester Times Co., 1915).
10. Gott, *History of the Town of Rockport*.
11. Shannon Ryan, *Fish Out of Water: The Newfoundland Salt Fish Trade, 1814–1914* (St. John's, Newfoundland: Breakwater Books Ltd., 1986).
12. *Cape Ann Weekly Advertiser*, December 13, 1867.
13. Smith, *Fisheries of Cape Ann*, pp. 38–39.
14. *Gloucester Telegraph*, April 13, 1850.
15. Three excellent sources on Gloucester schooners: Howard I. Chapelle, *The American Fishing Schooners*; Joseph Garland, *Down to the Sea: The Fishing Schooners of Gloucester*; and Dana A. Story, *The Shipbuilders of Essex*.
16. George H. Proctor, *The Fishermen's Memorial & Record Book* (Gloucester, MA: Proctor Brothers, 1873).
17. Jon Littleton, *Jottings of a Summer Codfishing Trip* (September 26, 1879).
18. Originals in collection of Cape Ann Museum.
19. Original in collection of Cape Ann Museum.
20. Gordon Thomas, "The Portuguese of Gloucester," *Gloucester Magazine*, Vol. 2, No. 3.
21. Cecile Pimental, *The Mary P. Mesquita: Rundown at Sea* (C. Pimental, 1998), p. 28.
22. Eugene L. Alves, *Portuguese Folklore in Gloucester* (mimeo, June 1954, copy available at Cape Ann Museum).
23. James B. Connolly, *American Fishermen* (New York: W. W. Norton & Co., 1961), p. 29.
24. Based on material from the Jeff and Gordon Thomas Oral History Collection.
25. This narrative is based on information provided by Jeff's son, Gordon Thomas, and is taken from the Jeff and Gordon Thomas Oral History Collection.
26. Stephen A. White, "The Arichat Frenchmen in Gloucester: Problems of Identification and Identity," *New England Historical Genealogical Society*, 1976.
27. Jeff and Gordon Thomas Oral History Collection.

28. *The Fisherman*, May 1895, pp. 1–2.

29. Jeff and Gordon Thomas Oral History Collection.

30. Ibid.

31. Original of diary in collection of Cape Ann Museum.

32. Log of *Concord* written by Alex D. Bushie, prepared by his great-granddaughter Joanne M. [Burke] Gallant of Gloucester and made available to the author by his great-grandson Johann Diego Arnorsson of Iceland.

33. Roselle Mercer, "The Evolution of the Codfish-Cake," *Leslie's Weekly*, August 28, 1902, p. 197.

34. W. M. P. Dunne, *Thomas F. McManus and the American Fishing Schooners* (Mystic, CT: Mystic Seaport Museum, Inc., 1994), pp. 198–99.

35. Story, *The Shipbuilders of Essex*, p. 165.

36. Dunne, *Thomas F. McManus and the American Fishing Schooners*.

37. Found at the Cape Ann Museum Library.

38. Jeff and Gordon Thomas Oral History Collection.

39. Thanks to Justin Demetri for his comments on the Italians moving into Gloucester.

40. John Morris tape with Carl's son, Carl Jr., fall 2001.

41. Mystic Seaport Museum Oral History Interview of Robert Merchant, by Virginia Jones, April 1, 1978. Recorded at Stonington, CT.

42. Jeff and Gordon Thomas Oral History Collection.

43. From Gorton's of Gloucester archives.

44. R. H. Gibson and Maurice Prendergast, *The German Submarine War, 1914–1918* (Annapolis, MD: U.S. Naval Institute Press, 2003), p. 308.

45. Material taken in part from Jeff and Gordon Thomas Oral History Collection.

46. *St. John's Daily News,* August 22, 1918.

47. Jeff and Gordon Thomas Oral History Collection.

48. Ibid.

49. Nina Columbia McCarthy Flag, Jeff and Gordon Thomas Oral History Collection.

50. Oral History, Jennie Auditore, Sawyer Free Library, Gloucester.

51. Jeff and Gordon Thomas Oral History Collection.

52. Ibid.

53. Story, *The Shipbuilders of Essex*, p. 217.

54. Michael Wayne Santos, *Caught in Irons: North Atlantic Fishermen in the Last Days of Sail* (Danvers, MA: Rosemont Publishing & Printing Corp., 2002), p. 91.

55. Gordon Thomas, Jeff and Gordon Thomas Oral History Collection.

56. Howard I. Chapelle, *The American Fishing Schooners, 1825–1935* (New York: W. W. Norton & Company, 1973), p. 300.

57. Gordon Thomas, *Fast & Able: Life Stories of Great Gloucester Fishing Vessels* (Beverly, MA: Commonwealth Editions, 2002), pp. 188–89.

58. Material provided by Gordon Thomas, Jeff and Gordon Thomas Oral History Collection.

59. Material drawn in part from Gordon Thomas and Harry Eustis conversation, Jeff and Gordon Thomas Oral History Collection.

60. Captain Charlton L. Smith, "The International Fiasco," *Atlantic Fisherman*, III, No. 10, November 22, 1922, pp. 5, 6, 14.

61. Material drawn from Gordon Thomas and Harry Eustis conversation in 1982, Jeff and Gordon Thomas Oral History Collection.

62. For more on *Columbia*, see Dana A. Story, *Hail Columbia! The Rise and Fall of a Schooner* (Barre, MA: Barre Publishing, 1970).

63. Jeff and Gordon Thomas Oral History Collection.

64. Ibid.

65. Ibid.

66. October 25, 1927, letter to Captain Ben Pine from Mrs Arthur Firth, Shelburne, N.S. From Ben Pine Collection, Cape Ann Museum.
67. From Ben Pine Collection, Cape Ann Museum.
68. OH-48 Gordon Thomas Recording, Sawyer Free Library, Gloucester, 1978.
69. John Morris Tape, November 2002.
70. U.S. Census, Historical Statistics of the United States, Colonial Times to 1970.
71. Material from Jeff and Gordon Thomas Oral History Collection.
72. View of Gordon Thomas, Jeff and Gordon Thomas Oral History Collection.
73. John R. Arnold, *Earnings of Fishermen and of Fishing Craft, Work Material No. 31*, Office of National Recovery Administration, Industry Studies Section, January 1936. Document available at the Franklin D. Roosevelt Library.
74. Carl Olsen (John Morris tapes 2001, 2005).
75. Jeff Thomas, interview with John Morris, September, 3, 2006.
76. Jeff Thomas, grandson of his namesake, John Morris tape collection, October 2003.
77. Mercedes Daley Lee, John Morris tape collection, 2002.
78. These quotes from Gerda, Romaine, Jack, and Harriet come from taped sessions between 1999 and 2002, John Morris tape collection.
79. John Morris tape with Captain Billy Shields, one of Gloucester's last living dorymen, 2001.

APPENDIX A

Glossary of Terms

Aft   Near the stern (rear) of the vessel.

Aloft   To be at the masthead on lookout for, say, a school of fish or a dory.

Anchor davits   Spars used as cranks to hoist the flukes of the anchor to the top of the bow without damaging the hull of the schooner. In the 1800s and 1900s they were made of iron; earlier they were made of wood and were called cat heads.

Anticosti grounds   (48°N, 061°W) Fishing grounds that sit above the Anticosti Platform (or shelf) that surrounds Anticosti Island, at the mouth of the St. Lawrence River. The prime fishing area lies to the east of the island. The primary fish species were cod and halibut.

Auxiliary   Sailing schooner with an engine.

Awash   When the seawater is level with and flowing across the deck of the schooner.

Bacalieu Bank   (48°N, 052.8°W) North-western extension of the Grand Banks, lying off Bacalieu Island, which is 3½ miles off the northern tip of the Conception Bay area of Newfoundland's Avalon Peninsula (note that the word "bacalieu" is Portuguese for salt cod). The bank extends about 30 miles to sea. By the 1890s, vessels in Gloucester's northern salt-cod and flitched-halibut fleets would sometimes make trips to this bank, beginning in March (or later) after the winter's ice had receded. Such trips would be made in conjunction with more northern voyages to the shelf areas of eastern Labrador.

Bailer   Wooden and metal scoop used to get the water out of a dory.

Bait   In dory trawling, the primary bait was herring, but baits also included squid, salted clams, capelin, mackerel, and a few other species of fish.

Baiting up   Cutting bait and putting hooks on the gangings that come off the trawl line.

Ballast   Rock and pig iron carried in the bottom of the vessel to provide draft and stability.

Banker   Schooner fishing off Newfoundland on the Grand Banks.

Banquereau (Quero Bank)   (44.5°N, 058°W) A fishing bank to the east of Nova Scotia on the Scotian Shelf, about 90 miles off the commercial fishing port of Canso, and separated from it by three smaller banks (Canso, Middle, and Missaine), 555 miles from Gloucester. The bank has a somewhat rectangular shape, 120 miles long by 47 miles at the widest spot. On the eastern portion of the bank, in an area called the Rocky Bottom, the depth is about 16 fathoms; elsewhere it ranges from 18 to 50 fathoms. Halibut and cod were the primary fish on this bank. One of the more interesting fishing spots on Quero was the "Stone Fence." Much gear was lost there because of the trees—some soft, some petrified—that lay below. They had a pinkish color, and in later years were brought up by the otter trawler fishermen.

Bare poles   Schooner condition when under way with no sails set—a precautionary measure taken in severe winds.

Barrel of fish   A barrel of fish (of, for example, herring, mackerel, shad, halibut fins) usually weighed 200 pounds.

Bay of Island grounds   (48.6°N, 058°W) Bay located on the west coast of Newfoundland, encompassing a 137-square-mile area. The bay consists of an outer large bay with 12 islands and three major arms reaching into Newfoundland in an easterly direction. The arms are known as North, Middle, and Humbers. It was a major herring fishing spot, beginning with the arrival of the herring in October and ending in late January, when the bay was closed by ice floes coming south from Belle Isle Strait (an area 90 miles long and 10 to 17 miles wide that had an extended period of ice cover).

Beam   Breadth of vessel's hull; the maximum breadth.

Beams   Structural timber stretching across the vessel, supporting the decks and bracing the sides; bound to the hull by knees.

Beckets   Twenty-five-and-one-half-inch ropes threaded through holes in the cap rails at either end of the dory; used in hoisting and lowering the dory to and from the schooner.

Bend a sail   Put on a new sail; attaching it with lines to mast and spas. The head of the sail is laced to the gaff, the foot to the boom.

Big four   Another name for the Gorton-Pew Fisheries Company; references the four companies that came together when Gorton-Pew was incorporated.

Big hook   A term used in Gloucester for the halibut fishery.

Bight   Slack portion of a rope—for example, a loop.

Bloater   Term used in grading mackerel. Bloaters were mackerel of the best quality; not mutilated; measuring not less than 14 inches from the extremity of the head to the notch, or fork, of the tail; free from rust, taint, or damage. There were no more than 150 bloaters to a barrel.

Booby hatch   A sliding cover, or hatch, that must be pushed away or raised to allow passage. Booby hatches were raised about 22 inches and covered the fish hatches on winter fishing trips; some had sliding companionways.

Boom   A horizontal spar, running fore or aft; the forward end is attached to the mast at the gooseneck. The foot of the sail is bent onto the boom.

Bosun (boatswain, bos'n)   Crew member who keeps up repairs on the schooner's rigging, sails, and hull.

Bowsprit   Spar projecting forward of the bow in order to increase the usable area for sails without having to increase the hull's length.

Breeze   A strong wind. The average fisherman seldom dignified even the most furious gale with any other name than "breeze."

Broken trip   To return from a trip with only a few fish.

Broker   To come home with no money or "share" from a trip.

Brown's (Browns) Bank   (42°N, 060°W) South of Cape Sable, Nova Scotia (40 miles off the coast), and north of Georges Bank, about 230 miles due east of Gloucester. Brown's and Georges banks are separated by a 15-mile-wide gully, the North (or Northeast) Channel. The bank is 63 miles wide and about 43 miles from north to south; depth of water ranges from 20 fathoms to 75 fathoms. The bottom is largely coarse sand, gravel, pebbles, and rocks. Cod, halibut, and haddock were the principal commercial species. Quite a few of the schooners fishing Georges Bank did some of their fishing on this bank.

Bull eyes   Type of mackerel, also known as a "hard head." They weigh about one-half pound each. The name comes from the hardness of the fish's head when split with a mackerel splitting knife. The fish is of a different genus than the more common mackerel landed by the Gloucester fleet, and did not appear every year in great numbers. A small fish, it took a good deal of work, as a full barrel would contain 600 to 1,100 salted fish.

Bulwarks (bullwords)   Solid wood sides, standing 24 to 27 inches above the schooner's deck and topped by the rail. They helped to prevent a man from falling overboard and to keep the seas from washing over the deck.

Bunks   Wooden beds on either side of a schooner's forecastle (or focsle) built in tiers, with a maximum of three to a tier.

Cape North and North Bay   (47°N, 061°W) Fishing grounds on the undersea shelf north of Cape Breton Island, Nova Scotia, and south of St. Paul Island and the Magdalen Islands; 590 miles from Gloucester. The shelf begins 4 to 15 miles from land and extends west for about 15 to 20 miles; it is part of the Arcadian Platform, which extends from the Gaspé Peninsula in the west (between New Brunswick and the St. Lawrence River), goes around Prince Edward Island, and ends to the north of Cape Breton Island in the east. The depth of water on the banks is about 65 to 100 fathoms and the bottom is rocky. Gloucester vessels came to these grounds for cod, hake, and mackerel: Cod trips were made just after the winter ice fields receded to the north; the mackerel fleet arrived in late spring (usually May) and stayed for a couple of months, and this fleet could be quite large. By the 1900s, the mackerel fleet visited North Bay only at the end of the season, as summer ended and fall began.

Cape Shore   Fishing grounds about 20 miles off the coast of Nova Scotia and about 300 miles from Gloucester at the southern end. A major mackerel fishing area in May and June as the fish migrated up the coast.

Capelin (caplin)   A slender, elongated bait fish that swims in Arctic regions; typically 5 to 8 inches long, reaching a maximum of 10 inches. It is a silvery, translucent fish, with green above, silvery sides, and white beneath.

Catchee   A young boy working on a schooner. Term comes from the boy "catching" the painter as dories returned with their catch. The catchee assisted wherever he could; launching and retrieving the dories, fishing off the side of the vessel when the dories were away, doing odd jobs for the skipper, assisting the cook, and even taking the wheel. Eventually the catchee would go out in a dory with an experienced hand, and with the passage of time he became a full member of the crew.

Caulking   Material (often oakum) used to seal the wooden seams of a hull to make it watertight.

Chain plate   Steel plate attached to the sides of the hull, with dead-eyes in the upper end, by which the lower end of the standing rigging (for example, the shroud) is attached to the hull.

Close hauled   The sails are trimmed in tight to enable the vessel to sail as close to the wind as possible.

Cod   A fish with three dark dorsal fins on its back, two dark anal fins, a square tail, and a lateral line running along the side of a heavy and tapered body. Its most distinguishing feature is its chin barbel, a whisker-like sensory organ under the large mouth. In the 1800s and early 1900s, the typical fish was 30 inches or more in length, and a small fish weighed in at 15 pounds, while a larger fish could weigh 75 to 190 pounds, and 55- to 75-pound fish were not uncommon.

Comber   A large, curling wave; a deepwater wave whose crest is pushed forward by a strong wind, much larger than a whitecap.

Come in "deep"   To return after an especially good trip, the craft so laden with fish that water laps at the scuppers.

Companionway   The entrance from the deck to the cabin below; also the wooden covering over the staircase to the cabin.

Counter   Part of the stern of the vessel that overhangs and projects aft of the sternpost over the waterline, past the rudder stock.

Crosstrees (spreaders)   Small spars set perpendicular to the mast, to spread the angle of the shrouds from mast to deck.

Cusk   Member of the cod family, distinguished by a single dorsal fin; generally weighing between 5 and 20 pounds (although it can grow to over 60 pounds), and ranging between 18 and 35 inches long (although it can grow to over 40 inches). Cusk prefer hard, rough, or rocky bottoms and are found in moderately deep (40 to 250 fathoms), cold waters. These solitary fish were not abundant and so were an incidental by-product, rather than a primary catch.

Cutwater   Forward part of a ship's prow that first cuts the water as she sails.

Davits   Crank-like devices used for raising and lowering boats to and from the deck.

Dead reckoning   Determining the position of the schooner based on an educated guess, given the last known position and the conditions of the wind, tide, and so on.

Derelict   Vessel abandoned at sea.

Dip   To set one's trawl.

Dogfish   A voracious small shark (5 to 10 pounds, 2 to 3 feet), predator of a number of commercial species of fish—known to attack cod, hake, and haddock when hooked on a trawl line, although more typically feeding on herring, squid, and shrimp. It has an elongated, slender body, a pointed head, and is slate gray to brownish on top. In the 1800s and early 1900s, it had little commercial value.

Dory cradle   Fixed wooden rest on which to stack dories on a schooner.

Dory plug   A birch plug placed in the floor of the dory before launching; when removed the plug's hole permitted water to drain from the dory when on deck. A manila line was passed through the center of the plug strap, and this protruded from the bottom of the dory; it could be used as a grip by the doryman should his boat be overturned and he find himself in the water.

Dory tackle   Block and tackle, with hook that came down from the schooner's fore and aft rigging to grab the bow and stern dory beckets and raise the dory over the side of the schooner; also used to recover the dory from the water after fishing.

Draft (draught)   Depth of the hull in the water, measured from the underside of the lowest part of the keel to the waterline.

Driver   Schooner skipper who drove his vessel hard in heavy winds in order to be the first to get his fare back to port, to catch the market, to get the best price for his fish.

Drogue   A bucket, canvas bag, or coils of cable set out for use as a sea anchor; it slowed the vessel and offered directional stability.

Dropping out of the sling   Launching a dory over the side of the schooner, the slings were the fore and aft lines used to hoist the heavy dory out of the nest and move it over the side.

Dryers   Drying rooms fitted with flakes on which salted cod were placed for processing; there were pipes through which hot air was distributed in the room at the right temperature.

Even keel   A vessel sitting upright at her waterline, with no list to either port or starboard.

Fathom   Measurement used in relation to depth of water: 1 fathom is 6 feet.

Fish flakes   Raised wooden platforms on which salted cod (or other salted fish) was exposed to the sun to dry and become ready for market.

Fish pens   Boarded-off areas on the deck of the schooner into which the dorymen placed their catch for sorting prior to processing and storage.

Fisherman   A sailing fishing schooner from the East Coast of the United States.

Fitting out   Stowing on board the schooner all the supplies required for a trip, including sails, rigging, food, gear, ice, salt, wood, coal, water, and so on.

Flemish Cap or Outer Bank   (47°N, 045°W) The easternmost extension of the Grand Banks. Depth is generally from 60 to 160 fathoms, and some areas are as shallow as 8 fathoms. The bottom is smooth and is composed of rock, sand, pebbles, and gravel. Predominant fish are cod and halibut, and the area was normally visited by a small number of dory fishermen from April to August. Icebergs were common, and at other times of the year the weather was too rough to permit an effective fishery. At the western edge of the Flemish Cap, a 35-mile-wide trough known as Flemish Pass separates the Flemish Cap from the Grand Bank.

Fletching (flitching) halibut   Filleting halibut at sea. The fish's head is cut off and the bones are removed from the body, which is then cut into fillets and salted.

Flying fishermen   Schooners launched from 1887 to 1930. The schooner *Gertrude L. Thebaud*, last of the new class of fast, beautiful schooners, was launched in 1930.

Flying set   Plan by which a schooner launched her dories across the fishing grounds: The schooner set a straight course across the grounds, and as she passed down this course, dories were released one after another, going out 1 to 2 miles, at a right angle from the flying set line.

Fore-and-aft sail   Sail that can take the wind from either side of the sail, depending on the direction of the wind.

Foremast   One of two masts on the typical fore-and-aft-rigged schooner; nearer the bow and generally shorter than the mainmast.

Foresail   The principal sail on the foremast; on a Gloucester schooner, this sail would have been gaff-rigged with a triangular fore-topsail above it.

Freeboard   Measurement of hull showing between the waterline and the deck; minimum vertical distance from the surface of the water to the gunwale. Ships with very little freeboard are harder to sink.

Fresh fishing   To bring in fresh fish preserved in ice, stored in pens in the hold of the schooner.

Full trip   To return to the harbor from a trip with the hold full of fish.

Funk Island Bank (or the Funks)   (49.8°N, 053°W) On the Newfoundland Shelf some 40 miles northeast of Cape Freels, on Newfoundland's north coast, about 120 miles north-northwest of Bacalieu Island, and 180 miles southeast of the Belle Isle Strait. The bank runs 125 miles north–south and 70 miles east–west, with a depth of no more than 110 fathoms. To the south and west of the bank is the deeper trough known as the Funk Island Deep, while to the northwest are two of the banks of the Labrador Shelf—the Belle Isle Bank and Hamilton Bank. The Labrador Current flows southward over the bank, bringing down cold surface waters and ice floes, and in the 1890s and later, vessels in Gloucester's northern salt-cod and flitched-halibut fleets would visit the Funks from March onward, after the winter's ice had receded.

Furl   To lower a sail and roll it up snugly to a yard or boom.

Gaff   A spar that extends diagonally upward from the side of the mast to which the head of a four-sided fore-and-aft sail is attached. On a Gloucester fishing schooner, both the foresail and mainsail were gaff-rigged.

Gaff hook   Curved, sharp iron hook at the end of a wooden pole, used to lift fish from the sea into the dory, and from the dory into the schooner. In lifting the fish to the deck, one doryman worked in the dory, the other on the deck.

Gale   Unprecise term for winds stronger than a breeze (say 30 knots) and less than a hurricane (say 80 knots). For the men from Gloucester, a typical gale would have been a 50- to 60-knot wind.

Galley   Schooner's kitchen.

Ganging   At fixed intervals (2 to 10 feet) along the trawl line, baited fishhooks were attached to lighter ganging lines that came off of the trawl (one hook per ganging—pronounced "ganjing").

Gear   The encompassing name for ropes, blocks, tackle, tools, and so on.

Georges Bank   (41°N, 067°W) The grounds most frequented by the Gloucestermen, south of Brown's Bank and separated from it by the Northeast Channel; 150 miles from Gloucester, primarily to the east of Cape Cod and Nantucket Shoals (41°N, 069.5°W), about 80 miles to shore. The area between the shoal area on the western side of Georges Bank and Nantucket Shoals is known as the South Channel (or Great South Channel, or the Channel). The greatest length north–south is 150 miles, and the greatest width is 98 miles; the depth of the water ranges from 2 to 50 fathoms, with a bottom that is chiefly sand. On the western part of the bank are a number of productive shoals, known as the East Shoal, North Shoal, Southwest Shoal, and Cultivator. The Southwest Shoal was the largest, 2½ by 15 miles, centered at 41.39°N, 067.48°W. This shoal had some of the shallowest water, about 2 fathoms in spots. The primary commercial species on Georges were haddock, cod, and mackerel, and in the early years, halibut.

Gloucestermen   Term used for both a fishing man and a fishing schooner from Gloucester.

Go adrift   To break away from the anchor when on the fishing grounds.

Gob-stick   A two-pronged stick used by the dorymen to hit and thereby stun fish as they were brought into the dory; also used to remove a deeply set hook from the mouth of a fish.

Gooseneck   Metal fitting attaching a boom to the mast.

Grand Bank   (45°N, 050°W) One of the component parts of the Grand Banks, lying about 90 miles to the southeast of Newfoundland and 1,200 miles northeast of Gloucester; about 260 miles long by 220 miles wide at its longest and narrowest points. The bottom varies both in depth and material. The principal fish species sought by the Gloucestermen were halibut and cod. The other smaller component parts of the Grand Bank continue to run to the west and include Green, St. Pierre's, and Burgeo banks. St. Pierre's, the most westerly of the Grand Banks, became a favorite spot for the Gloucester halibut fishermen. It has a total area of 5,310 square miles, is relatively flat, and is covered by coarse sand, except in one rocky area on the western edge. Halibut Channel lies to the east and separates St. Pierre's from Green Bank. Halibut tended to be found on the eastern slope of the bank and along the Laurentian Channel to the south. The slopes were characterized by coarse sand, with halibut typically found at depths greater than 110 fathoms. The southwest corner of the bank had a high level of tidal mixing and was a very productive fishing area.

Green fish　Split and salted fish that is not dried; weighs about three times as much as the final dried fish.

Ground fish　Fish that swim at the bottom of the ocean, such as haddock, hake, halibut, cusk, and pollock (although at times of the year, pollock school near the surface).

Gulf of Maine Banks　Grand Manan, Jeffreys Ledge, Stellwagen.

Gulf of St. Lawrence Banks　Anticosti, Bay of Chaleur, Magdalen, Orphan, Strait of Belle Isle.

Gully　(44°N, 059°W) Deep passage that lies between Banquereau and Sable Island on the Scotian Shelf. Extends in a west-northwest and east-southeast direction north of Sable Island, cutting into the Scotian Shelf, but turns southward at its eastern end and continues down between the eastern end of the Western Bank and the southwest prong of Banquereau; about 60 by 20 miles, with a depth of about 65 to 145 fathoms; the bottom consists of rocks, gravel, sand, and mud. It was an important halibut fishing ground, with the halibut more frequently found on the slopes leading into the Gully. The Scotian Shelf, into which the Gully cuts, has an average depth of about 50 fathoms, while the Gully is about 550 fathoms deep.

Gunwale　Thick plank or rail on top of a vessel's side planking.

Gurdy　A hand crank that fits across the bow of a dory and is used to haul in the trawl lines that are loaded with fish.

Gurry　Refuse (viscera, skin, head, bones) produced during preparation for splitting or salting fish; thrown away or later used to make other products (for example, glue and animal feed).

Gurry-kid　In the early years of dory fishing, bait was stored aboard the schooner in a gurry or bait box (known as the gurry-kid). It was placed between the cabin trunk (in the aft section of the vessel) and the aft hatch. It was as wide as the trunk, with room to walk around it on larger vessels, and maple cutting boards fit into a fixed piece of grooved wood on the sides of the box's top. Later, gear was stored in the gurry-kid.

Gutter　One of three men who worked on raw fish on board the boat, preparing them to be salted. The first man, who removed the head and split open the body of the fish, was the "throater"; the second man, who removed the viscera, was the "gutter"; and the third man, who split the fish and removed part of the backbone, was the "splitter."

Haddock　Bottom-feeding, purplish-gray ground fish of the cod family, with three dorsal fins and two anal fins, including a prominent first dorsal fin that is high and pointed. Has a small chin barbel and a black lateral line along the side of the body, interrupted at the shoulder by a black blotch, which is often called the "devil's thumbprint." Haddock could be 48 inches long (although 12- to 24-inch fish were common) and average 2 to 5 pounds. They were found at depths of 25 to 100 fathoms, in cool waters, on a bottom composed of sand, gravel, pebbles, and broken seashells. In the 1800s and early 1900s, they were distributed on banks from the mid-Atlantic coast of the United States to the Great Banks off of Newfoundland, with a major concentration on Georges Bank. Prior to refrigeration, they were less popular than cod because cod were better suited to salting, and for this reason the Gloucestermen concentrated on the haddock fishery in the colder months from November through April. The principal grounds for this fresh fishery were 75 to 300 miles from Gloucester, including Cashes, La Have, and Georges banks.

Hake　Several species of bottom-feeding fish, closely related to cod, with two dorsal fins and one anal fin. They include white hake, red hake, and silver hake (whiting). They have a long, slender body and a flattened head; are darkish gray in color and 2 to 3 feet in length; weigh 2 to 3 pounds; and migrate seasonally to cooler waters, although the silver hake prefer warmer waters. They were found from the mid-Atlantic coast to Newfoundland, and were a secondary fish brought in by Gloucestermen from trips to Georges Bank.

Halibut   Bottom-feeders, the largest of the flounder family of flat fish, growing to 300 to 350 pounds and 8 feet 3 inches long, although the medium-sized fish were the most commercially desirable. The fish were first graded by color, the so-called white bringing a higher price than the gray (even though in a retail store there would be no difference in the price charged the consumer); and second by size, the so-called large halibut ("whale"—85 pounds or more) getting the lowest price per pound.

Handline   Fishing line dropped over the side of the schooner (or later from the side of a dory launched from a schooner) to the bottom below. It consisted of up to 900 feet of strong twine on the hand reel, with conical lead weights and two short gangings, each with a baited hook.

Haven't a shot in the locker   When a seaman's funds ran low in the nineteenth century, he would use the saying "I haven't a shot in the locker" to indicate his destitute state.

Hawkins' Spot   (44.47°N, 56.40°E) One hundred fifty miles east of Sable Island, on the westerly end of St. Peter's Bank.

Head sail   Fore-and-aft sail forward of the foremast (for example, a jib).

Heave to   To trim a vessel's sails and rudder so that she is controlled but does not move ahead.

Heel   To lean over to one side (for a vessel).

Helm   The wheel controlling the rudder.

Helmsman   Man who steers the schooner, or any other vessel.

Herring   Migratory fish that tends to feed in the upper water column and remains in certain areas for extended periods. Feeds on plankton, securing its food through gill rakes. Grows to a mature length of 10 inches or more, with a maximum size of about 16 inches and 2.6 pounds. In the 1800s and early 1900s, they were exploited as both a source of bait and a food fish.

High flier   Windward buoy of a dory's trawl line. Typically it was a 100-pound nail keg with a wooden shaft, of which 24 inches was exposed on one side and 12 inches on the other; the short end was weighted down with metal to ensure that the long end stood up in the water. A second stick, of the same diameter, approximately 4 to 4½ feet long, was fastened to the shaft that protruded from the top of the keg; this carried the high flier flag, with the dory's number in black against a white or buff background.

Highland Light   Lighthouse on Cape Cod, north of Chatham, and 48 miles south of Eastern Point, Gloucester.

Highliner   Skipper or schooner with the highest total catch for the season in each fishery line (for example, the halibut highliner).

Hogshead   Large cask holding 63 gallons or about 440 pounds of salt; two barrels make one hogshead.

Hogshead of salt   On salt-bank codfishing trips, each vessel would go to sea with 400 hogsheads of salt, with the rule of thumb being that a vessel would bring in 1,000 pounds of fish to each hogshead of salt (that is, 675 to 800 pounds of dry fish, or 2,300 to 2,700 pounds of green fish).

Hold   Large compartment in the hull of the schooner in which the fish were stored.

Hove down   Vessel knocked over (but likely to come back up) by a squall or breaking seas.

Jogging   Keeping the schooner lying almost head to the wind, making about 2½ points leeway to 5 points headway, so that it would stay on a spot.

Keel   Fore-and-aft timber at the very bottom of the hull that supports the frame of the vessel; composed of several pieces of timber running the length of the vessel.

Kench   Deep bin in which fish were salted while at sea. The fish were thoroughly washed and then slipped down a canvas chute into the hold, where two men worked, salting and kenching, or piling, the fish in the hold. This had to be done carefully in order to prevent the fish from spoiling, which would mean sorting the good and bad out and re-kenching later in the trip.

Kick boards   Boards that divided the floor of a dory into separate areas to hold the fish as they were taken off the trawl line. The kickboards were removed so the dories could be stacked, one inside the other on the schooner's deck.

Knocked down   Vessel heeled over by a squall or breaking sea to such an extent that she is unable to recover.

Knot   Measurement of speed: 1 knot = 1 nautical mile; 1 nautical mile is 1 mile of latitude (or 6,082 feet, 1,852 meters); a statute mile is 5,280 feet, 1,609.34 meters.

La Have bank   (43°N, 064°W) Bank lying about 45 miles east of the southern tip of Western Bank about 60 miles off the Nova Scotian coast, 295 miles from Gloucester. It has two portions—the eastern part runs north–south for 39 miles, while the western part extends east–west for 35 miles. The bottom is largely coarse gravel, pebbles, and rocks; and the depth is from 40 to 50 fathoms. Cod, haddock, hake, and halibut were the principal commercial species.

Lay   Crewmember's, skipper's, or owner's share of a trip's profits (see also "one-fourth/one-fifth lay" and "share").

Lee side   Of the two sides of a vessel, the one farther away from the wind.

Lighter   Beamy, flat-bottomed boat used to transport cargo between a large vessel that cannot tie up to the dock and the shore.

Mackerel   Migratory fish that runs in schools all along the coast. It has a torpedo shape, two dorsal fins, and a forked tail. Typically 12 to 19 inches long and weighs from less than 1 pound to over 2 pounds. The mackerel was graded as follows: tacks or spikes under ½ pound; tinkers ½ to 1 pound; small 1 to 1½ pounds; medium 1½ to 2¼ pounds; and large over 2¼ pounds.

Mainmast   Of the two masts on a Gloucester schooner (a few schooners had three masts), the taller, and the one closer to the stern, stepped just aft of mid-ship.

Mainsail   Principal sail on the mainmast; on a Gloucester schooner, this sail would have been gaff-rigged with a triangular main-topsail above it.

Make a set   To set the trawl line from a dory.

Maritime Provinces   Collective name of the eastern Canadian Provinces of New Brunswick, Nova Scotia, and Prince Edward Island. The term "Atlantic Provinces" came in after Newfoundland joined Canada at a point after this story is over.

Masthead   Top of a mast.

Midships   Point along a vessel's length at which the beam is widest.

Mollycods   Name for men whose trawls tangle and refuse to play out; also known as tommycods or nincompoops, according to the degree of the mess.

Mug-up   To take a break from working in order to get something to drink and eat.

Newfoundland Shelf Banks   Bacalieu, Burgeo, Flemish Cap, Funk Island, Grand, Green, St. Pierre's, Whale, Whale Deep.

Nippers   Doughnut-shaped pieces of material that fit over the fingers of the doryman to protect his hand as he pulled in the trawl line.

One-fourth/one-fifth lay   Method of payment by which the owner furnished the vessel, receiving one-fourth or one-fifth the value of each trip made. The crew—after paying for the fishing gear, or for the use of it, and all food supplies, bait, and ice—had the balance divided equally among them (except the cook, who generally received something extra).

Outward bound   A schooner sailing for the fishing grounds.

Painter   Eighteen-foot manila rope spliced into the bow of the dory; used to tow the dory; thrown on deck of the schooner to pull the dory in alongside; left to stream behind the dory when moving to serve as a lifeline should a man fall overboard. Dories had both bow and stern painters.

Plug strap   See "Dory plug."

Pollock   Member of the cod family, generally 4 to 15 pounds (a reach 40 pounds), and 2 feet to 3 feet 6 inches long. Olive green to greenish brown on top, with a forked tail and a straight lateral line on the side. Migrates seasonally to cooler waters and found from Virginia to Labrador; generally in large, fast-swimming schools, frequenting all depths and feeding on large zooplankton and fish such as herring.

Port   Left side of a vessel when looking forward.

Pulpit   Stand on the forward end of the bowsprit, where the man who handled the lance in swordfishing took up his position.

Purse-seine   Large net used to capture a school of fish by two boats—a larger rowed seine-boat (about the same size as a whaleboat) and a dory. It was closed at the bottom by means of a line that passed through rings attached to the bottom of the net: to "purse a seine" was to close the net.

Quarter   Aft sections of a vessel on either side of the centerline; in a Gloucester schooner, aft of the shrouds and forward of the stern.

Quero Bank   See "Banquereau."

Quintal   A 112-pound measure, applied to dry salt fish that are ready for market; sometimes written and pronounced "kentle."

Raft   Very large school of mackerel.

Rang the bell   Crew member's share for a trip was $100 or more.

Redfish   Bottom-feeding member of the rockfish family, from orange-red to deep-scarlet color above and paler hues below, up to 20 inches long. A tall, rather thin fish with a big head and mouth, and big eye; also known as ocean perch or rosefish. Found at depths of 55 to 276 fathoms in cold waters. Distributed throughout the Gulf of Maine and the southern edge of Georges Bank, as well as on the Scotian and Grand banks.

Reef   To reduce the size of a sail by tying up part of it with reef points, short pieces of line permanently set in the sail. Decreases the amount of sail the wind can affect.

Riding sail   Small, fixed sail used to help the vessel maintain its direction with little or no movement; could be used when at anchor or in heavy seas; also known as a trysail.

Rigging   Items used to attach sails and spars to a vessel.

Rips fishing   The Rips codfishing fleet started in the nineteenth century and continued into the twentieth century. The season began

in April, with the return of the cod in great numbers, and ended with the return of winter weather to the grounds. As late as 1907, 45 craft participated in this fishery. In the early days, vessels caught fish as they drifted over the grounds, rather than at anchor as did the regular Georgesmen; the ground first fished was the Rips, part of Nantucket Shoals. Gradually the fleet shifted over to Georges, until most fished on this ground. The technique was the same as that used by the Georges handliners, except that they used cockles as bait and fished while under way. The fish they brought in was of a fine quality and brought top prices. In earlier years, the fish was salted; later it was iced and brought to T-Wharf for fancy prices. Over time this fleet replaced the Georges handline fleet; in 1907 there were only 10 Georges handliners left. The Georges handliners, however, went out in the winter, while the Rips fleet did not.

Sable Island   Twenty-five-mile-long sandy island 180 miles southeast of Halifax, Nova Scotia; the only portion of the Sable Island Bank (see Western Bank) above sea level. Two long sandbars, resembling big hooks, run under the water about 32 miles on each side of the island; known as the "Graveyard of Ships."

Sail carrier   Skipper who has the ability and courage to carry a lot of sail in an increasing breeze in order to get to the Banks or port more quickly than his competitors.

St. Pierre's   Farthest west of the Grand Banks, lying to the north of the Laurentian Channel and south of St. Pierre and Miquelon islands. It was a favorite spot for the Gloucester halibut fishermen. It has a total area of 5,310 square miles. It has a flat bottom surface and is covered in most areas by coarse sand, except a rocky area on the western edge of the bank. Halibut Channel lies to the east of St. Pierre's and separates the bank from Green Bank. Halibut tended to be found on the eastern slope of the bank (to the east) and along the Laurentian Channel to the south, at a depth of more than 110 fathoms. The southwest corner of the bank had a high level of tidal mixing and was a very productive fishing area.

Salt fishing   Fish are dressed and salted while at sea and stored in the schooner's hold until landed in port. Once a fish is on board, one man removes the head and splits open the body; the second removes the viscera; and a third splits the fish and removes part of the backbone. In fishing parlance, these men are known as throaters, gutters, and splitters. Each fish loses about 40 percent of its weight by the process. Thus, 100,000 pounds of raw, green fish will translate into 60,000 pounds of salted.

Scandinavian grounds Another name used in Gloucester for the Cape Shore fishing grounds off of the east coast of Nova Scotia; also applied to a small ground 18 miles south-southwest from Shelburne, Nova Scotia.

Schooner Fore-and-aft-rigged sailing vessel with two or more masts, where the mast nearest the stern is the largest. Almost all Gloucester schooners had two masts.

Scotian Shelf Banks Off the east coast of Nova Scotia, extending about 60 miles to sea, with an average depth of about 50 fathoms. Among the named fishing banks on the shelf are the following: Baccaro, Banquereau, Brown's, Canso, Emerald, German, Gully, La Have, Middle, Missaine, Roseway, Sable Island, Sambro, Scatteri, and Western.

Scup Fish that was first sought by Gloucester's Italian fishermen in the 1930s; also known as a "porgy." It has a dull silvery color, with a white belly. Scup average 14 to 16 inches and 1 to 2 pounds; their range is from southern New England to North Carolina.

Scuppers Deck-level holes through vessel's sides (bulwarks) that allow water to drain from the deck over the side.

Seine-boat A 38- to 40-foot rowed boat, carried on seining schooners, usually on the port side deck; it carried the seine net on the fishing grounds and worked with a dory to encircle a school of fish with the net.

Set To lay the trawl line from a dory.

Settle up The owner or master of the schooner pays off the crew at the end of the trip.

Shadowing When a school of mackerel shadows, it flashes up to the surface for a minute and then dives out of sight.

Share Money paid to crew, skipper, and owner from the profit of a trip (see also "lay" and "one-fourth/one-fifth lay").

Sheet Line attached to a sail and used to adjust its position relative to the wind.

Shoal Area of shallow water.

Shrouds Ropes or wires that support the mast on either side. On a Gloucester schooner, the shrouds went from the top of the mast to the spreaders or crosstrees, and then to the chainplates.

Silver bellies Small, blue-back herring used as baitfish.

Skate Three-hundred-foot section of tarred cotton trawl line used in ground fishing, onto which the baited gangings were placed at fixed intervals; six skates made up a standard 1,800-foot trawl line.

Skinners Men who use knives to take the skin off the fish during processing.

Slob ice Type of broken ice that could close off a bay, but through which a vessel could be sailed in a fair wind.

Sounding Lowering a lead attached to a line (marked in fathoms) over the side of a vessel to get the depth of the water; also used to describe the measurement obtained.

Sou'wester Oilskin hat with an oversized brim at the back.

Splitter See "Gutter."

Spring and summer fare In the early part of the nineteenth century, the coast and Banks vessels usually made two trips a year, called the spring and summer fare.

Squall Sudden, violent, short-lived windstorm usually accompanied by rain.

Stanchion Upright wooden post that supports the schooner's bulwarks or railings.

Starboard Right side of the vessel when looking forward.

Steaker A codfish weighing from 15 to 25 pounds.

Stock (or gross stock) Gross proceeds from the sale of the catch. Under the share system, the stock would be divided as follows: some covered costs, some to the owner and skipper, and the rest to the crew. Depending on the formula, the crew might get from 30 to 40 percent of the gross stock, but it depended on the size of the catch, the price received for the fish, and the amount of equipment lost on the fishing grounds.

Straight halibuter   Men who went north in search of halibut to the Grand Bank, Quero Bank, La Have Bank, Bacalieu Bank, and so on, as opposed to the Georges halibuters, who brought in a more mixed catch from the nearby Georges Bank.

Street clothes   Set of clean clothes stored on board until a fisherman reached port; also known as off clothes.

Striker   Man who stands in the pulpit of the vessel to throw the harpoon at the swordfish.

T-Wharf   Major fresh-fish landing pier in Boston before the second decade of the twentieth century; it lay beside Long Wharf.

Tack   To turn the bow of the vessel through the wind so that the breeze will come to the opposite side.

Thole pins   Wooden pins placed in twos in the gunwales of a dory; the oar was placed between the two pins for rowing; equivalent of the modern-day oar locks.

Throater   See "Gutter."

Thumb the hat   One means by which watches were set on the schooner. One man pulls off his hat, and each crew member places a thumb and finger on the rim. The captain commences at random, counts eleven, placing his hand on a finger, and each time he utters a numeral, stops at that number. The first person at whom he stops has the first watch, and immediately takes himself to the wheel to stand his turn of steering. Meanwhile, the counting goes on, the men dropping out as their watches are assigned to them.

Thwart   Dory's removable seat used by the dorymen when rowing.

Tillies Bank   Off Eastern Point and Thatcher's Island; 10 miles long east to west, 5 miles wide, with a rocky bottom; it runs toward Jeffreys and is frequented in spring and fall by cod, haddock, hake, and pollock.

Ton   Term used to measure a vessel's carrying capacity, as in gross tons or net tons. Each ton represents 100 cubic feet. The volume (in cubic feet) of the vessel's interior divided by 100 gives her gross tonnage. This is measured exactly by including all spaces, without any exception, below the deck as well as all permanently enclosed spaces on the deck. The net tonnage is supposed to represent the actual capacity of a vessel for carrying cargo. It is obtained by subtracting from the gross tonnage the cubic contents of all spaces that cannot be used for cargo (for example, the crew's quarters, engine rooms, and so on).

Topmast   Mast above another mast (for example, fore-topmast).

Transient skipper   Member of the crew to whom the captain would assign the responsibility, on a onetime basis, to command the vessel when the permanent captain remained ashore.

Trawl   Buoyed line, worked from a dory, brought to the bottom by small anchors; could extend for up to one mile along the bottom of the sea.

Trawl tub   Half barrel used to hold the coiled trawl line with the attached gangings and hooks.

Trough   Valley between two waves.

Turtle position   Overturned dory in the water.

Waist   Central part of a vessel; portion of the deck between the focsle and quarterdeck.

Waterline   Horizontal line on the vessel's hull indicating the level of water when the vessel is trim and in ballast.

Western Bank (or Sable Island Bank) (43.5°N, 060°W) About 80 miles off Cape Breton Island and the eastern part of Nova Scotia, below Banquereau, on the Scotian Shelf; 475 miles from Gloucester. To the south are the La Have Ridges and La Have Bank. One hundred fifty-six miles long by 76 miles wide (including the Middle Ground); on its eastern side is Sable Island. Depth ranges from 18 to 60 fathoms and the bottom is mostly sandy. Cod and halibut were the principal commercial fisheries.

Wet all her bait   To use all of a schooner's bait when on a trip.

Wet all her salt   On a salt-bank trip, catching so much cod that all of the salt on board is used to preserve the catch.

Windlass   Mechanical device used to lower or raise heavy objects, such as the anchor. Consists of a cylinder around which a rope or chair is wound or unwound by a crank or lever. On the Gloucester schooner, it was set up near the bow.

Windward side   Side of the vessel nearest to the wind; also known as the weather side.

Wristers   Tubes of knitted wool that went from the middle of the forearm down and over the palm of the hand; the lower end had a hole for the thumb.

APPENDIX B

Fish Landings*: 1808–1939[1]

Year	Gloucester Landings	Total Landings	Cod: Fresh and Salt	Haddock/ Other Fresh	Halibut: Fresh and Flitched	Hake and Cusk	Mackerel: Fresh and Salt	Pollock	Herring: Fresh, Salt, and Frozen
1808– 1820							None– 0.05		
1821							0.4		
1822							0.6		
1823							0.7		
1824							1.3		
1825							1.9		
1826							2.3		
1827	13.0[2]						3.3		
1828			7.2[3]				6.8		
1829							7.5		
1830							10.3		
1831							14.0		
1832							8.1		
1833							9.1		
1834							12.3		
1835							9.7		
1836							8.8		
1837							6.7		
1838							4.9		
1839							2.1		
1840							1.8		
1841							1.7		
1842							3.1		
1843							3.3		
1844			3.2				3.5		
1845							9.7		
1846							8.3		
1847	21.6		7.1		3.5	0.7	9.4	0.9	
1848							10.7		
1849							9.1		

*Landings are in millions of pounds of fish.

Year	Gloucester Landings	Total Landings	Cod: Fresh and Salt	Haddock/ Other Fresh	Halibut: Fresh and Flitched	Hake and Cusk	Mackerel: Fresh and Salt	Pollock	Herring: Fresh, Salt, and Frozen
1850							10.0		
1851							16.3		
1852	25.9		9.5		4.5	0.6	10.7	0.6	
1853							7.2		
1854	20.0		11.0		0.4		8.6		
1855							14.6		
1856							13.3		
1857							11.0		
1858							9.2		
1859	29.2		12.8		4.5		11.9		??
1860							19.6		
1861							18.1		
1862							23.2		
1863							30.8		
1864							31.0		
1865	44.4		12.7	0.4			28.3		3.0
1866							22.8		
1867							22.8		
1868							15.1		
1869	64.8		28.0	8.0		2.8	18.0		8.0
1870	59.7		26.8	7.5		3.4	19.6		2.4
1871	71.8		35.8	8.3		3.9	20.8		3.0
1872	72.7		43.0	8.0		3.0	14.2		4.5
1873	81.6		51.5	9.0		2.8	17.3		1.0
1874	97.4		53.2	11.0	3.8	3.4	24.0		2.0
1875[4]	96.1		50.5	12.0	9.7	4.6	10.6	1.1	7.6
1876	103.9		47.6	11.0	14.5	4.5	19.1		7.2
1877					16.1		9.8		
1878					11.4		11.1		
1879	91.1[5]		40.1	12.6	13.2	6.6	9.7	2.9	6.0
1880	93.5		49.5		9.1		25.9		9.0
1881	109.3		46.7	0.3	8.7	1.0	32.8	0.6	19.2
1882	110.5		42.9	2.0	7.8	2.0	34.0	2.4	19.4
1883	105.3		55.0	2.1	7.3	0.8	21.7	1.2	17.2
1884	129.0		55.3	1.0	9.0	0.6	44.7	3.3	15.1
1885	123.1		54.0	0.9	10.0	0.9	31.5	5.9	19.9
1886	105.6		52.7	1.2	12.2	1.0	10.8[6]	1.9	25.8
1887	99.7		58.2	0.6	9.0	1.7	9.0	2.1	19.1

Year	Gloucester Landings	Total Landings	Cod: Fresh and Salt	Haddock/ Other Fresh	Halibut: Fresh and Flitched	Hake and Cusk	Mackerel: Fresh and Salt	Pollock	Herring: Fresh, Salt, and Frozen
1888	79.4		48.8	0.1	7.7	0.4	4.7	1.2	16.5
1889	75.8		44.3	1.5	7.2	1.0	2.3	1.7	17.8
1890	78.1		45.6	1.3	6.9	3.2	3.2	1.5	16.4
1891	96.9		44.0	4.6	7.8	15.1	5.6	3.5	16.3
1892	111.3	140.5	49.8	4.8	8.2	13.7	6.7	3.0	25.1
1893	99.3	137.2	41.7	3.7	8.9	12.9	8.3	2.9	20.9
1894	98.8	145.3	42.0	6.2	9.6	14.1	6.2	1.0	19.7
1895	86.2		45.2	4.3	8.7	5.7	4.3	0.5	17.5
1896	112.0	152.0	48.0	20.0[7]	10.0		16.0		18.0
1897	75.0	115.3	32.7	3.7	10.4	10.9	2.5	2.1	13.0
1898	104.1		32.3	46.7[8]	10.1		2.7		12.3
1899	121.0	127.0	62.0	8.7	7.5	9.1	4.1	3.3	19.6
1900	106.1	136.4	46.4	5.4	7.3	6.9	16.1	3.0	14.4
1901	106.4	146.4	52.4	6.5	4.6	5.7	13.6	2.2	17.6
1902	102.2	142.2	46.4	4.9	4.9	6.5	7.5	4.9	20.0
1903	87.8	124.7	37.8	5.2	3.8	7.3	8.7	5.0	16.7
1904	103.5	135.3	34.1	7.3	2.7	15.5	5.7	7.3	23.2
1905	112.4	157.1	29.4	13.7	2.8	20.4	5.7	17.6	15.8
1906	93.8	128.1	26.9	14.1	4.0	10.6	2.6	7.3	23.0
1907	109.9	148.9	31.9	6.1	3.9	14.6	6.5	16.8	23.2
1908	96.7	129.3	36.2	8.4	3.7	16.2	4.3	7.2	19.0
1909	88.4	124.7	45.3	4.4	3.7	3.1	3.6	5.9	15.9
1910	89.7	121.9	37.3	4.9	3.2	5.9	0.7	9.4	21.9
1911	91.4								
1912	84.0		36.8	11.4	2.7	8.1	1.6	9.6	6.3
1913	69.9		27.8	9.0	3.8	7.0	1.8	11.2	2.7
1914	78.5	123.7	27.9	11.9	2.6	8.8	3.3	9.0	11.2
1915	111.0		24.1[9]	10.3	2.9	8.2	5.6	11.4	12.5
1916	132.3		21.2[10]	4.7	1.8	2.4	6.7	10.4	14.7
1917	131.0		27.9[11]	2.8	0.9	1.4	7.0	9.1	18.3
1918	144.2[12]		32.6	8.7	0.6	1.2	4.3	16.2	20.7
1919	120.4[13]		31.0	16.1	0.3	1.6	6.2	18.5	8.1
1920	75.8[14]		28.9	8.7	0.1	2.1	13.6	4.9	1.2
1921	54.8		20.4	8.3	0.4	4.6	4.8	4.8	1.9
1922	59.3[15]		23.0	11.6	0.16	3.2	4.3	3.0	3.5
1923	62.1[16]		24.9	10.6	0.22	3.5	6.3	3.4	3.6
1924	65.1[17]	107.0	26.1	11.2	0.2	2.9	4.9	3.1	4.1
1925	80.5[18]		30.0	14.5	0.3	2.4	11.9	2.5	2.9

Year	Gloucester Landings	Total Landings	Cod: Fresh and Salt	Haddock/ Other Fresh	Halibut: Fresh and Flitched	Hake and Cusk	Mackerel: Fresh and Salt	Pollock	Herring: Fresh, Salt, and Frozen
1926	72.4[19]		34.9	8.7	.04	1.6	10.4	1.6	4.6
1927	89.3	111.9							
1928	41.8								
1929	53.4								
1930	47.3								
1931	24.9								
1932	25.3								
1933	21.7								
1934	40.1								
1935	51.2								
1936	59.1	83.9[20]	8.2	4.0	0.04	0.9	7.4	17.2	0.6
1937	46.2	63.3	9.5	4.2	0.04	2.1	1.5	9.6	0
1938	62.9[21]	83.0	10.4	3.8	0.1	2.0	4.7	9.0	0
1939	79.1	99.8[22]	5.8	6.1	0.01	2.5	3.3	8.6	0

Note: This chart was compiled from annual published reports of the landings of the Gloucester fleet. In the early years, relatively complete estimates are available for the mackerel catch, but only incidental information on the landings of the other fish species. Thus, prior to the mid-1860s, the absence of an estimate in the chart should not be construed as the absence of a catch. Later, as Gloucester emerged as the center of the American fisheries, the local papers noted the catch of the Gloucester fleet on a daily basis—no matter where the fish were landed. This information was then summarized during the course of the year, the newspapers tracking the current performance of the fleet against what had happened in the recent past.

[Note: Canadian and N.F. vessels landed an additional 4.9 million pounds in 1913 and 13.6 million pounds in 1914 at Gloucester.]

[Note: Canadian and N.F. vessels landed an additional 13.1 million pounds in 1915 at Gloucester, while an additional 8.7 million pounds of fish arrived by rail.]

[Barrel conversions: Barrel mackerel = 200 pounds; barrel herring = 200 pounds; barrel shad = 200 pounds; barrel halibut fins = 200 pounds; quintal = 112 pounds.]

[1] Sources: Fishermen's own book, yearly reports in Gloucester newspapers, *The Fisherman*, *The Fisherman's Memorial and Record Book*.

[2] The first recorded total catch of fish at Gloucester, recorded in the *Gloucester Telegraph*.

[3] Babson, p. 571. This was the high point in this era, and dropped to insignificant portions after that with the advent of the mackerel fishery.

[4] 1875 was the first year for which there is a detailed and reasonably correct statement of the receipt of fish for Gloucester. The data were gathered by the *Cape Ann Weekly Advertiser* from all of the fishing firms, although the data reported here come from another source and are reported in order to maintain continuity over the years.

[5] The fresh-fish component came to 44 million pounds. Included in the total are 8.1 million pounds of mixed and other fish.

[6] The mackerel disappeared, and the 14-year period of reciprocity with Canada ended. This period went from 1871 to 1885. The decrease in total landings in the next few years was due thus to the decline in the mackerel catch, not to the abandonment of the reciprocity agreement. In 1887, Congress passed a law that virtually prohibited southern mackerel fishing over the next few years; no seining operations were permitted prior to June 1.

[7] Unspecified mixed fresh fish.

[8] Unspecified mixed fresh fish.

[9] Salt fish by rail, add in an additional 8.7 million pounds; add in an additional 13.1 million pounds that was not a product of American fisheries.

[10] An additional 28.3 million pounds that was not a product of American fisheries.

[11] 32.2 million pounds that was not a product of American fisheries; an additional 13.3 million pounds came in by rail.

[12]27.1 million pounds that was not a product of American vessels; an additional 22.9 million pounds came in by rail.

[13]25.7 million pounds that was not a product of American vessels; an additional 23.4 million pounds came in by rail.

[14]7.9 million pounds came in by rail.

[15]10.7 million pounds that was not a product of American vessels.

[16]Included were 1.2 million pounds of cured fish, 0.4 million pounds of miscellaneous fish, 0.4 million pounds of foreign salt mackerel, and 8.2 million pounds of foreign cured fish.

[17]Included were 2.9 million pounds of cured fish, 1.3 million pounds of miscellaneous fish, 1.1 million pounds of foreign salt mackerel, and 9.6 million pounds of foreign cured fish. One-third of the catch of the Gloucester fleet was landed at Gloucester.

[18]10.6 million pounds that was not a product of American vessels.

[19]7.2 million pounds that was not a product of American vessels.

[20]Included were 24.5 million pounds of imported salt whole fish. There were also 2.6 million pounds of whiting and 17.1 million pounds of redfish.

[21]Included were 29.4 million pounds of redfish, 1.6 million pounds of whiting, and 19.1 million pounds of imported salt fish.

[22]Included were 20.7 million pounds of imported salt fish, 44.4 million pounds of redfish, and 3.8 million pounds of whiting.

APPENDIX C

The Gloucester Fleet, Vessels Lost, Men Lost at Sea

Year	Size of Fleet	New Vessels Added to Fleet—Total and From	LOST Vessels Men	LEFT Widows Children
1693	*8 total craft.* 6 sloops, 1 boat, and 1 shallop. One or more involved in transporting Cape Ann wood to Boston.			
1700	*70 to 80 large fishing vessels*			
1716			5 Ves 20 Men	
1721	*49 fishing vessels, 245 crew* (recorded at Canso, Nova Scotia)			
1722			1 Ves 5 Men	
1741	*70 schooners* (almost all in Grand Bank fishery for codfish)			
1765	*146 total vessels—5,530 gross tons, 888 crew.* 120 were fishing vessels (average of 37.5 tons), carrying 560 fishermen.			
1766			9 Ves 40 Men	
1770	*145 total vessels.* 70 to 80 went to the Grand Bank; about 70 fished for cod, hake, and pollock on the ledges near the Massachusetts coast.			
1775	*146 vessels—5,530 gross tons (average of 37.9 tons).* 80 in Gloucester; 70 in Sandy Bay [Rockport]. 75 schooners Banks fishing from Gloucester for cod, the boats in Sandy Bay engaged in the shore fishery, averaging 160 quintals of fish each. The cured fish were generally exported to Europe and the West Indies.			
1783	*60 vessels*			
1786–1790	*160 total codfishing vessels, 3,600 gross tons (average of 22.5 tons)*			
1789	*50 fishing vessels*			
1790	*43 total vessels—4,018 tons.* 23 schooners, 7 sloops, 4 ships, 9 brigs			
1801	*42 total vessels—5,635 gross tons.* 55 boats in Gloucester Harbor, 23 under 20 tons. 61 boats in Sandy Bay, 57 under 20 tons. 62 boats in Annisquam, 54 under 20 tons. 9 in town and West Parish, all under 20 tons. Of these vessels and boats, 43 were of only 6 tons burden. There were at least 9 vessels over 100 tons burden.			

Year	Size of Fleet	New Vessels Added to Fleet—Total and From	LOST Vessels Men	LEFT Widows Children
1803	*3 schooners and 200 boats*			
1804	*208 total vessels, but only 8 over 30 tons (average of 15.6 tons)—3,240 gross tons.* 200 Chebacco boats averaging 15 tons each.			
1808			3 Ves 10 Men	
1811	*185 total vessels—6,080 gross tons.* 153 vessels belonging to Gloucester in the coasting trade and cod fishery; 73 over 20 tons; 80 under 20 tons. Total tons = 3,278. The average craft was 8.7 years old. One vessel was added to the coasting and cod fleet in 1811. In 1810, 8 were added. Other craft: 1 ship, 6 brigs, 17 schooners, 8 sloops.			
1816	*261 total craft in Gloucester (average 30.7 tons)—8,014 gross tons.* 143 schooners of 20 tons or larger in fishing and coastal trade, including 6 larger vessels (average 86.7 tons) in the intercountry merchant trade. Excluding the 6, the remaining 137 schooners were 30.4 tons on average, with large numbers of these craft coming in between 20 and 25 tons burden. 41 of the schooners were less than 2 years old. 108 smaller boats of less than 20 tons (average 14.7 tons). 10 brigs (average 174.0 tons).			
1818	*304 total craft in Gloucester (average 33.0 tons)—10,040 gross tons.* 177 schooners of 20 tons or larger (average 36.7 tons). 43 of the schooners less than 2 years old (average 40.1 tons burden). 126 smaller boats of less than 20 tons (average 14.7 tons). 10 brigs (average 168.7 tons).			
1822	*249 total craft in Gloucester (average 36.2 tons)—9,009 gross tons.* 162 schooners of 20 tons or larger (average 32.8 tons). 11 of the schooners less than 2 years old. 67 smaller boats of less than 20 tons (average 14.5 tons). 13 brigs (average 159 tons), 3 ships (average 232 tons), and 7 large merchant schooners (average 94.1 tons).			
1824	*222 total craft in Gloucester (average 39.3 tons)—8,734 gross tons.* 152 schooners of 20 tons or larger (average 33.7 tons). 16 schooners less than 2 years old. 55 smaller boats of less than 20 tons (average 15.8 tons). 11 brigs (average 177.8 tons), 3 ships (average 232 tons), and one large merchant schooner (87 tons).			
1825	*194 schooners and sloops—6,499 gross tons.* [Note: this estimate does not fit into the numbers for 1825 and 1826.] 154 schooners over 20 tons and 40 boats less than 20 tons.			
1826	*291 total craft in Gloucester (average 45.9 tons)—13,356 gross tons.* 187 schooners of 20 tons or larger (average 38.8 tons). 49 of the schooners less than 2 years old (average 46.8 tons). 38 smaller boats of less than 20 tons (average 15.2 tons). 8 brigs (average 205.9 tons), 5 ships (average 247 tons), and 4 large merchant schooners (average 85.5 tons).			

Year	Size of Fleet	New Vessels Added to Fleet—Total and From	LOST Vessels Men	LEFT Widows Children
1827	*235 total vessels belonging to the district of Glouces-ter (average 55.6 tons)—13,073 gross tons.* 5 ships, 1 barque, 7 brigs, 46 coastal schooners, 14 ships, 165 fishing schooners.The fishing schooners come in at 6,376 tons (average 39 tons). Twenty-three individu-als (firms) owned from 2 to 5 schooners. Those with 3 or more were S. W. Brown (5), W. Blanchford (3), F. Davis (3), S. Caswell (4), E. Merchant (3), W. Mans-field (3), G. Steele (3), F. Tarr (3). It was also common for fathers, sons, brothers, and cousins to be in the trade each owning one vessel. By number of persons with same last name we find the following owner-ship: Allen (2), Babson (3), Blanchford (3), Brown (2), Chard (2), Davis (2), Fears (2), Friend (2), Giles (2), Lane (4), Langsford (2), Merchant (2), Mansfield (2), Norwood (4), Pool (7), Parsons (3), Sargent (2), Steele (2), Stevens (2), Tarr (8), Wheeler (2), Wonson (3), Woodbury (4), Young (3). In total, there were 127 individuals listed as owners of a Gloucester fishing schooner; 103 owned only one vessel.			
1828	*214 total vessels in the port of Gloucester of 20 or greater tons—12,041 gross tons.* 201 schooners of 20 tons or larger (8,133 tons, for an average of 40.5 tons). 5 ships (1,235 tons), 7 brigs (1,389 tons), and 6 schooners were registered in the foreign trade. 1 barge (264 tons),14 sloops, and 39 schooners were registered in the coasting trade. 164 schooners were enrolled in the cod fishery (for part of the year). There were also 68 smaller boats (average 15 tons). The size of the fleet had recently enlarged, stimulated in part by an 1819 act of Congress that gave a bounty to vessels engaged in the cod fisheries; 33 of the schooners were less than 2 years old.			
1829			1 Ves 1 Man	
1830			3 Ves 8 Men	
1832	*256 total craft in Gloucester (average 51.3 tons)—13,135 gross tons, 800 crew.* 213 schooners of 20 tons or larger (average 43.8 tons). 13 schooners less than 2 years old (average 47.3 tons). 26 smaller boats of less than 20 tons (average 15.7 tons). 11 brigs (average 212.5 tons), 3 ships (average 201.3 tons), 1 barque (264 tons), and 2 large merchant schooners (average 92.0 tons). 6,500 tons in fishing.		1 Ves 4 Men	
1833	*307 total vessels in Gloucester—15,271 gross tons.* 6 ships, 7 brigs, 240 schooners, 24 sloops, 30 boats. 190 were employed in the fisheries and 117 in the foreign and coasting trade.		1 Ves 0 Men	
1834			1 Ves 4 Men	
1835			0 Ves 2 Men	

Year	Size of Fleet	New Vessels Added to Fleet— Total and From	LOST Vessels Men	LEFT Widows Children
1836			1 Ves 4 Men	
1837	*221 total vessels in Gloucester engaged in the cod and mackerel fishery—9,824 gross tons, 1,580 crew, average of 7 men per boat*		5 Ves 26 Men	
1838	*247 total vessels in Gloucester engaged in the cod and mackerel fishery—11,000 gross tons*		4 Ves 4 Men	
1839			2 Ves 4 Men	
1840			2 Ves 8 Men	
1841			2 Ves 8 Men	
1842			3 Ves 0 Men	
1843			3 Ves 11 Men	
1844	*130 fishing schooners in Gloucester—6,500 gross tons, 850 crew, average of 6.5 men per boat*		3 Ves 7 Men	
1845	*334 total vessels in Gloucester (157 in Gloucester, 103 in Rockport, 58 in Annisquam, 16 in Manchester)— average vessel 49.5 tons, 16,524 gross tons, 1,850 crew. 242 schooners (127 in Gloucester, with 220 engaged in fishing), 22 sloops, 11 brigs, 1 ship, 58 boats.*		5 Ves 16 Men	
1846	*357 total vessels in Gloucester—17,548 gross tons, 1,850 crew, average of 6.6 men per boat.* (Engaged in fishing, 279 total vessels—220 schooners, 58 boats, 1 sloop.) [In Gloucester, 176 vessels (147 schooners, 17 boats), in Annisquam 60 (36 schooners, 20 boats).] 1 ship, 8 brigs, and 1 schooner registered in the foreign trade. 3 brigs, 80 schooners, 21 sloops, and 3 boats enrolled in the coasting trade. 161 schooners, 1 sloop, 55 boats enrolled in the fisheries (16,528 tons). Total of 263 schooners, 26 sloops, 9 brigs, 1 ship, 58 boats.	35 Tot 24 Ess 11 Glou	7 Ves 15 Men	
1847	*287 total vessels in Gloucester—12,354 gross tons, 1,867 crew, average of 6.5 men per vessel*	34 Tot 24 Ess 10 Glou	3 Ves 8 Men	
1848			0 Ves 0 Men	
1849	*Number of vessels unspecified; tonnage is for vessels in Gloucester—17,000 gross tons*		6 Ves 43 Men	
1850	*184 total vessels in Gloucester—13,531 gross tons. 166 schooners, 4 barques, 12 brigs, and 2 sloops.*		4 Ves 39 Men	
1851	*241 total vessels in Gloucester—2,326 crew*		10 Ves 40 Men	
1852	*357 total vessels—19,143 gross tons, 2,650 crew, average of 7.4 men per vessel*	55 Tot 45 Ess 9 Glou	13 Ves 40 Men	

Year	Size of Fleet	New Vessels Added to Fleet—Total and From	LOST Vessels Men	LEFT Widows Children
1853			3 Ves 23 Men	
1854	*282 vessels in cod and mackerel fisheries—19,734 gross tons, 2,820 crew, average of 10.0 men per vessel*	48 Tot 33 Ess 11 Glou	4 Ves 26 Men	
1855			7 Ves 21 Men	
1856			6 Ves 2 Men	
1857	*301 total vessels—13,882 gross tons, 3,658 crew*	28 Tot 20 Ess 6 Glou	5 Ves 15 Men	
1858			7 Ves 42 Men	
1859	*322 total vessels in Gloucester, 301 in fishing—23,882 gross tons (manned by 3,434 men and 134 boys [total of 3,568], average of 11.1 men per vessel)*		6 Ves 70 Men	9 Wid 13 Chl
1860			7 Ves 75 Men	17 Wid 15 Chl
1861			16 Ves 52 Men	11 Wid 19 Chl
1862			19 Ves 162 Men	74 Wid 153 Chl
1863	*305 schooners in Gloucester—24,316 gross tons.*		10 Ves 6 Men	3 Wid 1 Chl
1864			13 Ves 85 Men	31 Wid 58 Chl
1865	*378 total vessels in Gloucester (measures in accordance with new tonnage law)—25,170 gross tons, 4,700 crew, average of 13.2 men per vessel.* 365 schooners over 20 tons, 53 boats under 20 tons, and 13 sloops (300 schooners Gloucester, 35 Rockport, 28 Annisquam, 1 Essex).		8 Ves 11 Men	4 Wid 6 Chl
1866	*478 total vessels in Gloucester—22,534 gross tons.* 417 schooners, 50 boats, and 11 sloops. [369 in Gloucester, 39 in Annisquam, 1 in Essex, and 69 in Rockport] *The outfitters/owners included:* Morton C. Alexander (1) Joseph L. Andrews & Sons (5) Andrew G. Arey (1) D. C. Babson & Co (7) Philip Babson (1) E. A. Brazier (1) David A. Brown (1) Samuel W. Brown (10) Chas. A. Chadwick (1) Wm. W. Chard (1) Clark, Rust & Co. (7) Collins & Clark (6) E. Crowell II (1) Cunningham & Pool (9) Dennis & Ayer (7) George Dennis & Co. (7) Dodd, Tarr & Co (12) Andrew W. Dodd (1) Thomas Douglas (1) Nath'l Duley (1) Walter W. Fault (1) Peter Foley (1) Charles Friend & Co. (5) Joseph Friend & Son (11)		15 Ves 26 Men	10 Wid 18 Chl

Year	Size of Fleet	New Vessels Added to Fleet—Total and From	LOST Vessels Men	LEFT Widows Children
1866 (cont.)	William H. Friend & Bro. (6) George Gardner (1) George Garland (7) Abraham Gerring (1) G. Griffin & Son (6) J. T. Harvey (1) Samuel Haskell (3) Chas. M. Ingersoll (1) William G. Kiff (1) George W. Lane (1) Samuel R. Lane & Son (7) William Lane (1) Archibald Linniken (1) Francis Locke (1) Alfred Low & Co. (12) John Low Jr. & Son (12) James Mansfield & Sons (10) Knud Markuson (1) Geo. J. Marsh (1) Wm. H. Matthews (1) William McKenzie (8) Addison Merchant (6) M. C. Morgan (1) Nelson & Day (2) George Norwood & Son (9) James Parker (1) Charles Parkhurst (11) David Parkhurst (5) Wm. Parsons II and Co. (11) John Perkins & Co. (10) Pettingell & Bro. (9) Pettigrell & Wallace (7) John Pew & Son (16) George F. Plummer (1) George Poole (1) Patrick Powers (1) Addison Procter (2) Joseph O. Procter (11) Amos N. Rackliffe (1) Fitz E. Riggs (15) H. Roberts (1) John B. Roberts (1) Thomas Rorke (1) Rowe & Smith (10) Wm. Saunders (1) Daniel Sayward (5) Epes Sayward Jr. & Co. (4) Shute & Merchant (11) William H. Smith (1) George Steele (10) G. L. Stevens (1) Dodd & Tatt & Co. (12) Samuel Tarr (1) Jonathon Taylor (1) Samuel Upham (1) Henry Uson (1) Leonard Alen (1) James Wheeler Jr. (1) John F. Wonson & Co. (13) William C. Wonson (7) William H. Wonson III (1)			
1867	*527 total vessels in Gloucester fleet fishing for mackerel and cod—25,540 gross tons*	46 Tot	13 Ves 66 Men	20 Wid 36 Chl
1868	*490 total vessels in Gloucester—26,184 gross tons. 423 total fishing vessels (384 Gloucester, 39 Annisquam).*	47 Tot 29 Ess 5 Glou	4 Ves 40 Men	19 Wid 46 Chl
1869	*510 total vessels (317 owned by 42 firms)—24,891 gross tons, 6,120 crew, average of 12 men per vessel.* 435 fishing vessels in Gloucester. *Outfitters/owners with three or more vessels in Gloucester:* Joseph L. Andrews & Sons (14) D. C. & H. Babson (9) Brown Brothers (7) Edward E. Burnham (9) Clark & Somes (7) John J. Clark (4) Dennis & Ayer (8) George Dennis & Co. (8) Dodd, Tarr & Co. (16) Charles Friend & Co. (6) George Friend & Co. (3) Joseph Friend (18) Lemuel Friend & Co. (4) William H. Friend (4) George Garland (11) Gerring & Douglas (6) G. Griffin & Son (5) Samuel Haskell Jr. (7) John Harvey (3) Samuel Haskell (5) Samuel Lane & Bros. (7) Alfred Low & Co. (5) David Low & Co. (10) John Low & Son (14) Maddocks & Co. (13) James Mansfield & Sons (12) William McKenzie (8) Nelson & Day (6)		16 Ves 67 Men	24 Wid 35 Chl

Year	Size of Fleet	New Vessels Added to Fleet—Total and From	LOST Vessels Men	LEFT Widows Children
1869 (cont.)	George Norwood & Son (10) Charles Parkhurst (11) John Perkins & Co. (9) Pettingell & Cunningham (5) John Pew & Son (17) George W. Plummer (3) Solomon Pool (10) Joseph O. Procter (9) Rowe & Jordan (7) Saunders, Hunter & Co. (5) Daniel Sayward (7) Epes Sayward & Co. (5) Shute & Merchant (11) Smith & Gott. (8) Smith & Oakes (4) Stanwood & Leighton (7) George Steele (10) Tarr Brothers (3) Walen & Wonson (6) John F. Wonson & Co. (14) William C. Wonson (6)			
1870	*567 total vessels in Gloucester—29,022 gross tons, 6,084 crew, average of 10.7 men per vessel. 506 schooners, 12 sloops, 1 steamer, 48 boats. 426 schooners in Gloucester Harbor, 31 in Annisquam, 46 in Rockport, 1 in Essex, and 2 in Manchester.* *Outfitters/owners with three or more vessels in Gloucester:* D. C. & H. Babson (11) Brown Brothers (7) George Brown & Co. (6) Clark & Somes (7) John J. Clark (6) Sargent S. Day (6) Dennis & Ayer (12) George Dennis & Co. (7) Dodd, Tarr & Co. (13) Charles Friend (4) George Friend Jr. & Co. (5) Joseph Friend (14) Lemuel Friend & Co. (3) Sydney Friend & Bros. (7) William H. Friend (4) John H. Gale (2) George Garland (10) Gerring & Douglas (8) John T. Harvey (4) Samuel Haskell Jr. (5) Francis W. Homans (3) Samuel Lane & Bros. (6) Leighton & Co. (12) Alfred Low & Co. (2) David Low & Co. (16) Maddocks & Co. (12) James Mansfield & Sons (13) Hardy McKenzie (9) Knowlton McKenzie & Co (4) Charles Parkhurst (10) David Parkhurst (3) William Parsons II & Co. (13) Perkins Bros. (8) Pettingell & Cunningham (8) John Pew & Son (16) George W. Plummer (2) Solomon Pool (9) Joseph O. Procter (12) Rowe & Jordan (9) Daniel Sayward (9) Epes Sayward Jr. & Co. (6) Shute & Merchant (12) Smith & Gott. (11) Smith & Oakes (6) George Steele (7) Tarr Brothers (5) Walen & Allen (12) Walen & Wonson (4) Aaron D. Wells (4) John F. Wonson & Co. (14) William C. Wonson (8)	39 Tot 24 Ess 2 Glou	13 Ves 97 Men	26 Wid 45 Chl
1871	*548 total vessels in Gloucester district—28,560 gross tons, 5,900 crew, average of 10.8 men per vessel. 465 schooners.*		20 Ves 140 Men	48 Wid 98 Chl
1872	*538 total vessels in Gloucester—28,794 gross tons, 5,500 crew, average 10.5 men per vessel. 479 schooners (391 Gloucester, 28 Annisquam, 49 Rockport, 1 Essex, 1 Manchester), 11 sloops, 41 boats, 3 steamers. 389 in fishing, 63 in foreign trade, 85 coasting, 1 yacht.*	15 Tot 7 Ess 4 Glou	12 Ves 63 Men	21 Wid 32 Chl

Year	Size of Fleet	New Vessels Added to Fleet—Total and From	LOST Vessels Men	LEFT Widows Children
1872 (cont.)	*Outfitters/owners with three or more vessels:* D. C. & H. Babson (9) Brown Brothers (6) George Brown & Co. (5) Clark & Somes (13) Dennis & Ayer (12) George Dennis & Co. (6) Dodd, Tarr & Co. (13) Eldridge & Stetson (10) Fernald, Sargent & Co. (4) George Friend & Co. (5) Joseph Friend (10) Sidney Friend & Bro. (6) William H. Friend (5) George Garland (9) Gerring & Douglas (8) Samuel Haskell Jr. (7) Knowlton & Horton (2) Samuel Lane & Brothers (5) Leighton & Co. (17) Alfred Low & Co (3) D. Low & Co. (12) Maddocks & Co. (11) James Mansfield & Sons (12) McKenzie, Hardy & Co. (9) Charles Parkhurst (7) Wm. Parsons II & Co. (13) Perkins Bros. (7) Pettingell & Cunningham (7) John Pew & Son (19) Solomon Pool (8) Joseph O. Procter (12) Rowe & Jordan (8) Daniel Sayward (7) Sayward Bros. (5) Shute & Merchant (12) Smith & Gott. (13) Smith & Oakes (6) George Steele (9) Walen & Allen (9) Leonard Walen (4) Aaron D. Wells (5) John F. Wonson & Co. (14) William C. Wonson (7)			
1873	*521 total vessels in Gloucester—28,982 gross tons, 5,500 crew, average of 9.7 men per vessel. 422 vessels in fishing at 21,517 tons, 95 in coasting, 3 in foreign trade, 1 yacht.*		31 Ves 174 Men	47 Wid 47 Chl
1874	*497 total vessels—29,000 gross tons, 5,200 crew, average of 10.5 men per vessel. 438 schooners (387 Gloucester, 12 Annisquam, 35 Rockport, 2 Essex, 2 Manchester). 12 sloops, 36 boats, 9 steamers, 2 yachts. Outfitters/owners with three or more vessels:* D. C. & H. Babson (10) Clark & Somes (10) Dennis & Ayer (14) George Dennis & Co. (7) George Friend & Co. (4) Joseph Friend (6) Sidney Friend & Bro. (10) George Garland (8) Gerring & Douglas (10) Samuel Haskell Jr. (6) Harvey Knowlton Jr. (3) Samuel Lane & Bro. (7) Leighton & Co. (19) David Low & Co. (10) Maddocks & Co. (11) James Mansfield & Sons (11) McKenzie, Hardy & Co. (7) Charles Parkhurst (6) Wm. Parsons II & Co. (12) Perkins Bros. (10) Pettingell & Cunningham (9) John Pew & Son (20) Pool & Cunningham (9) Joseph O. Procter (14) Rowe & Jordan (12) Daniel Sayward (5) Sayward Bros (5) Shute & Merchant (13) Smith & Gott. (17) Smith & Oakes (6) George Steele (9) James A. Stetson (3) James G. Tarr & Bro. (11) Walen & Allen (10) Leonard Walen (4) John F. Wonson & Co. (15) William C. Wonson (6)	31 Tot 19 Ess 11 Glou	10 Ves 68 Men	18 Wid 37 Chl

Year	Size of Fleet	New Vessels Added to Fleet—Total and From	LOST Vessels Men	LEFT Widows Children
1875	*499 total vessels in Gloucester district—30,134 gross tons, 5,100 crew, average of 10.2 men per vessel.* 419 comprise the Gloucester Harbor fleet. 437 schooners (393 in Gloucester), 3 yachts, 12 sloops, 7 steamers, 40 boats. 39 Gloucester Harbor firms engaged in the fishing business. *Outfitters/owners with two or more vessels:* D. C. & H. Babson (11) Clark & Somes (11) Dennis & Ayer (13) George Dennis & Co. (9) George Friend & Co. (5) Joseph Friend (5) Sidney Friend & Bro. (11) Gerring & Douglas (9) George Garland (8) Benj. Haskell & Sons (3) Samuel Haskell Jr. (5) Harvey Knowlton Jr. (2) David Low & Co. (12) Samuel Lane & Bros (8) Leighton & Co. (18) Maddocks & Co. (11) James Mansfield & Sons (10) McKenzie, Hardy & Co. (7) George Norwood & Son (7) John Pew & Son (21) Pettingell & Cunningham (5) Joseph O. Procter (13) Pool & Cunningham (9) Perkins Bros (10) Wm. Parsons II & Co. (12) Rowe & Jordan (12) Daniel Sayward (5) Sayward Bros (5) Smith & Oakes (7) Shute & Merchant (13) Smith & Gott (16) George Steele (10) James A. Stetson (2) James G. Tarr & Bro. (13) William C. Wonson (5 John F. Wonson & Co. (12) Leonard Walen (3) Walen & Allen (12)	48 Tot 25 Ess 14 Glou	16 Ves 123 Men	21 Wid 22 Chl
1876	*522 total vessels in Gloucester on June 30—34,580 gross tons.* This represents an apparent increase of 19 vessels, but it is accounted for largely by 12 vessels bringing in salt cargos that took out temporary papers, which were surrendered when they left port to take permanent register for a foreign voyage. 414 vessels are in fishing (22,408 tons), 100 vessels are in coasting (11,123 tons), 5 in foreign trade (867 tons), and 3 yachts (182 tons). *Outfitters/owners with three or more vessels:* D. C. & H. Babson (12) Clark & Somes (11) William B. Coombs (1) Cunningham & Thompson (10) Dennis & Ayer (13) George Dennis & Co. (8) George Friend & Co. (4) Joseph Friend (9) Sidney Friend & Bro. (14) George Garland (8) Benj. Haskell & Sons (11) Samuel Haskell Jr. (5) Harvey Knowlton Jr. (3) Samuel Lane & Bro. (8) Andrew Leighton (17) David Low & Co. (13) Maddocks & Co. (11) James Mansfield & Sons (10) McKenzie, Hardy & Co. (7) George Norwood & Son (7) Charles Parkhurst (6) Wm. Parsons II & Co. (14) Perkins Bros. (9) Pettingell & Cunningham (5) John Pew & Son (19) Joseph O. Procter (12) Rowe & Jordan (13) Daniel Sayward (5) Sayward Bros. (5) Shute & Merchant (13)	34 Tot 15 Ess 10 Glou	27 Ves 212 Men	54 Wid 112 Chl

Year	Size of Fleet	New Vessels Added to Fleet—Total and From	LOST Vessels Men	LEFT Widows Children
1876 (cont.)	Smith & Gott (18) Smith & Oakes (6) George Steele (12) James A. Stetson (1) James G. Tarr & Bro. (16) Walen & Allen (14) Leonard Walen (6) John F. Wonson & Co. (13) William C. Wonson (5)			
1877	*516 total vessels in Gloucester—31,445 gross tons, 5,300 crew, average of 10.3 men per vessel.* 417 engaged in fisheries, 8 engaged in foreign trade, 97 in the coasting trade, and 3 in yachting.	27 Tot	8 Ves 39 Men	10 Wid 21 Chl
1878	*524 total vessels in Gloucester—31,694 gross tons.* 419 fishing vessels and 466 total vessels in Gloucester. 429 schooners (390 Gloucester, 6 Annisquam, 28 Rockport, 3 Essex, 2 Manchester), 13 sloops, 73 boats, 6 steamers, 3 yachts. *Outfitters/owners with three or more vessels:* D. C. & H. Babson (12) Clark & Somes (12) William B. Coombs (5) Cunningham & Thompson (12) Dennis & Ayer (13) George Dennis & Co. (9) George Friend & Co. (1) Joseph Friend (6) Sidney Friend & Bro. (13) George Garland (3) Benj. Haskell & Sons (7) Samuel Haskell Jr. (6) Harvey Knowlton Jr. (9) Samuel Lane & Bro. (8) Andrew Leighton (18) David Low & Co (11) Maddocks & Co. (9) James Mansfield & Sons (6) McKenzie, Hardy & Co. (7) George Norwood & Son (9) Charles Parkhurst (6) Wm. Parsons II & Co. (13) Perkins Bros. (9) Pettingell & Cunningham (6) John Pew & Son (20) Joseph O. Procter (12) Rowe & Jordan (14) Daniel Sayward (3) Sayward Bros. (5) Shute & Merchant (10) Smith & Oakes (7) Sylvannus Smith (14) George Steele (11) James A. Stetson (1) James G. Tarr & Bro. (15) Walen & Allen (12) Leonard Walen (4) Wonson Bros. (10) John F. Wonson & Co. (13) William C. Wonson (8)	12 Tot 4 Ess 6 Glou	13 Ves 33 Men	8 Wid 20 Chl
1879	*490 total vessels—29,045 gross tons, 5,253 crew, average of 10.7 men per vessel.* 445 vessels in fishing (91 fishing here but owned out of town). Of vessels in fishing: 338 Gloucester, 11 Rockport, 81 other NE ports,15 British Provinces. 104 vessels followed Georges all season and 82 followed the Banks cod and halibut fisheries all season (another 32 vessels made one trip). 25 vessels made Bay of St. Lawrence trips for mackerel. About 100 vessels in the shore mackerel fishery. 47 vessels in the shore fishery. Half a score made herring trips to Newfoundland.	6 Tot	32 Ves 266 Men	92 Wid 222 Chl
1880	*475 total vessels in Gloucester—27,456 gross tons.* 442 in fishing. 374 schooners (342 Gloucester, 4 Annisquam, 26 Rockport, 1 Manchester), 12 sloops, 81 boats, 5 steamers, 3 yachts. 97 vessels engaged in coasting trade.	5 Tot 2 Ess 2 Glou	7 Ves 55 Men	11 Wid 18 Chl

Year	Size of Fleet	New Vessels Added to Fleet—Total and From	LOST Vessels Men	LEFT Widows Children
1880 (cont.)	*Outfitters/owners with three or more vessels:* Atlantic Halibut Company (9) D. C. & H. Babson (11) Clark & Somes (11) Cunningham & Thompson (10) Dennis & Ayer (13) George Dennis & Co. (8) Joseph Friend (7) George Garland (3) Samuel Haskell (6) Harvey Knowlton (5) Samuel Lane & Bros. (7) Andrew Leighton (14) Benjamin Low (10) Maddocks & Co. (7) James Mansfield & Sons (6) McKenzie, Hardy & Co. (6) Benj. Montgomery (3) George Norwood & Sons (5) William Parsons II & Co. (11) Wm. H. Perkins Jr. (5) Pettingell & Cunningham (5) John Pew & Son (20) Joseph O. Procter Jr. (10) Rowe & Jordan (12) Sayward Brothers (5) Shute & Merchant (9) Smith & Oakes (6) Sylvannus Smith (13) George Steele (12) James G. Tarr & Bro. (15) Leonard Walen (3) Michael Walen & Son (10) Wonson Brothers (11) John F. Wonson & Co. (14) William C. Wonson & Co. (6)			
1881	*437 total fishing vessels in Gloucester district—4,142 crew.* Gloucester had 343 vessels: 17 belonged elsewhere but fished from Gloucester during the year, and 77 made one or more trips during the year. Of the men in the fleet, 1,460 in the Georges fleet, 284 in Western Bank, 360 Grand Bank, 330 fresh halibut, 1,120 mackereling, 430 shore, 125 dory fishing, 24 trap fishing.	17 Tot 11 Ess 2 Glou	9 Ves 55 Men	6 Wid 15 Chl
1882	*483 total vessels in Gloucester—27,810 gross tons.* 382 schooners (353 Gloucester, 25 Rockport, 4 Annisquam), 14 sloops, 2 yachts, 6 steamers, 79 boats. *Outfitters/owners with two or more vessels:* Atlantic Halibut Company (7) Daniel Allen & Son (7) D. C. & H. Babson (11) Cunningham & Thompson (10) George Clark & Co. (9) William B. Coombs (2) Dennis & Ayer (12) George Dennis & Co. (6) Joseph Friend (5) George Garland (2) Samuel Haskell (6) Samuel Lane & Brothers (5) Andrew Leighton (14) Benjamin Low (9) Maddocks & Co. (6) James Mansfield & Sons (6) John S. McQuin & Co. (3) McKenzie, Hardy & Co. (5) Benj. Montgomery (4) George Norwood & Sons (7) Oakes & Foster (6) William Parsons II & Co. (10) Pettingell & Cunningham (4) John Pew & Son (19) Joseph O. Procter Jr. (9) Rowe & Jordan (13) Sayward Brothers (5) Shute & Merchant (9) Sylvannus Smith (12) George Steele (11) James G. Tarr & Bro. (16) George J. Tarr (3) Leonard Walen (3) Michael Walen & Son (8) Wonson Brothers (3) William C. Wonson & Co. (5)	43 Tot 25 Ess 8 Glou	15 Ves 115 Men	50 Wid 113 Chl

Year	Size of Fleet	New Vessels Added to Fleet—Total and From	LOST Vessels Men	LEFT Widows Children
1883	508 *total vessels in Gloucester—31,675 gross tons.* 418 schooners, 2 yachts, 11 sloops, 5 steamers, 72 boats. Of the schooners, 391 in Gloucester, 25 in Rockport, 1 in Annisquam, 1 in Lanesville. *Outfitters/owners with two or more vessels:* Daniel Allen & Son (8) D. C. & H. Babson (11) Burnham & Parsson (6) George Clark (8) William B. Coombs (2) Cunningham & Thompson (11) Dennis & Ayer (13) George Dennis (3) Thomas Douglass (3) Walter M. Fault (2) Joseph Friend (5) George Garland (2) Benjamin Haskell & Sons (11) Samuel Haskell (5) Samuel Lane & Bros. (8) Timothy A. Langsford (3) Andrew Leighton (16) Benjamin Low (11) Maddocks & Co. (5) James Mansfield & Sons (7) John H. McDonough (4) McKenzie, Hardy & Co. (5) John S. McQuin & Co. (2) Benj. Montgomery & Son (4) George Norwood & Son (8) Oakes & Foster (7) Will. Parsons II & Co. (11) Pettingell & Cunningham (5) John Pew & Son (19) Pool & Gardner (7) Joseph O. Procter Jr. (12) Rowe & Jordan (14) Sayward Brothers (7) Shute & Merchant (6) Joseph Smith (3) Sylvannus Smith (12) Benjamin H. Spinney (3) George Steele (12) James G. Tarr & Bro. (14) George J. Tarr & Son (4) Leonard Walen (3) Michael Walen & Son (7) Wonson Brothers (5) John F. Wonson & Co. (17) William C. Wonson & Co. (4)	46 Tot 30 Ess 6 Glou	17 Ves 209 Men	40 Wid 68 Chl
1884	501 *total vessels in Gloucester district—30,827 gross tons.* Rockport 51, Lanesville 5, Bay View 2, and Annisquam 2. 415 schooners; 52 schooner boats; 3, three-masted schooners; 18 sloops; 4 boats; 1 pinky; 2 yachts; 1 steam water boat; 1 steam lighter; 2 steam tugs; 1 steamboat. 48 fishing firms fit out 350 schooners and boats; the balance of the fishing fleet is comprised of vessels whose masters are the owners. *Outfitters/owners with two or more vessels:* Daniel Allen & Son (8) D. C. & H. Babson (11) Burnham & Presson (5) George Clark & Co. (8) William B. Coombs (2) Cunningham & Thompson (10) Dennis & Ayer (12) George Dennis & Co. (8) Thomas Douglass (3) Walter M. Falt (2) Joseph Friend (5) Benjamin Haskell & Sons (16) Samuel Haskell (4) Samuel Lane & Brothers (8) Andrew Leighton (16) T. A. Langsford & Son (3) Benjamin Low (13) Maddocks & Co. (5) James Mansfield & Sons (5) John S. McQuin (3) John H. McDonough (3) Mergerson & Blatchford (3) McKenzie, Hardy & Co (6) B. Montgomery & Son (4) George Norwood & Sons (8) Oakes & Foster (7) William Parsons II & Co. (10) Pettingell & Cunningham (5)	28 Tot 19 Ess 3 Glou	16 Ves 131 Men	50 Wid 66 Chl

Year	Size of Fleet	New Vessels Added to Fleet— Total and From	LOST Vessels Men	LEFT Widows Children
1884 (cont.)	John Pew & Son (19) Poole & Gardner (10) Wm. H. Perkins Jr. (4) Joseph O. Procter Jr. (11) Rowe & Jordan (15) Sayward Brothers (7) Shute & Merchant (4) Joseph Smith (3) Benjamin H. Spinney (3) Sylvannus Smith (13) George Steele (11) Stockbridge & Co. (3) James G. Tarr & Bro. (12) Leonard Walen (3) Michael Walen & Son (6) Wonson Brothers (4) John F. Wonson & Co. (16) Walter W. Wonson & Bro. (8) William C. Wonson & Co. (6)			
1885	*511 total vessels in Gloucester—31,800 gross tons.* 452 in Gloucester Harbor. 433 schooners (403 Gloucester, 26 Rockport, 1 Annisquam, 3 Lanesville, 1 Bay View), 11 sloops, 59 boats, 4 steamers, 2 yachts, 3 steam sloops. *Outfitters/owners with three or more vessels:* Daniel Allen & Son (10) D. C. & H. Babson (10) John Chisholm (3) George Clark & Co. (8) Cunningham & Thompson (10) Dennis & Ayer (11) George Dennis & Co. (9) Joseph Friend (5) Samuel Haskell (3) Samuel Lane & Brothers (9) Andrew Leighton (17) T. A. Langsford & Son (7) Benjamin Low (13) Maddocks & Co. (5) James Mansfield & Sons (5) Margeson & Blatchford (3) John H. McDonough (3) McKenzie, Hardy & Co. (6) Benj. Montgomery & Son (4) George Norwood & Sons (7) Oakes & Foster (7) William Parsons II & Co. (9) Wm. H. Perkins Jr. (1) Pettingell & Cunningham (6) John Pew & Son (19) Pool & Gardner (10) David S. Presson (4) Joseph O. Procter Jr. (11) Rowe & Jordan (15) Sayward Brothers (7) Shute & Merchant (4) Joseph Smith (4) Sylvannus Smith (13) George Steele (12) Stockbridge & Co. (4) James G. Tarr & Bro. (12) Leonard Walen (3) Michael Walen & Son (6) Wonson Brothers (4) John F. Wonson & Co. (16) Walter W. Wonson & Bro. (7) William C. Wonson & Co. (5)	20 Tot 15 Ess 3 Glou	13 Ves 35 Men	5 Wid 13 Chl
1886	*487 total vessels in Gloucester—30,584 gross tons.* 428 in Gloucester Harbor. 410 schooners (382 Gloucester, 26 Rockport, 1 Annisquam, 1 Lanesville—average size 70.6 tons), 10 sloops, 57 boats, 8 steamers, 2 yachts. *Outfitters/owners with two or more vessels:* Daniel Allen & Son (9) James S. Ayer (11) D. C. & H. Babson (8) John Chisholm (3) George Clark & Co. (8) William B. Coombs (2) Cunningham & Thompson (9) George Dennis (7) Joseph Friend (5) B. Haskell & Sons (3) Samuel Haskell (3) Thomas Hodge (5) Samuel Lane & Brothers (8) Andrew Leighton (17) T. A. Langsford & Son (6)	16 Tot 10 Ess 3 Glou	27 Ves 136 Men	22 Wid 50 Chl

Year	Size of Fleet	New Vessels Added to Fleet—Total and From	LOST Vessels Men	LEFT Widows Children
1886 (cont.)	Benjamin Low (13) Maddocks & Co. (5) James Mansfield & Sons (5) J. Edward Margeson (3) John H. McDonough (3) McKenzie, Hardy & Co. (6) James S. McQuin (2) Benj. Montgomery & Son (3) Edward Morris (3) George Norwood & Son (7) Oakes & Foster (7) William Parsons II & Co. (8) Wm. H. Perkins Jr. (1) Pettingell & Cunningham (6) John Pew & Son (18) Pool, Gardner & Co. (7) Joseph O. Procter Jr. (10) Reed & Gamage (8) Rowe & Jordan (15) Sayward Brothers (7) Shute & Merchant (3) David B. Smith (3) Joseph Smith (4) Sylvannus Smith & Co. (12) Benjamin H. Spinney (3) George Steele (12) James G. Tarr & Bro. (14) Leonard Walen (3) Michael Walen & Son (6) C. F. Wonson & Co. (6) John F. Wonson & Co. (16) William C. Wonson & Co. (5) Wonson Bros. (3)			
1887	*475 total vessels in Gloucester—30,716 gross tons.* 419 in Gloucester Harbor. 401 schooners (376 Gloucester, 24 Rockport, 1 Annisquam, 1 Lanesville—average size 74.0 tons), 10 sloops, 54 boats, 7 steamers, 2 yachts, 1 barge. *Outfitters/owners with two or more vessels:* Daniel Allen & Son (9) James S. Ayer (10) D. C. & H. Babson (10) George Clark & Co. (8) William B. Coombs (1) Cunningham & Thompson (11) George Dennis (7) Joseph Friend (5) Samuel Haskell (3) Thomas Hodge (5) Samuel Lane & Brothers (8) Andrew Leighton (18) T. A. Langsford & Son (7) Benjamin Low (11) Maddocks & Co. (6) James Mansfield & Sons (5) John H. McDonough (3) McKenzie, Hardy & Co. (5) Benj. Montgomery & Sons (3) George Norwood & Son (6) Oakes & Foster (7) William Parsons II & Co. (8) William H. Perkins Jr. (1) Pettingell & Cunningham (6) John Pew & Son (18) Pool, Gardner & Co. (7) Joseph O. Procter Jr. (9) Reed & Gamage (7) Rowe & Jordan (14) Sayward Brothers (6) Shute & Merchant (4) David B. Smith (5) Sylvannus Smith & Co. (12) Benjamin H. Spinney (3) George Steele (9) James G. Tarr & Bro. (14) Leonard Walen (3) Michael Walen & Son (5) C. F. Wonson & Co. (5) John F. Wonson & Co. (15) William C. Wonson & Son (5) Wonson Bros. (2)	20 Tot 12 Ess 5 Glou	17 Ves 123 Men	25 Wid 59 Chl
1888	*471 total vessels in Gloucester—30,254 gross tons.* 416 in Gloucester Harbor. 386 schooners, 10 sloops, 64 boats under 20 tons, 8 steamers, 2 yachts, 1 barge.	18 Tot 13 Ess 3 Glou	14 Ves 73 Men	6 Wid 16 Chl

Year	Size of Fleet	New Vessels Added to Fleet—Total and From	LOST Vessels Men	LEFT Widows Children
1889	*466 total vessels in Gloucester—30,630 gross tons.* 376 schooners (354 Gloucester, 20 Rockport, 1 Lanesville), 1 Annisquam, 13 sloops, 65 boats, 10 steamers, 2 yachts, 1 barge. *Outfitters/owners with two or more vessels:* Daniel Allen & Son (10) James S. Ayer (7) D. C. & H. Babson (8) George Clark & Co. (7) Cunningham & Thompson (10) George Dennis (6) Joseph Friend (9) Samuel Haskell (3) Hodge & Pool (12) Wm. H. Jordan (13) Samuel Lane & Brothers (8) T. A. Langsford & Son (8) Andrew Leighton (17) Benjamin Low (10) James Mansfield & Sons (5) Wm. N. McKenzie & Co. (3) Benj. Montgomery & Son (3) George Norwood & Sons (5) Oakes & Foster (7) Henry A. Parmenter (5) William Parsons II & Co. (7) Wm. H. Perkins (2) Pettingell & Cunningham (6) John Pew & Son (16) Pool & Gardner (10) Joseph O. Procter Jr. (7) Reed & Gamage (9) Sayward Brothers (7) Shute & Merchant (1) David B. Smith (9) Sylvannus Smith (13) George Steele (9) James G. Tarr & Bro. (16) Michael Walen & Son (6) C. F. Wonson & Co. (7) John F. Wonson & Co. (15) William C. Wonson & Co. (2)	24 Tot 18 Ess 3 Glou	18 Ves 86 Men	13 Wid 48 Chl
1890	*470 total vessels—33,421 gross tons.* 393 enrolled in fishing.	22 Tot 16 Ess 4 Glou	20 Ves 86 Men	7 Wid 13 Chl
1891	*481 total vessels in Gloucester—32,242 gross tons.* 383 schooners (365 Gloucester, 16 Rockport, 1 Lanesville, 1 Annisquam), 13 sloops, 69 boats, 10 steamers, 5 yachts, 1 barge. *Outfitters/owners with three or more vessels:* Daniel Allen & Son (11) James S. Ayer (7) Albert P. Babson (4) D. C. & H. Babson (8) George Clark & Co. (7) John Chisholm (11) S. V. Colby (3) Cunningham & Thompson (12) George Dennis (5) Wm. H. Gardner & Co. (15) Samuel Haskell (3) Hodge & Pool (23) Wm. H. Jordan (14) Samuel Lane & Brothers (11) T. A. Langsford & Son (9) Andrew Leighton (13) Benjamin Low (10) James Mansfield & Sons (3) Wm. N. McKenzie & Co (2) Benj. Montgomery & Son (1) Edward Morris (7) George Norwood & Sons (5) Oakes & Foster (7) Henry A. Parmenter (6) William Parsons II & Co. (7) Wm. H. Perkins (1) Pettingell & Cunningham (6) John Pew & Son (16) Joseph O. Procter Jr. (5)	40 Tot 28 Ess 11 Glou	17 Ves 78 Men	21 Wid 58 Chl

Year	Size of Fleet	New Vessels Added to Fleet— Total and From	LOST Vessels Men	LEFT Widows Children
1891 (cont.)	Reed & Gamage (8) Sayward Brothers (6) Shute & Merchant (1) David B. Smith (16) Sylvannus Smith (12) George Steele (10) James G. Tarr & Bro. (14) Michael Walen & Son (8) C. F. Wonson & Co. (7) John F. Wonson & Co. (17) William C. Wonson & Co. (1)			
1892	*474 total vessels in Gloucester—32,795 gross tons. 426 in Gloucester Harbor. 387 schooners (369 Gloucester, 16 Rockport, 1 Annisquam, 1 Lanesville—average size 79.9 tons), 11 sloops, 61 boats, 10 steamers, 4 yachts, 1 barge.* *Outfitters/owners with two or more vessels:* Daniel Allen & Son (9) James S. Ayer (7) Albert P. Babson (4) D. C. & H. Babson (8) George Clark & Co. (7) John Chisholm (13) S. V. Colby (3) Cunningham & Thompson (12) George Dennis (5) Gardner & Parsons (13) Samuel Haskell (3) Hodge & Pool (21) Eli Jackman & Co. (2) Wm. H. Jordan (15) Samuel Lane & Brothers (8) T. A. Langsford & Son (9) Benjamin Low (11) John H. McDonough (2) Wm. N. McKenzie & Co. (2) Benj. Montgomery & Son (2) Edward Morris (13) George Norwood & Son (3) Oakes & Foster (7) Parmenter & Co. (7) William Parsons II & Co. (7) Wm. H. Perkins (2) Pettingell & Cunningham (6) John Pew & Son (16) Sayward Brothers (6) Shute & Merchant (1) David B. Smith (16) Sylvannus Smith (14) George Steele (11) James G. Tarr & Bro. (14) Michael Walen & Son (8) John F. Wonson & Co. (18) William C. Wonson & Co. (1) Wonson Bros. (1)	28 Tot 19 Ess 9 Glou	12 Ves 46 Men	4 Wid 8 Chl
1893	*483 total vessels in Gloucester—33,927 gross tons. 432 in Gloucester Harbor. 383 schooners (360 Gloucester, 15 Rockport, 1 Lanesville, 1 Annisquam—average size 81.0 tons), 11 sloops, 52 schooner boats, 21 sloop boats, 73 boats, 11 steamers, 5 yachts, 1 barge.* *Outfitters/owners with two or more vessels:* Daniel Allen & Son (9) James S. Ayer (7) Albert P. Babson (4) John Chisholm (14) George Clark & Co. (6) Cunningham & Thompson (13) Davis Bros. (5) George Dennis (3) Gardner & Parsons (14) Augustus G. Hall (3) Hodge & Pool (19) Eli Jackman & Co. (2) William H. Jordan (16) Samuel Lane & Bros. (8) T. A. Langsford & Son (8) Andrew Leighton (2) Benjamin Low (12) Jerome McDonald (4) Wm. N. McKenzie & Co. (2) Benj. Montgomery & Son (2) Edward Morris (11) Oakes & Foster (7) Parmenter & Co. (7) William Parsons II & Co. (7) William H. Perkins (2) Pettingell & Cunningham (6) John Pew & Son (16) Sayward Brothers (5) Shute & Merchant (1) David B. Smith (18)	32 Tot 23 Ess 8 Glou	19 Ves 72 Men	13 Wid 30 Chl

Year	Size of Fleet	New Vessels Added to Fleet—Total and From	LOST Vessels Men	LEFT Widows Children
1893 (cont.)	Sylvannus Smith (14) George Steele (15) James G. Tarr & Bro. (14) Michael Walen & Son (8) John F. Wonson & Co. (18) William C. Wonson & Co. (2) Wonson Bros. (1)			
1894	*463 total vessels in Gloucester—32,628 gross tons.* 416 in Gloucester Harbor, including 341 schooners. 359 schooners (341 Gloucester, 15 Rockport, 1 Annisquam, 1 Lanesville—average size 77.2 tons), 3 three-masted schooners, 11 sloops, 48 schooner boats, 24 sloop boats, 11 steamers, 6 yachts, 1 barge. *Outfitters/owners with two or more vessels:* Daniel Allen & Son (7) James S. Ayer (6) Albert P. Babson (4) John Chisholm (13) George Clark & Co. (5) Cunningham & Thompson (18) Davis Bros. (6) Gardner & Parsons (13) Augustus G. Hall (3) Loring B. Haskell (6) Hodge & Pool (18) Eli Jackman & Co. (3) William H. Jordan (14) Samuel Lane & Brothers (9) T. A. Langsford & Son (7) Andrew Leighton (2) Benjamin Low (9) Jerome McDonald (4) Wm. N. McKenzie & Co. (2) Benj. Montgomery & Son (2) Edward Morris (9) Oakes & Foster (6) Hugh Parkhurst & Co. (3) Parmenter & Co. (6) William Parsons II & Co. (9) William H. Perkins (2) Pettingell & Cunningham (6) John Pew & Son (17) Sayward Brothers (5) Shute & Merchant (1) David B. Smith (18) Sylvannus Smith & Co. (13) George Steele (15) James G. Tarr & Bro. (12) Michael Walen & Son (11) Charles F. Wonson & Co. (4) John F. Wonson & Co. (18) William C. Wonson & Co. (2) Wonson Bros. (1)	24 Tot 15 Ess 7 Glou	30 Ves 137 Men	50 Wid 54 Chl
1895	*445 total vessels in Gloucester—32,010 gross tons.* 400 in Gloucester Harbor, including 333 schooners. 348 schooners (333 Gloucester, 13 Rockport, 1 Annisquam, 1 Lanesville—average size 82.9 tons), 2 three-masted schooners, 13 sloops, 38 schooner boats, 25 sloop boats, 11 steamers, 6 yachts, 2 barges. *Outfitters/owners with two or more vessels:* Daniel Allen & Son (7) James S. Ayer (6) Albert P. Babson (3) John Chisholm (8) Cunningham & Thompson (16) Davis Bros. (9) Gardner & Parsons (14) Nathaniel Greenleaf (2) Augustus G. Hall (3) Loring B. Haskell (11) Hodge & Pool (18) Eli Jackman & Co. (3) William H. Jordan (14) Samuel Lane & Brothers (10) T. A. Langsford & Son (8) Andrew Leighton (2) Benjamin Low (7) Jerome McDonald (4) Wm. N. McKenzie & Co. (2)	9 Tot 8 Ess	13 Ves 125 Men	19 Wid 86 Chl

Year	Size of Fleet	New Vessels Added to Fleet— Total and From	LOST Vessels Men	LEFT Widows Children
1895 (cont.)	Benj. Montgomery & Son (2) Edward Morris (4) Oakes & Foster (6) Hugh Parkhurst & Co. (4) Parmenter & Co. (6) William Parsons II & Co. (8) William H. Perkins (1) Pettingell & Cunningham (6) John Pew & Son (16) Sayward Brothers (5) Shute & Merchant (1) David B. Smith (17) Sylvannus Smith & Co. (13) George Steele (15) James G. Tarr & Bro. (11) Michael Walen & Son (10) Charles F. Wonson & Co. (4) John F. Wonson & Co. (17) William C. Wonson & Co. (2) Wonson Bros. (1)			
1896	*435 total vessels in Gloucester district—30,767 gross tons.* 365 vessels in fishing. 342 schooners, 1 three-masted schooner, 12 sloops, 37 schooner boats, 23 sloop boats, 11 steamers, 5 yachts, 4 barges. (51 Grand Banks dory trawling, 79 on Georges Bank, 95 fresh fishing, 3 salt, and 42 in fresh-halibut fishing, 15 mackerel seining, 20 frozen herring trade, 60 yet to fit out.)	6 Tot	14 Ves 88 Men	15 Wid 41 Chl
1897	*401 total vessels in Gloucester—31,352 gross tons.* 312 schooners (297 Gloucester, 13 Rockport, 1 Annisquam, 1 Lanesville), 1 three-masted schooner, 11 sloops, 30 schooner boats, 31 sloop boats, 11 steamers, 1 yacht, 4 barges. *Outfitters/owners with three or more vessels:* Daniel Allen & Son (6) James S. Ayer (6) Albert P. Babson (1) John Chisholm (10) Cunningham & Thompson (16) Davis Brothers (12) Gardner & Parsons (11) Loring B. Haskell (12) Hodge & Pool (17) Wm. H. Jordan (13) W. A. King & Co. (4) Samuel Lane & Brothers (11) T. A. Langsford & Son (4) Andrew Leighton (2) Benjamin Low (6) Jerome McDonald (4) Wm. N. McKenzie & Co. (1) Benj. Montgomery & Son (2) Oakes & Foster (6) Hugh Parkhurst & Co. (4) Parmenter & Co. (6) William Parsons II & Co. (4) John Pew & Son (14) Sayward Brothers (5) David B. Smith (18) Sylvannus Smith (12) George Steele (13) James G. Tarr & Bro. (11) Michael Walen & Son (9) Chas. F. Wonson & Co. (4) John F. Wonson & Co. (14) William C. Wonson & Co. (1)	5 Tot 3 Ess 2 Glou	11 Ves 63 Men	7 Wid 24 Chl
1898	*382 total vessels in Gloucester district.* 280 schooners (264 Gloucester, 14 Rockport, 1 Annisquam, 1 Lanesville), 2 three-masted schooners, 10 sloops, 19 schooner boats, 33 sloop boats, 14 steamers, 1 yacht, 4 barges.	5 Tot	20 Ves 97 Men	29 Wid 74 Chl

Year	Size of Fleet	New Vessels Added to Fleet—Total and From	LOST Vessels Men	LEFT Widows Children
1898 (cont.)	*Outfitters/owners with three or more vessels:* James E. Bradley (6) Albert P. Babson (1) John Chisholm (9) Cunningham & Thompson (15) Davis Brothers (11) Gardner & Parsons (9) Augustus G. Hall (2) Loring B. Haskell (10) Hodge & Pool (13) Wm. H. Jordan (10) Samuel Lane & Brothers (11) T. A. Langsford & Son (2) Andrew Leighton (2) Jerome McDonald (3) Benj. Montgomery & Son (2) Oakes & Foster (6) Hugh Parkhurst & Co. (4) Parmenter & Co. (6) William Parsons II & Co. (3) John Pew & Son (14) Sayward Brothers (5) David B. Smith (20) Sylvannus Smith (13) George Steele (10) James G. Tarr & Bro. (11) Michael Walen & Son (10) John F. Wonson & Co. (14) William C. Wonson & Co. (1)			
1899	*350 total vessels in Gloucester—27,042 gross tons.* 308 in Gloucester Harbor. 254 schooners (240 in Gloucester Harbor), 1 three-masted schooner, 9 sloops, 32 schooner boats, 34 sloop boats, 13 steamers, 3 yachts, 4 barges.	16 Tot 13 Ess 3 Glou	17 Ves 68 Men	15 Wid 37 Chl
1900	*351 total vessels in Gloucester—28,076 gross tons, 4,140 crew, average of 11.8 men per vessel.* 251 schooners (237 Gloucester, 13 Rockport, 1 Lanesville), 1 three-masted schooner, 9 sloops, 34 schooner boats, 32 sloop boats, 15 steamers, 2 sloop yachts, 1 schooner yacht, 6 barges. *Outfitters/owners with three or more vessels:* James E. Bradley (4) John Chisholm (10) Cunningham &Thompson (12) Davis Brothers (13) Gardner & Parsons (9) John Gleason Jr. (5) Loring B. Haskell (5) Thomas Hodge (3) Wm. A. King & Co. (3) Samuel Lane & Brothers (9) T. A. Langsford & Son (3) Est. of Andrew Leighton (2) Jerome McDonald (4) Orlando Merchant (10) Benj. Montgomery & Son (2) Oakes & Foster (5) Hugh Parkhurst & Co. (4) John Pew & Son (16) Samuel G. Pool & Sons (8) Sayward Brothers (2) David B. Smith (21) Sylvannus Smith (12) Lemuel Spinney (2) George Steele (8) James G. Tarr & Bro. (13) Michael Walen & Son (8) John F. Wonson & Co. (12) William C. Wonson & Co. (1)	23 Tot 15 Ess 6 Glou	9 Ves 53 Men	18 Wid 24 Chl
1901	*353 total vessels in Gloucester—28,366 gross tons.* 248 schooners, 2 auxiliary schooners, 1 three-masted schooner, 10 sloops, 28 schooner boats, 38 sloop boats, 17 steamers, 3 schooner yachts, 1 sloop yacht, 5 barges. 194 vessels over 20 tons.	27 Tot 15 Ess 12 Glou	9 Ves 47 Men	9 Wid 21 Chl

Year	Size of Fleet	New Vessels Added to Fleet—Total and From	LOST Vessels Men	LEFT Widows Children
1902	*362 total vessels in Gloucester—29,209 gross tons.* 252 schooners, 6 auxiliary schooners, 1 three-masted schooner, 9 sloops, 30 schooner boats, 41 sloop boats, 1 auxiliary schooner boat, 1 auxiliary sloop boat, 13 steamers, 2 schooner yachts, 1 sloop yacht, 5 barges. 194 vessels over 20 tons.	28 Tot	10 Ves 82 Men	30 Wid 52 Chl
1903	*364 total vessels in Gloucester—28,892 gross tons.* 255 schooners (244 Gloucester, 12 Rockport, 1 Lanesville, 1 Manchester—average size 96.4 gross tons), 5 auxiliary schooners, 1 three-masted schooner, 8 sloops, 33 schooner boats, 43 sloop boats, 12 steamers, 1 auxiliary sloop boat, 1 auxiliary schooner boat, 2 schooner yachts, 1 sloop yacht, 2 barges. *Outfitters/owners with two or more vessels:* James E. Bradley (4) John Chisholm (12) Cunningham & Thompson (16) Davis Brothers (14) Gardner & Parsons (10) John Gleason Jr. (7) Thomas Hodge (3) Wm. A. King & Co. (5) Samuel Lane & Brothers (10) T. A. Langsford & Son (2) Jerome McDonald (7) Orlando Merchant (14) Benj. Montgomery & Son (1) Oakes & Foster (3) Hugh Parkhurst & Co. (6) John Pew & Son (15) Samuel G. Pool & Sons (5) Sayward Brothers (2) David B. Smith (26) Samuel P. Smith (6) Sylvannus Smith & Co. (13) Lemuel E. Spinney (3) George Steele (8) James G. Tarr & Bro. (12) Michael Walen & Son (7) John F. Wonson & Co. (13) Charles C. Young (2)	20 Tot 11 Ess 9 Glou	9 Ves 73 Men	14 Wid 40 Chl
1904	*350 total vessels in Gloucester—28,021 gross tons.* 239 schooners (227 Gloucester, 1 Lanesville, 5 Rockport, 6 Pigeon Cove), 5 auxiliary schooners, 1 three-masted schooner, 8 sloops, 32 schooner boats, 45 sloop boats, 12 steamers, 1 auxiliary schooner boat, 1 schooner yacht, 2 sloop yachts, 3 barges. *Outfitters/owners with two or more vessels:* James E. Bradley (4) John Chisholm (11) Cunningham & Thompson (17) Davis Brothers (13) Gardner & Parsons (10) John Gleason Jr. (7) Thomas Hodge (3) Wm. A. King & Co. (5) Samuel Lane & Brothers (10) T. A. Langsford & Son (2) Jerome McDonald (7) Orlando Merchant (14) Benj. Montgomery & Son (1) Oakes & Foster (3) Hugh Parkhurst & Co (6) John Pew & Son (15) Pinkham & Foster (2) Samuel G. Pool & Sons (4) David B. Smith (27) Samuel P. Smith (5) Sylvannus Smith & Co. (13) Lemuel E. Spinney (3) George Steele (8) James G. Tarr & Bro. (10) Michael Walen & Son (6) John F. Wonson & Co. (10) Charles C. Young (3).	14 Tot 6 Ess 7 Glou	11 Ves 32 Men	5 Wid 21 Chl

Year	Size of Fleet	New Vessels Added to Fleet—Total and From	LOST Vessels Men	LEFT Widows Children
1905	*341 vessels in Gloucester—26,330 gross tons.* 224 schooners, 5 auxiliary schooners (total 229 schooners —219 Gloucester, 1 Lanesville, 3 Rockport, 6 Pigeon Cove), 1 three-masted schooner, 6 sloops, 30 schooner boats, 45 sloop boats, 14 steamers, 10 gasoline boats, 1 schooner yacht, 1 sloop yacht, 2 gasoline yachts, 2 barges.	17 Tot 7 Ess 6 Glou	10 Ves 21 Men	7 Wid 16 Chl
1906	*321 vessels in Gloucester—25,480 gross tons.* 205 schooners (193 Gloucester, 5 Rockport, 6 Pigeon Cove, 1 Lanesville), 7 auxiliary schooners, 1 three-masted schooner, 6 sloops, 24 schooner boats, 23 gasoline boats, 14 steamers, 3 barges.	10 Tot 4 Ess 3 Glou	7 Ves 27 Men	8 Wid 20 Chl
1907	*287 total vessels in Gloucester—22,416 gross tons.* 180 schooners, 5 auxiliary schooners, 1 three-masted schooner, 6 sloops, 22 schooner boats, 28 sloop boats, 30 gasoline boats, 13 steamers, 3 barges. Gorton-Pew owned 50 vessels.	5 Tot 3 Ess 1 Glou	8 Ves 30 Men	16 Wid 34 Chl
1908	*273 total vessels in Gloucester—21,864 gross tons.* 169 schooners, of sail and auxiliary (165 Gloucester, 4 Rockport, 5 Pigeon Cove, 1 Lanesville—average size 105 gross tons), 10 auxiliary schooners, 1 three-masted schooner, 17 schooner boats, 22 sloop boats, 35 gasoline boats, 10 steamers, 6 sloops, 3 barges. *Outfitters/owners with two or more vessels:* James E. Bradley (3) John Chisholm (9) Cunningham & Thompson (16) Davis Brothers (11) Gardner & Parsons (5) Gorton-Pew Fisheries Co. (46) Thomas Hodge (1) Wm. A. King & Co (3) Lane & Smith (5) Fred C. Langsford (1) Jerome McDonald (7) Orlando Merchant (12) Benj. Montgomery & Son (1) Hugh Parkhurst & Co (8) Samuel G. Pool & Sons (4) Manuel Simmons (7) Samuel P. Smith (5) Sylvannus Smith & Co. (11) Lemuel E. Spinney (3) Michael Walen & Son (5) Charles C. Young (2)	6 Tot 5 Ess 1 Glou	5 Ves 68 Men	16 Wid 95 Chl
1909	*275 total vessels of 5 gross tons or larger—22,000 gross tons, 3,926 crew, average of 14.2 men per vessel.* 167 schooners, 6 auxiliary schooners, 29 schooner boats, 4 sloops, 40 sloop boats, 10 gasoline boats, 3 tugs, 2 steamboats, 1 steam lighter, 1 sloop yacht. *Outfitters/owners with two or more vessels:* James E. Bradley (3) John Chisholm Corp. (10) Cunningham & Thompson (12) Davis Bros. (9) John C. Foster (2) Gardner & Parsons (4) John Gleason (2) Gorton-Pew Fisheries Co. (48) Lovell Hines (2) William A. King (2) Orlando Merchant (11) Jerome McDonald (7)	3 Tot 1 Ess 1 Glou	4 Ves 47 Men	16 Wid 49 Chl

Year	Size of Fleet	New Vessels Added to Fleet—Total and From	LOST Vessels Men	LEFT Widows Children
1909 (cont.)	Oakes & Foster (3) Charles O'Neill (2) Hugh Parkhurst (8) Charles M. Parsons (2) Samuel G. Pool & Son (3) Thomas E. Reed (3) Manuel Simmons (8) Lemuel E. Spinney (2) Sylvannus Smith & Co. (11) Arthur L. Story (2) Manuel Viator (2) Michael Walen & Son (4) George W. Wright (2) [Note: 102 vessels were individually owned.]			
1910	*181 total vessels of 20 tons or more. Gorton-Pew owned 47 vessels.*	9 Tot 2 Ess 4 Glou	1 Ves 26 Men	8 Wid 24 Chl
1911	*242 total vessels in Gloucester and Rockport of 5 gross tons or larger—18,697 gross tons. 151 schooners, 4 auxiliary schooners, one gasoline schooner, 3 sloops, 34 sloop boats, 2 auxiliary sloop boats, 8 schooner boats, 9 gasoline boats, 47 gasoline screws, 4 steam tugs, 1 steamboat, 1 steam lighter, 1 sloop yacht. Outfitters/owners with three or more vessels:* James E. Bradley (3) John Chisholm Corp. (8) Cunningham & Thompson (15) Fred L. Davis (11) John Foster (3) Gorton-Pew (56) Orlando Merchant (11) Jerome McDonald (3) Hugh Parkhurst (8) Manuel S. Sears (3) Manuel Simmons (9) Lemuel Spinney (2) Sylvannus Smith & Co. (11) The Story Co. (3)		7 Ves 62 Men	2 Wid 40+ Chl
1912	*222 total vessels in Gloucester and Rockport of 5 gross tons or larger—18,684 gross tons. 125 individuals and firms owned one or more vessels. 130 schooners, 4 auxiliary schooners, 3 sloops, 5 sloop boats, 1 auxiliary sloop boat, 3 schooner boats, 8 gasoline boats, 81 gasoline screws, 10 steam screws, 1 gasoline auxiliary, 1 gasoline sloop boat, 1 steam lighter, 2 steam tugs, 1 steam boat, 1 sloop yacht. Outfitters/owners with two or more vessels:* Raymond Balter (2) John F. Barrett (2) James E. Bradley (2) John Chisholm Corp. (9) A. Cooney (2) Cunningham & Thompson (15) Fred L. Davis (10) John Foster (3) Gorton-Pew (47) Orlando Merchant (10) Jerome McDonald (3) Charles Nelson (2) Hugh Parkhurst (6) Manuel S. Sears (3) Manuel Simmons (4) Lemuel Spinney (2) Sylvannus Smith & Co. (9) The Story Co. (7) Manuel F. Viator (2) Michael Walen & Son (3)	10 Tot 5 Ess 1 Glou	10 Ves 45 Men	13 Wid 17 Chl
1913	*251 total vessels in Gloucester and Rockport of 5 tons or larger—18,670 gross tons. 247 in Gloucester. Outfitters/owners with three or more vessels:* John Chisholm Corp. (9) Cunningham & Thompson (15) Fred L. Davis (10) Gorton-Pew (48) Orlando Merchant (10) Jerome McDonald (3) Hugh Parkhurst (6) Sylvannus Smith & Co. (9) The Story Co. (6)	6 Tot 3 Ess	13 Ves 26 Men	2 Wid 2+ Chl

Year	Size of Fleet	New Vessels Added to Fleet—Total and From	LOST Vessels Men	LEFT Widows Children
1914	*233 total vessels in Gloucester and Rockport of 5 tons or larger—15,568 gross tons. 227 in Gloucester. Outfitters/owners with three or more vessels:* John Chisholm Corp. (9) Cunningham & Thompson Corp. (15) Fred L. Davis (8) Gorton-Pew (43) Jerome McDonald (3) Orlando Merchant (10) The Story Co. (6) Sylvannus Smith & Co. (7).	4 Tot 1 Ess 2 Glou	8 Ves 28 Men	2 Wid 0 Chl
1915	*210 total vessels in Gloucester—13,137 gross tons. Owners with three or more vessels:* John Chisholm Corp. (7) Cunningham & Thompson Corp. (12) Fred L. Davis (8) John C. Foster (3) Gorton-Pew (38) William H. Jordan (8) Jerome McDonald (1) Orlando Merchant (1) Manuel Domingoes (3) Manuel Simmons (4) The Story Co. (6) Sylvannus Smith & Co. (6).	6 Tot 5 Ess 1 Glou	5 Ves 31 Men	7 Wid 23 Chl
1916	*225 vessels in Gloucester—16,151 gross tons.* 88 schooners, 3 auxiliary schooners, 1 gasoline schooner, 2 sloops, 3 sloop boats, 1 gasoline sloop boat, 2 schooner boats, 6 gasoline boats, 92 gasoline screws [note: these were schooners], 7 steam tugs, 1 steam boat, 3 steam lighters, 7 steam screws, 2 oil screws, 5 barges. 105 individuals or firms owned vessels, 84 a single craft. *Outfitters/owners with two or more vessels:* Atwood & Payne (2) Jacob P. Barrett (2) John Chisholm Corp. (7) Cunningham & Thompson Corp. (12) A. Cooney (2) John Dahlmar (2) Fred L. Davis (8) John C. Foster (3) Isaac H. Frost (2) Gorton-Pew (44) Lovett Hines (3) William H. Jordan (8) Manes Fishing (8) J. M. Marshall (3) Master Mariners Towboat (2) Merrimack River Towing (3) Charles Nelson (2) Thomas E. Reed (3) Rockport Granite (9) Manuel Simmons (2) Lemuel Spinney (2) The Story Co. (7) Sylvannus Smith & Co. (6)	6 Tot 4 Ess 2 Glou	6 Ves 33 Men	
1917	*224 vessels in Gloucester—16,228 gross tons.* 85 schooners, 3 auxiliary schooners, 2 sloops, 4 sloop boats, 1 gasoline sloop boat, 1 diesel oil screw, 1 gasoline sloop boat, 2 schooner boats, 4 gasoline boats, 95 gasoline screws, 1 oil screw, 7 steam tugs, 1 steamboat, 3 steam lighters, 8 steam screws, 5 barges. 107 individuals or firms owned vessels, 83 a single craft. *Outfitters/owners with three or more vessels:* Jacob P. Barrett (2) Chisholm Corp. (7) A. Cooney (2) Cunningham & Thompson Corp. (12) John Dahlmar (2) Fred L. Davis (7) John C. Foster (3) Isaac H. Frost (2) Gorton-Pew (44) Lovett Hines (3)	10 Tot 9 Ess	1 Ves 20 Men	6 Wid 15 Chl

Year	Size of Fleet	New Vessels Added to Fleet—Total and From	LOST Vessels Men	LEFT Widows Children
1917 (cont.)	William H. Jordan (8) Manes Fishing (8) J. M. Marshall (3) Master Mariners Towboat (2) Merrimack River Towing (3) Charles Nelson (2) Thomas E. Reed (3) Rockport Granite (9) Manuel Simmons (2) Lemuel Spinney (2) The Story Co. (7) Sylvannus Smith & Co. (5) M. Walen & Sons (2)			
1918	*215 total vessels in Gloucester of 5 tons or larger—15,235 gross tons. 208 in Gloucester. 107 individuals or firms owned one or more vessels.* *Outfitters/owners with two or more vessels:* Atwood & Payne (3) J. F. Barrett (2) John Chisholm Corp. (6) A. Cooney (3) J. A. Dahlmar (2) Fred L. Davis (7) Gorton-Pew (65) Frank H. Hall (2) Wm. H. Jordan (7) Eugene Lafond (2) Lufkin & Tarr (2) J. M. Marshall (4) Auguste Oulette (2) Ben Pine (4) The Story Co. (7) Sylvannus Smith (2) M. Walen and Sons (4) [Note: Gorton's purchased the fleet of Cunningham & Thompson Corp.]	13 Tot 4 Ess	14 Ves 25 Men	10 Wid 27 Chl
1919	*200 total vessels in Gloucester district of 5 tons or larger—13,987 gross tons. 197 in Gloucester. 99 individuals or firms owned one or more vessels.* *Outfitters/owners with two or more vessels:* Atwood & Payne (3) Cape Ann Cold Storage (3) John Chisholm Corp. (4) A. Cooney (2) Liborio Curcuru (3) J. A. Dahlmar (2) Fred L. Davis (7) Gorton-Pew (52) Lovett Hines (2) Wm H. Jordan (8) John Lafound (2) Lufkin & Tarr (2) J. M. Marshall (7) Charles Nelson (2) Auguste Oulette (2) Ben Pine (5) Story Co. (7) Lewis Tarr (2) Peter Tysver (2) M. Walen and Sons (4) [Note: Gorton's transferred registry of a number of craft to Canada once fish could come down in Canadian bottoms.]	9 Tot 6 Ess 1 Glou	22 Men	
1920	*196 total vessels of 5 tons or more in Gloucester district—13,205 gross tons. 177 in Gloucester. 220 vessels including those smaller than 5 tons, resulting in 15,153 gross tons for all vessels. 180 were in fishing (12,172 tons), still 25 sailing vessels. 100 individuals or firms owned one or more vessels.* *Outfitters/owners with two or more vessels:* E. R. Brigham (2) Cape Ann Cold Storage (3) John Chisholm Corp. (4) A. Cooney (2) John A. Dahlmar (2) Fred L. Davis (8) Gorton-Pew (38) Interstate Fish Corp. (8) Wm. H. Jordan (6) John Lafond (2) Lansford & Pine (7) Lufkin & Tarr (2) J. M. Marshall (6) Wm. B. McDonald (2) Auguste Oulette (2) Lemuel E. Spinney (2)	5 Tot 3 Ess	10 Men	

Year	Size of Fleet	New Vessels Added to Fleet—Total and From	LOST Vessels Men	LEFT Widows Children
1920 (cont.)	Peter Tyser (2)　　United Fisheries (4) M. Walen and Sons (4)			
1921	*214 vessels including those smaller than 5 tons— 18,216 tons. 176 were in fishing (12,560 tons), still 22 sailing vessels.*	4 Tot 1 Ess	10 Men	
1922	*196 total vessels of 5 tons or larger in Gloucester and Rockport—15,406 gross tons. 189 in Gloucester. 103 individuals or firms owned one or more vessels. Outfitters/owners with two or more vessels:* Jacob P. Barrett (2)　　W. W. Campbell (2) John Chisholm Corp. (4) Liborio Curcuru (3) Gospe C. Contrino (2)　Cunningham & Thompson (2) John A. Dahlmer (2)　　Fred L. Davis (9) Gorton-Pew (29)　　　Interstate Fish Corp. (5) International Petroleum (2) Wm. H. Jordan (4)　　Eugene Lafond (3) Langsford and Ben Pine (10) Lufkin & Tarr (3)　　J. M. Marshall (7) Master Mariners Towboat (2) Merrimack River Towing (2) Joseph P. Mesquita (2)　Newburyport Fisheries (2) Auguste Oulette (2)　　Perkins & Corliss (2) Thomas Reed (2)　　Rockport Granite Co. (9) The Story Co. (3)　　H. P. Wenneberg (2) M. Walen and Sons (3)	3 Tot 3 Ess	23 Men	
1923	*176 total vessels of 5 tons or larger in Gloucester and Rockport—13,113 gross tons. 169 in Gloucester. 98 individuals or firms owned one or more vessels. Outfitters/owners of two or more vessels:* John Chisholm Corp. (3) John A. Dahlmar (2) Fred L. Davis (7)　　Gorton-Pew (34) Interstate Fish Corp. (9) Wm. H. Jordan (4) Langsford and Ben Pine (10) Lufkin and Tarr Vessels (4) J. M. Marshall (7)　　Auguste Oulette (2) Perkins & Corliss (2)　John Silveria (2) H. P. Wenneberg (2)　M. Walen and Sons (3)	3 Tot 1 Ess	6 Ves 24 Men	
1924	*149 total vessels in Gloucester fleet of 5 tons or larger, of which 137 were in fishing—9,768 gross tons. 127 powered with gasoline or oil engines. Six were non-powered, relying on sails alone. Average age of vessel 15 years. Average value per vessel $16,100. 86 individuals or firms owned one or more fishing vessels. Owners of two or more fishing vessels:* John Chisholm Corp. (3) Fred L. Davis (5) Philip Giammanco (2)　Gorton-Pew (21) Wm. H. Jordan (4)　　Langsford and Ben Pine (6) J. M. Marshall (7)　　United Fisheries (4) Edward J. Weiderman (2) M. Walen and Sons (3)	4 Tot 3 Ess		
1925	*149 total vessels in Gloucester and Rockport of 5 tons or larger—11,224 gross tons. 146 in Gloucester. 86 individuals or firms owned one or more vessels. Owners of two or more vessels:*	10 Tot 6 Ess	31 Men	

Year	Size of Fleet	New Vessels Added to Fleet—Total and From	LOST Vessels Men	LEFT Widows Children
1925 (cont.)	Albert Arnold (2) John Chisholm Corp. (4) Fred L. Davis (4) Gorton-Pew (24) William H. Jordan Vessels Co. (4) Langsford and Pine (12) J. M. Marshall (7) Frank C. Pearce (2) Marcus Sears (2) J. F. Silveria (2) Herman Tysver (2) United Fisheries (3) Ed. J. Weiderman (3) M. Walen and Sons (3)			
1926	*144 total vessels in Gloucester of 5 tons or larger, none listed for Rockport—11,480 gross tons.* Of these vessels, 23 built in last 5 years, none of which were owned by Gorton-Pew Fisheries. 90 individuals or firms owned one or more vessels. *Owners of two or more vessels:* Albert Arnold (2) John Chisholm Corp. (4) Fred L. Davis (4) Gorton-Pew (21) William H. Jordan Vessels Co. (4) Wm. LaFond (2) Langsford and Ben Pine (9) J. M. Marshall (6) Benjamin Pine (2) M. S. Sears (2) Herman Tysver (2) United Fisheries (5) Ed J. Weiderman (3) M. Walen and Sons (3)	9 Tot 7 Ess	8 Men	
1927	*141 vessels in Gloucester of 5 tons or larger—10,702 gross tons, 1,716 crew.* Average age of vessel 15⅓ years; 13 vessels over 60 tons. 92 individuals or firms owned one or more vessels. *Owners of two or more vessels:* Ray Adams (3) Albert Arnold (2) John Chisholm Corp. (4) Cleaves & Pine (2) Fred L. Davis (4) Gorton-Pew (20) William H. Jordan Vessels Co. (2) Langsford and Ben Pine (7) J. M. Marshall (4) Manual F. Roderick (2) Matthew S. Sears (3) Gerry Shoares (3) United Fisheries (3) M. Walen and Sons (2) John C. Williams (2)	8 Tot 7 Ess	41 Men	
1928	*157 vessels in Gloucester of 5 tons or larger—11,665 gross tons.* 99 individuals or firms owned one or more vessels. *Owners of two or more vessels:* Ray Adams (4) Albert Arnold (2) John Chisholm Corp. (6) Henry R. Clattenberg (2) Barbara Curcuru (2) Cleaves & Pine (2) Fred L. Davis (4) Gorton-Pew (20) William H. Jordan (2) Simon A. Landry (2) Langsford and Ben Pine (7) J. M. Marshall (4) Manual F. Roderick (2) Matthew S. Sears (2) Frontero Scola (2) Gerry Shoares (3) United Fisheries (7) United Sail Loft (2) John C. Williams (2)	10 Tot 5 Ess	1 Man	
1929	*181 vessels in Gloucester—13,165 gross tons, 1,943 crew.* 125 vessels in fishing. 117 individuals or firms owned one or more vessels.	16 Tot 11 Ess	9 Men	

Year	Size of Fleet	New Vessels Added to Fleet—Total and From	LOST Vessels Men	LEFT Widows Children
1929 (cont.)	*Owners of two or more vessels:* Ray Adams (6) Reuben Cameron (2) John Chisholm Corp. (5) Henry R. Clattenberg (2) Benjamin Curcuru (2) Fred L. Davis (3) Gorton-Pew (20) Simon A. Landry (2) Langsford and Ben Pine (6) J. M. Marshall (3) Joseph Mesquita (2) Matthew S. Sears (2) Frontero Scola (3) Paul Scola (2) Gerry Shoares (3) Jeremiam Smurrage (2) United Fisheries (11) United Sail Loft (2) John C. Williams (2)			
1930	*187 vessels in Gloucester of 5 tons or greater—13,150 gross tons, 2,058 crew.* 196 vessels including those smaller than 5 tons, 13,817 tons 163 were in fishing (10,894 tons), 9 sailing vessels. 159 were fishing vessels (11,090 gross tons), 25 in freight and towing, 12 yachts. *Owners of two or more vessels:* John Chisholm Corp. (5) Fred L. Davis (3) Gorton-Pew (20) Langsford and Ben Pine (6) J. M. Marshall (3) M. S. Sears (2) United Fisheries (11) This increase in the fleet was due largely to the addition of 18 newly constructed vessels, representing the last great cycle of launchings from the Essex shipyards (note that 11 of these vessels came from an Essex yard).	18 Tot 7 Ess	5 Men	
1931	*170 vessels in Gloucester—11,663 gross tons.* 188 vessels including those smaller than 5 tons, 13,294 tons —154 were in fishing (10,495 tons), 8 sailing vessels. 111 individuals or firms owned one or more vessels. *Owners of two or more vessels:* John Chisholm Corp. (4) L. B. Curcuru (12) Fred L. Davis (1) Gorton-Pew (15) Langsford and Ben Pine (4) J. M. Marshall (2) M. S. Sears (2) United Fisheries (9)	7 Tot	9 Men	
1932	*170 vessels of 10 tons or larger—10,931 gross tons.* 141 were in fishing (9,175 tons), 3 sailing vessels.	2 Tot	6 Men	
1933	*171 vessels of 10 tons or larger—1,420 gross tons.* 140 were in fishing (9,413 tons), 3 sailing vessels.		11 Men	
1934			3 Men	
1935	*134 vessels of 10 tons or larger—9,224 gross tons.* 95 individuals or firms owned one or more vessels. *Owners of two or more vessels:* John F. Barrett (2) John Chisholm Corp. (2) L. B. Curcuru (10) John Dahlmar (2) Lemuel Firth (2) Gorton-Pew (11) Simon A. Landry (2) J. M. Marshall (2) Alex J. McDonald (2) John Piscitello (2) Gerry Shoares (4) United Fisheries (7) Central Wharf & Vessels (3) R. R. Wonson (2)		13 Men	

Year	Size of Fleet	New Vessels Added to Fleet—Total and From	LOST Vessels Men	LEFT Widows Children
1936	*146 vessels in Gloucester fishing fleet of 5 tons or larger—9,571 gross tons*		9 Men	
1937	*149 vessels in Gloucester fishing fleet of 5 tons or larger—9,618 gross tons*		8 Men	
1938	*146 vessels in Gloucester fishing fleet of 5 tons or larger—9,648 gross tons*		10 Men	
1939	*132 vessels in Gloucester fishing fleet of 5 tons or larger—8,500 gross tons.*		3 Men	
1940	*132 vessels in Gloucester fishing fleet of 5 tons or larger—8,380 gross tons*		3 Men	

Note: This chart contains several types of information. For vessels lost, deaths, and surviving family members, I relied largely on the annual reports found in the local papers—checking the annual summaries against the ongoing reports during the course of the year. For early years, prior to the nineteenth century, I relied on specific historical accounts of known disasters. This process may underestimate the full loss of life. For example, in 1918 my great uncle Macker Simmons died many months following a fall from the rigging of a schooner, and his death was not included in the annual summary for 1918. The information on the size of the fleet is drawn largely from annual reports of all Gloucester vessels. Beginning in the mid-1800s the local papers published a detailed list of the fleet. Later, annual compendiums came out (either by the Gloucester Master Mariners or the Boston-based Fishermen of the Atlantic). I have purchased about 30 of the later volumes and have secured photocopies of all extant lists of Gloucester vessels going back to the 1840s, reaching out to libraries and private collections in the United States and Canada. From these sources, I was able to create the information on the fleet: number and types of vessels, tonnage, ownership, crew, and where the vessels came from. Note, in earilier years, I report on numbers available in the historical record.

Photo Credits

Oil painting on front cover: *Fog on the Grand Banks* by Jeff Weaver, 2007. Gloucester, Massachusetts; www.jeffweaverfineart.com. Reproduced by permission.

p. ii: John Morris collection. Drawn by M. J. Burns, *Harpers Weekly*, 1885, courtesy of Joseph E. Garland.

p. 4: John Morris collection.

p. 18: John Morris collection. *Leslie's* magazine, 1840s.

p. 22: John Morris collection.

p. 24: John Morris collection. From an unidentified weekly, c. 1857.

p. 36: John Morris collection.

p. 37: John Morris collection. From an 1864 German-language issue of *Leslie's Weekly.*

p. 43: John Morris collection.

p. 44: John Morris collection.

p. 53: John Morris collection. From Frederick William Wallace, "Life on the Grand Banks," *National Geographic*, July 1921.

p. 54: John Morris collection. From Frederick William Wallace, "Life on the Grand Banks," *National Geographic*, July 1921.

p. 57: Cape Ann Museum, Gloucester, Massachusetts.

p. 65: John Morris collection.

p. 66: John Morris collection.

p. 69: John Morris collection.

p. 75: John Morris collection.

p. 76: John Morris collection.

p. 85: John Morris collection. Photograph by W. A. Elwell, 1876, as part of album for Centennial Exposition in Philadelphia.

p. 89: John Morris collection.

p. 90: John Morris collection. Courtesy of Lillian Lund Files.

p. 108: John Morris collection. Courtesy of Jóhann Diego Arnórsson.

p. 111: John Morris collection. From Frederick William Wallace, "Life on the Grand Banks," *National Geographic*, July 1921.

p. 121: David Cox collection.

p. 124: David Cox collection.

p. 131: Jeff Thomas collection.

p. 135: David Cox collection.

p. 138: John Morris collection.

p. 139: John Morris collection.

p. 145: John Morris collection.

p. 147: John Morris collection. Photograph by Alden W. Flye.

p. 155: Cape Ann Museum, Gloucester, Massachusetts.

p. 156: Cape Ann Museum, Gloucester, Massachusetts.

p. 159: John Morris collection.

p. 166: David Cox collection.

p. 170: Jeff Thomas collection.

p. 173: Jeff Thomas collection.

p. 177: John Morris collection.

p. 184: John Morris collection.

p. 193: Cape Ann Museum, Gloucester, Massachusetts.

p. 196: Cape Ann Museum, Gloucester, Massachusetts.

p. 222: John Morris collection.

p. 226: John Morris collection. Courtesy of Lillian Lund Files.

p. 229: John Morris collection.

p. 232: Cape Ann Museum, Gloucester, Massachusetts.

p. 233: Cape Ann Museum, Gloucester, Massachusetts.

p. 249: John Morris collection.

p. 257: John Morris collection.

p. 260: John Morris collection. Courtesy of Lillian Lund Files.

p. 264: John Morris collection. Courtesy of Mercedes Daley Lee.

p. 265: John Morris collection. Courtesy of Lillian Lund Files.

p. 267: John Morris collection.

p. 268: John Morris collection.

p. 271: John Morris collection. *Esperanto*, photograph by Robert W. Phelps.

p. 276: Cape Ann Museum, Gloucester, Massachusetts.

p. 280: John Morris collection. From Frederick William Wallace, "Life on the Grand Banks," *National Geographic*, July 1921.

p. 289: Cape Ann Museum, Gloucester, Massachusetts.

p. 294: John Morris collection.

p. 296: Cape Ann Museum, Gloucester, Massachusetts.

p. 305: Cape Ann Museum, Gloucester, Massachusetts.

p. 313: John Morris collection.

p. 320: John Morris collection.

p. 323: John Morris collection, from *Fishing Gazette* magazine.

p. 343: John Morris collection.

p. 348: Cape Ann Museum, Gloucester, Massachusetts.

p. 350: Cape Ann Museum, Gloucester, Massachusetts.

p. 353: Cape Ann Museum, Gloucester, Massachusetts.

p. 354: Cape Ann Museum, Gloucester, Massachusetts.

p. 356: Cape Ann Museum, Gloucester, Massachusetts.

p. 358: Cape Ann Museum, Gloucester, Massachusetts.

p. 359: Cape Ann Museum, Gloucester, Massachusetts.

p. 360: Cape Ann Museum, Gloucester, Massachusetts.

p. 374: John Morris collection.

Index